# Praise for *The Organic Medicinal Herb Farmer*

"Jeff and Melanie Carpenter give me hope for the future of herb farming. In this book, they generously share very useful experience-based information and lessons learned, which will help young prospective herb farmers to avoid pitfalls and plan for an economically viable and appropriately scaled operation for sustainable production. Read this book before you start up!"

—JOSEF BRINCKMANN, vice president of sustainability,
Traditional Medicinals

"Finally, a book to recommend to the increasing number of organic farmers who are looking to grow medicinal herb crops. This inspiring handbook provides the quality and depth of information that only comes from years of first-hand experience. Having begun our own herb-growing and herbal-product manufacture thirty-five years ago, I am completely impressed with the comprehensive breadth of topics and business wisdom that the Carpenters have so generously shared."

—SARA KATZ, cofounder, Herb Pharm, and
board president, United Plant Savers

"Bullseye! Jeff and Melanie Carpenter nail it in addressing the needs of today's medicinal herb grower. In a friendly, easy-to-read style, the information in *The Organic Medicinal Herb Farmer* is straightforward and comprehensive, benefitting beginning and experienced farmers alike."

—RICHARD WISWALL, author,
*The Organic Farmer's Business Handbook*

"With this beautiful and informative book, Jeff and Melanie Carpenter share the knowledge they have gathered as they have realized their vision over the last fifteen years on their 10-acre farm in Vermont. *The Organic Medicinal Herb Farmer* serves as a training manual to support U.S. domestic production of high-quality medicinal herbs. A dirt-smudged copy should be within easy reach of every home gardener or farmer who grows—or wants to grow—medicinal plants."

—MICHAEL MCGUFFIN, president,
American Herbal Products Association (AHPA)

"Many of the medicinal herbs used in the herbal industry are still imported, even though we have ideal conditions as landowners and farmers to fulfill the growing demand. United Plant Savers' motto, 'conservation through cultivation,' is a way in which we can take demand off of wild harvested native medicinals, and also stimulate regional sources for a dynamic and growing market. This book could not be more timely. We need domestic herb farmers not only to supply a growing demand for herbal medicine but also as a critical component to ensuring an abundant supply of American medicinal plants for generations to come."

—SUSAN LEOPOLD, executive director, United Plant Savers

"*The Organic Medicinal Herb Farmer* rocks with practical insights for growing healing herbs and making a viable living. Locally grown medicine will be embraced by local food movements as more community herbalists get the word out. The 'Health Care Marketplace' we actually need today consists of more hard-working farm couples like Jeff and Mel Carpenter."

—MICHAEL PHILLIPS, author, *The Holistic Orchard*

"I highly recommend *The Organic Medicinal Herb Farmer* to all new and experienced growers of Western medicinal herbs. If you are an aspiring herb farmer, this is your book!"

—PEG SCHAFER, author, *The Chinese Medicinal Herb Farm*

"Drawing on their fifteen years of experience growing medicinal herbs commercially, Jeff and Melanie Carpenter have written the most comprehensive book available on growing, harvesting, drying, packaging, and selling medicinal herbs. Beginning farmers will find this book particularly useful with its detailed instructions on all aspects of herb farming, including field-site selection, cultural practices, tools, equipment, and business planning."

—JEANINE DAVIS, PhD, associate professor and extension specialist, Department of Horticultural Science, North Carolina State University

"Seasoned and novice growers alike will find a mother lode of information and wisdom packed into this gem of a book! Anyone interested in growing or using medicinal herbs will reap the benefits of Jeff and Mel's meticulous research and hard-won expertise in the field and marketplace. These savvy business people are stellar models of earth stewards making a right livelihood on the land. Reading *The Organic Medicinal Herb Farmer* will greatly help you along the same path!"

—NANCY PHILLIPS, author, *The Herbalist's Way*

# The Organic Medicinal Herb Farmer

The Ultimate Guide to Producing High-Quality Herbs on a Market Scale

Jeff Carpenter *with* Melanie Carpenter

Foreword by Rosemary Gladstar

Chelsea Green Publishing
White River Junction, Vermont

Figure 5-07 is reprinted from *The Organic Farmer's Business Handbook*, copyright © 2009 by Richard Wiswall, with permission from Chelsea Green Publishing (www.chelseagreen.com).

"Let it Grow," words by John Barlow, music by Bob Weir, copyright © 1973 by Ice Nine Publishing Co., Inc. Copyright renewed. All rights administered by Universal Music Corp. All rights reserved. Used by permission. Reprinted by permission of Hal Leonard Corporation.

Note to the reader: This book is intended to provide educational information to the reader on the covered subject. It is not intended to take the place of personalized medical counseling, diagnosis, and treatment from a trained health professional.

Project Manager: Patricia Stone
Developmental Editor: Makenna Goodman
Copy Editor: Eileen M. Clawson
Proofreader: Eric Raetz
Indexer: Linda Hallinger
Designer: Melissa Jacobson

Printed in the United States of America.
First printing April, 2015
10 9 8 7 6 5 4 3 2     16 17 18 19

**Our Commitment to Green Publishing**
Chelsea Green sees publishing as a tool for cultural change and ecological stewardship. We strive to align our book manufacturing practices with our editorial mission and to reduce the impact of our business enterprise in the environment. We print our books and catalogs on chlorine-free recycled paper, using vegetable-based inks whenever possible. This book may cost slightly more because it was printed on paper that contains recycled fiber, and we hope you'll agree that it's worth it. Chelsea Green is a member of the Green Press Initiative (www.greenpressinitiative.org), a nonprofit coalition of publishers, manufacturers, and authors working to protect the world's endangered forests and conserve natural resources. *The Organic Medicinal Herb Farmer* was printed on paper supplied by QuadGraphics that contains at least 10% postconsumer recycled fiber.

**Library of Congress Cataloging-in-Publication Data**
Carpenter, Jeff, 1969– author.
  The organic medicinal herb farmer : the ultimate guide to producing high-quality herbs on a market scale / Jeff Carpenter with Melanie Carpenter ; foreword by Rosemary Gladstar.
    pages cm
  Includes bibliographical references and index.
  ISBN 978-1-60358-573-6 (pbk.)—ISBN 978-1-60358-574-3 (ebook)
  1. Herb farming. 2. Herbs—Organic farming. 3. Herbs—Therapeutic use. 4. Herbs—Marketing. I. Carpenter, Melanie, author. II. Title. III. Title: Ultimate guide to producing high-quality herbs on a market scale.

  SB351.H5C295 2015
  635—dc23
                        2014049603

Chelsea Green Publishing
85 North Main Street, Suite 120
White River Junction, VT 05001
(802) 295-6300
www.chelseagreen.com

*This book is dedicated to*
*Lillian Marie Carpenter and Priscilla Gifford Carpenter.*

# Contents

# Foreword

*People need these plants. They want them live, dried, and fresh—for the medicine they make, the gardens they grow, and the classes they teach. As a result there is a growing need for medicinal herb farms, and there are many ways herb growers can participate in and become an integral part of this green movement.*
—THE ORGANIC MEDICINAL HERB FARMER

I love books! They can be so informative, practical, and inspiring. This book contains all three of these essential qualities. Furthermore, it is exceptionally well written and lavishly illustrated with gorgeous, pertinent photographs. Everything a great book should be, but it's the innovative and hopeful message to farmers about farming that moves me most. . . .

When my daughter, Melanie, and her husband, Jeff (who, rather than an "in-law," I fondly call my "son-of-the-heart"), first mentioned that they were thinking of writing a book about organic herb farming, I was at least as excited about the project as they were. I had no doubt it would be a practical, comprehensive, and useful resource for other farmers—and farmers to be. I also suspected it would contain the unique tools and talents that Jeff and Melanie both brought to their farming practices. In the past twenty or so years Jeff and Melanie have been "in the field" literally, learning through trial, error, and innovation about herbs and farming; shaping and defining what being "successful farmers" means to them. I knew a book written about herb farming, a subject they were both deeply invested in, would be exceptionally well done—a cut above—both because of their love of the subject and also because that's just the way they do things. When they asked me to write the foreword to their book, I was deeply honored.

Melanie and Jeff shared with me that this book is an introduction to medicinal herb farming and a resource for how to grow and process medicinal herbs for market but it is far more than an introduction. *The Organic Medicinal Herb Farmer* addresses everything that a person new to farming and just beginning the farming adventure, as well as the seasoned farmer seeking detailed information on growing herbs for the marketplace, would need to know to get started. The best of agricultural practices, including seed sowing, which herbs to grow, harvesting, and medicinal plant conservation, are combined with the less appealing—or as Melanie states, "the less sexy"—aspects of farming, such as business management, bookkeeping, and marketing, to give a comprehensive overview of what it takes to be a successful herb farmer in today's market.

While there are several excellent books on growing herbs, and some very good ones on farming, there are few books that combine the practical how-tos of organic herb farming and the ins and outs of the herbal industry with common-sense marketing skills—that is, how to sell your herbs once they've grown. This was the book, I'm sure, that Jeff and Melanie had wished they'd had available when they first got started. Woven amidst the necessary technical farming data, charts, yield-per-acre averages, and dry-to-wet herb ratios that make a book like this useful and practical are the personal tales and insights—the ups and downs and personal revelations of farming—which give richness and depth to any story. This "farmers tale" is the story of their successes and the stories of the crops that failed and the challenges they've met along the way, all powerful teachings to those considering farming. Between the seeding, weeding, and harvesting of the seasons,

Jeff and Melanie managed to write this book with the same spirit and integrity in which they live their lives. What shines most throughout the pages of this rather hefty tome is their heartful generosity, sharing what they've learned with others.

Since they were youngsters, both Jeff and Melanie demonstrated an early interest in plants. While still in high school, Melanie and her twin sister, Jennifer, started a small herbal business under my guidance and tutelage. It was our—their father's and my—attempt to teach them good work ethics and financial responsibility (they put their earnings into savings for their college fund), and also my personal desire to engage them in the study and practice of herbalism. Indeed, it worked! Both Jennifer and Melanie have an incredible work ethic and are keenly resourceful at managing money (they had more money saved up in their bank account when they graduated high school than I did in the first few years of running my own herbal business!) and, best of all, they both developed a lifelong love of herbs.

Jeff comes from a hearty sixth-generation Vermont farming family, which he's always proud to mention. I've often heard him say, "Farming is in my blood." And it is; he's a natural at it—and like most farmers, he does love a tractor. Just ask him why he really chose farming as a career! When Jeff first came to Sage Mountain, our herbal retreat center, fresh out of high school, the "wild" older brother of Melanie's best friend, I recognized immediately the exceptional person he was and saw in him the potential plant lover he would become. I was delighted when he signed up for my herbal apprentice program, and after the apprenticeship was over, because of his keen interest, I invited him to stay on the mountain for a few months longer to continue his study of herbs and also to help me with the chores. He fell in love with plants as well as in love with my daughter (though I think it was in the reverse order!), and those two love affairs have been a guiding light on his personal as well as professional path.

I have had the privilege of watching Jeff and Melanie mature not only as individuals and as caring, loving parents, but also as herbalists and farmers, into the savvy and successful business partners they've become over the past twenty years. They've had an impact not only on their local community and their home state of Vermont with their vision of organic herb farming but have also joined forces with other farmers throughout the country to become a strong voice for the cultivation of sustainable, high-quality organic herb production in the United States. Like other young farmers seeding the back-to-the-farm local food movement and supplying high-quality organic herbs to the burgeoning herbal renaissance, they have become "rock stars" of their communities.

In part because of the pioneering work of farming entrepreneurs such as Jeff and Melanie and their mentors Richard Wiswall (Cate Farm, Vermont), Todd Hardie (Caledonia Spirits, Vermont), and Andrea and Matthias Reisen (Healing Spirits, New York), many other farmers are creating successful herb businesses and are making a go of it financially. No longer an undesirable career, farming, and herb and vegetable farming specifically, is becoming a career of choice for young college graduates and visionaries seeking to make a difference in the world; experienced farmers wanting to grow more profitable and sustainable crops than wheat, corn, or soy; and even those urban dwellers seeking a healthier lifestyle for themselves and their families.

In an easy to understand, practical, and comprehensive manner, *The Organic Medicinal Herb Farmer* lays out the tools and step-by-step practices to help others become successful herb farmers. If you are trying to decide whether to farm medicinal plants exclusively or to add one to two herbal crops to your current inventory; or you wish to focus exclusively on common culinary herbs, or strictly medicinals, or at-risk native herbs; or you are trying to decide on the advantages of growing living potted plants versus harvested and dried herbs; or you wonder if it's wise to add a value-added product, you'll find the detailed information necessary to guide you through the process. This isn't necessarily information gleaned from other resources and passed along in a diluted format, but the direct experience of two farmers who have tried it all and are sharing the best of what they've learned with others.

However, no farmer farms alone. Farmers learn from and depend on one another and the community that supports them. Much like the fungal mycorrhizae that magically weaves plants together in a complex underground Internet system, Jeff and Melanie are part of a network of farmers who have created a thriving "aboveground" support system. These farmers share information and resources freely and generously with one another, often providing mentoring services and intern programs that help teach and revitalize the basic skills of farming that have been lost in this country over the past four decades because of the demise of the small family farm. This is cooperation and collaboration at its best, where each person's success is enhanced by the success of others. We are all cheering each other along.

Farming communities have traditionally been close, joined together by the community potluck, local Granges, churches, marriages, and family. The reality of today is that herb farmers are spread out across the country, and the Internet and phone have become the means of communication, the modern day Grange and potluck that binds them together. In the process of writing this book Jeff and Melanie visited and interviewed many other herbal farmers across the country. Their wisdom and insights, as well as practical farming and business advice, pepper the pages of this book, not only enriching it but also imparting a sense of the diversity of the herbal farming community.

Farming is a subject near and dear to me and still strokes tender heartstrings and childhood memories. I grew up on a small family farm in the ideal farming tillage of Sonoma County, California, in the 1950s and '60s. My father loved farming more than anything. He—and my entire family—worked from sunrise to sunset on that farm, milking the cows; planting and hauling hay; raising the calves, goats, and chickens; gardening; and canning. In spite of the tremendous amount of work required to run the farm, neither he nor anyone else in the family, for that matter, would ever have given up the farmer's life or the family farm.

Unfortunately, my father was farming at a time when agricultural practices were changing, and the small family farms and farmers were being systematically driven out of business, one by one, in old-time farming communities across the country, replaced by government-supported monocropping and factory farming. There were no options, no alternative crops or incentives offered to the small farmers to help them weather the hard times and stay in business. It was a sad day for my father, and for the other farmers in our community, when they were forced to sell their beloved dairy herds and farms and head off the farm to find work. In truth, it was a sorry day for this entire nation. For the past several decades we've witnessed the tragic effects of factory farming and monocropping on the landscape, in our water systems, in the food system, in our health care, and in our communities. I believe the health of a country is dependent on the health of the farming communities, the health of the soil, the food and medicine we grow, and the rich diversity of sustainable organic farming practices.

While I'm extremely proud of my two "children of the heart" for writing such a vital and important resource for farmers, it's not the primary reason I appreciate *The Organic Medicinal Herb Farmer*. It's because it carries a hopeful and pertinent message, provides detailed information and innovative tools and suggestions, and offers a roadway to success to the small family farm. This is the book I wish my father had had.

ROSEMARY GLADSTAR
HERBALIST AND AUTHOR
SAGE MOUNTAIN, VERMONT

# Preface

This book was conceived primarily as a result of hearing our family, friends, colleagues, students, and customers encouraging us for years to write a practical guide that seemed to be surprisingly absent from the vast library of agricultural books. Although many excellent books on growing herbs existed previously, few that we knew of seemed to fill the role of providing a comprehensive, common-sense "how-to" guide based on direct experience from both sides of the industry, the herbal product side and the organic herb farming side. Melanie and I had owned a successful herbal product company for years before selling it to buy the farm. That experience, coupled with fifteen years of commercially growing over fifty species of medicinal herbs on a small certified-organic herb farm has provided us with a good foundation of knowledge with which to share methods gleaned through our successes and challenges.

One of the biggest challenges we faced in the formative years of herb farming was a lack of good information about growing and processing medicinal herbs commercially in a region where growing seasons are short and conditions are often less than ideal. There were a few books on commercial herb production out there, but what we found was that the information contained in these books rarely applied to what we were trying to do, which was focusing on quality rather than quantity and keeping our costs down by innovating with the equipment and resources we could build, borrow, or afford to purchase rather than buying expensive equipment and planting large acreage. There are some great books on home-scale herb gardening, which were helpful with propagation and growing techniques for the plants we were focusing on. However, as far as production knowledge, we found ourselves stranded somewhere in the middle between the bigger, faster, more machinery-centric approach and the small cottage herb garden "hobby" approach. We basically learned most of what we know now through innovation, trial, and error, making plenty of mistakes along the way, balanced by intermittent revelations on how to make do with what we had, what we could build, or what we could afford to buy.

People often ask us how and why we got into medicinal herb farming, as opposed to another more traditional crop. Melanie's stepmother, Rosemary Gladstar, is an esteemed author and herbalist and is often referred to as "The Fairy Godmother of Western Herbalism." During Melanie's childhood Rosemary helped her cultivate the foundational knowledge and deep reverence for the plants that are at the core of Melanie's work today as a farmer, healer, and plant conservationist. I had the good fortune of being introduced to Melanie through my sister Janna, who explained to me that Melanie and her family were "into herbs and stuff." At the time I thought perhaps they liked to sprinkle lots of oregano on their pizza or something like that. That was what I thought until I made my first trip to Sage Mountain (Rosemary's home and herbal education center in Orange, Vermont) and had my first exposure to Rosemary and the magical world of medicinal plants and herbalism. That fateful day was the beginning of a relationship that would blossom into a marriage, both literally and figuratively.

Farming is in my blood. My paternal grandparents were both sixth-generation Vermont farmers. When I was young we often visited the dairy farm my grandmother grew up on in Randolph, Vermont, which is now owned by her nephew. I was completely enamored by the farmers who remained in the family, and when we visited the farms I wanted to be a part of that action. The tractors, the hay fields, the cows, the salt-of-the-earth farmers. As a child the agricultural landscape and lifestyle appealed to me in a way that was accessible, yet I had always felt a certain degree of separation.

Both my father and my grandfather moved away from the farms, chose careers in business, and seemed to view the remaining farmers in their family with a sense of pity because of the amount of work they had to perform and the sacrifices they had to make to survive economically. Their viewpoint was shared by many in their generation who had left their family farms for greener pastures. The feeling that I got from some members of my family, as well as others in the community, left me with a sense that although farms were cool to visit, there were certainly better career choices to be made.

Fortunately, after being lost for years I finally followed my heart and found my way back to the farm. Things certainly seem to have changed for the better in the last decade or so regarding people's attitudes about farmers and farming and the important role they play in our society and on our landscape. Right now, at least here in our region, where local, sustainable, organic agriculture is experiencing a kind of revolution, farmers are the "rock stars" of our communities, and in many cases, we are actually making a go of it financially. Gone seem to be the days when the "poor, hardscrabble dirt farmer" was not only often pitied for his or her meager existence but also often viewed as uneducated and unable to fill the mold of the economic ideal. Farmers have been and are becoming increasingly well educated, though not always in the traditional sense of education provided by colleges and universities. We are being educated by our experiences working with the plants and natural environment around us on an almost daily basis and through the knowledge shared by others doing similar work.

It is in the spirit of sharing knowledge that we have written this book. Melanie and I have always viewed our role in the medicinal herb industry in a cooperative sense rather than a competitive sense.

That may sound naive or idealistic according to the traditional business model, but the fact is we wouldn't be here without the collaboration and support of our colleagues and community, and we haven't yet come close to being able to meet the demand for high-quality, locally sourced organic medicinal herbs on our small farm. There are others like us doing similar work, but we need many more growers to join us in our efforts to help make the domestic bulk medicinal herb market become more viable. The United States currently imports a vast majority of the medicinal herbs used in the herbal products industry from foreign countries, and we feel that there is something really wrong with that fact. (We are currently forming a medicinal herb growers cooperative here in Vermont, and we encourage others to pool their talents and resources and follow suit.) Healing people, animals, and even plants themselves by utilizing plant medicine is returning to its former status as mainstream rather than alternative medicine, and we encourage others to enter into the field of medicinal herb farming.

*The Organic Medicinal Herb Farm* was coauthored by Melanie and me and was written primarily through the lens of our individual areas of expertise and responsibilities with the day-to-day farming operations. I wrote the majority of the chapters that focus on plant ecology, the herb industry, and the mechanical and agricultural aspect of our operation, and Melanie wrote the majority of the chapters that focus on business management, bookkeeping, marketing, harvesting, and plant conservation. We both contributed a great deal to each other's work as well.

This book is divided into two parts. Part one is a "how to" technical manual based primarily on our experiences in the medicinal herb industry and contains lots of experiential references discussing our trials, tribulations, and successes along the way. Interwoven within this narrative is knowledge we have gleaned from experts in the industry that we have spent time with or interviewed via phone or e-mail. Part two is composed of individual plant profiles detailing fifty species of medicinal herbs we have grown, processed, and marketed.

This book contains some terminology that may not be familiar to everyone. Some of the terminology is used interchangeably with other relatively obscure terminology, so here is a brief preemptive glossary to give you the reader some advance idea of what we are talking about when we use the following terms.

We refer to the crops that we produce as **medicinal herbs, medicinals, herbs, plants, species,** and **botanicals**. These terms are all used interchangeably.

When we speak of the medicinal properties of plants, we use the terms **bioactive compounds, medicinal constituents,** and **medicinal properties** interchangeably.

We refer to the plants that we grow and market in several different forms. **Dried herbs** refers to plants that were harvested and dehydrated before sale. **Fresh herbs** refers to plants that were harvested and sold or used in their freshly harvested form without being dehydrated. **Live plants** refers to live, potted nursery plants that we grow and sell. **Tea** refers to herbal tea blends that we produce and market but does not refer to the tea plant (*Camilla sinensis*), which does not grow in our region.

**Cultivate** means to grow something, as in "we cultivate medicinal herbs." It also means to work the soil as well as to remove weeds. **Till** means to work the soil in preparation for planting yet does not exclusively refer to working the soil with a rototiller, which is only one of many tools used to till soil.

**Phyto-** means plant based, as in **phytochemical, phytomedicinal,** or **phytogeographical**.

**Wildcraft** and **wild-harvest** are used interchangeably and refer to harvesting plants from the wild that were not purposely grown there.

The term **native** when referring to plants and their habitats is a fairly subjective term and can carry with it a certain degree of controversy. Since plants have been dispersed far and wide both naturally and anthropogenically since the dawn of humans, it can be challenging to determine exactly which species qualify as "native" and which ones were "introduced." We have chosen to use the term "native" when referring to plants on the North American continent that appear to have preexisted before European colonization. For plants on other continents we have chosen to use the term "native" to denote plants that have no recorded history of anthropogenic dispersal into the habitats that we refer to as their "native habitats."

Lastly, at the time of this writing, the herbal products industry is experiencing a great deal of flux regarding the FDA's regulation of dietary supplements (which herbal products qualify as) through the Dietary Supplement, Health, and Education Act (DSHEA) of 1994. Good Manufacturing Procedures (GMPs) are being mandated by the FDA in an attempt to regulate the production and sale of dietary supplements. In 2011, the Food Safety Modernization Act (FSMA) was also enacted by the FDA as part of a sweeping reform of our food safety laws.

As producers of medicinal herbs that may be categorized not only as herbal supplements under the DSHEA but may also be categorized as "produce" under the FSMA, we anticipate increased scrutiny and regulation in order to maintain FDA compliance. Since these laws are in their relative infancy, at this point, we still don't know for sure exactly what we will need to do (if anything) to become and/or remain compliant in the eyes of the FDA. Therefore, there are many procedures described in this book that we have utilized for years to produce our products ethically and legally that may need to be adapted in order to maintain compliance. We highly recommend that growers remain aware of FDA regulations concerning the manufacture and sale of bulk medicinal herbs and adapt their methods accordingly in order to maintain compliance in an affordable and manageable way.

JEFF CARPENTER
HYDE PARK, VERMONT

# Acknowledgments

Our deepest thanks and appreciation go to:

Rosemary Gladstar for opening the gate to the garden of healing plants and showing us the wonder within. We aspire to continue the legacy of knowledge, love of plants, and earth stewardship you have established and taught us.

Our parents and stepparents Donna Russo and Eric Seidel, Laura and Jeff Drury, Karl Slick, Rosemary Gladstar and Robert Chartier, and Don and Lynne Carpenter for their endless support, love, and inspiration.

Our dear family, circle of friends, and past and present employees, apprentices, and interns who have been instrumental from the beginning—believing in our dreams, working on the farm, and helping us keep our humor through it all.

Our grandmothers, grandfathers, and the generations before them whose lineage runs deep in our veins, forming the genesis of our lives and work.

Our spiritual teachers and mentors: Kip Roseman, Jan Sandman, Sandy Corcoran, and Trishuwa. The work you have done to help us form and maintain the connection between spirit and our work with the plants is immeasurable.

Our mentors Richard Wiswall, Todd Hardie, and Andrea and Matthias Reisen. You showed us that it is possible to make a living doing what you love, and then helped us figure out how to follow in your footsteps. We are forever grateful for the knowledge and inspiration you have shared.

Brian Hoogervorst, "the human combine," who has worked with us literally since day one and who has been the backbone of our farm crew for the last fifteen years. Thank you for the tons of herbs you have helped us plant, grow, harvest, and process and for always keeping it real.

Bethany Bond who worked endless hours on the book while simultaneously running the farm office. From taking pictures, creating the art log, formatting manuscripts, and being our in-house IT support, we wouldn't have met a single deadline without your help.

Alex Otto for contributing many hours of his scientific engineering background and knowledge to chapter 14 on drying and processing botanicals and for helping us fine-tune our own processing systems.

Micheal Phillips for sharing his knowledge regarding pest and disease control in plants and trees through his book *The Holistic Orchard*, through workshops we attended, and also via phone conversations. We also thank you for giving us insightful information from your own experience authoring books before we committed to our own writing endeavor.

Josef Brinckmann, Richo Cech, Jeanine Davis, Micheal Gillette, Matt Dybala, Jovial King, Katheryn Langelier, and Andrea and Matthias Reisen for making the time to be interviewed for our book and for sharing your knowledge with us and our readers.

Our green sisters, Kate Clearlight and Larken Bunce, for sharing many of your beautiful plant pictures that grace the pages of this book and Cornelius Murphy for creating the farm illustration on pages 38 and 39.

Margo Baldwin of Chelsea Green Publishing and Makenna Goodman, Patricia Stone, and Eileen Clawson for their guidance and help in shaping our ideas and inspiration into this book.

The Green Nations of Trees, Plants, and Fungi, our most powerful and beloved teachers and allies. The life force within you heals and sustains us all and we are forever in service and partnership.

Photograph by Alden Pellett

# PART ONE

# Growing and Processing Medicinal Herbs for Market

**Figure 1-1.** Zack Woods Herb Farm nestled in the heart of Vermont.

# CHAPTER 1

# Why Grow Medicinal Herbs?

The agricultural revolution is truly happening, especially where we live in Vermont. Nestled within these green rolling hills and mountains is a rich community of small farms, over five hundred of which are certified organic.[1] Today farmers are growing and producing a myriad of "alternative crops," from alpacas to cranberries, emus to mushrooms and even rice. Yet when asked what kind of farm we own and operate, there are often perplexed looks or awkward jokes when we say, "We grow medicinal herbs." Once it's established that we're *not* growing marijuana, the questions start and just don't stop. "What kind of herbs do you grow? Are they hard to grow? Which plants grow here? Where do you sell them? Is it profitable? Can we come to visit?" Many people don't realize herbs are used for medicinal purposes and that many of the pharmaceutical drugs they are familiar with were derived from plant compounds. They often think more of herbs as culinary ingredients than as healing agents. It is an entirely new concept for most people that medicinal herbs can grow successfully in Vermont, where the winters are cold and long and the summer sun variable. Growing local food makes sense to them, but growing local medicine sounds revolutionary.

With the explosive growth of the herbal products industry in the last twenty years or so, more people are becoming aware that herbs can increase their well-being and vitality. But until recently the public's focus has been primarily on therapeutic uses of medicinal herbs, with little attention paid to where the plants used in herbal products originated. For example, many people use echinacea extracts but have limited understanding about how the plant grows. People purchase chamomile tea to help relax, often with little thought about how it is grown, harvested, and whether it was sprayed with chemicals or irradiated during import. I am happy to report that increased attention to food safety and quality standards, the local food movement, and organic growing methods are changing this. The significant work of organizations such as the nonprofit United Plant Savers has also elevated awareness about sustainability for native wild plant populations that are at risk, due in part to overharvesting for the herb industry. Increasingly, individuals, manufacturers of herbal products, and health practitioners are demanding

**Figure 1-2.** *Echinacea purpurea* in full bloom. Photograph courtesy of Kate Clearlight

**Figure 1-3.** Fresh milky oats being prepared into a tincture.
Photograph courtesy of Kate Clearlight

and willing to pay a premium for high-quality, organically grown herbs. They are not inclined to settle for poor-quality or imported herbs that are irradiated, sprayed, and grown in unnatural, unsustainable, or unethical ways, even when they may be less expensive.

Despite these increased demands, commercial medicinal herb farming in the United States is still in its relative infancy. It is time to change that; we need to increase herb growing in every state and bioregion. Almost every species of medicinal herbs in common use today could be grown here in the United States, from Maine to Hawaii. These are fertile times for the industry, and it is time to entice more growers into the field and to encourage existing farmers to diversify their crops to include medicinal plants.

# A New Approach to Old-World Medicine

In order to understand where we are going, we must look first at where we've been. Herbalism has a

rich and fascinating history. While it is not unusual to hear herbalism referred to in the United States as "New Agey," the evidence of humans utilizing medicinal plants is very, very old. So old, in fact, that fifty-thousand-year-old Neanderthal skeletons recently excavated in Spain were determined to have the remains of two medicinally active plants, yarrow and chamomile, embedded in the plaque on their teeth. In northern Iraq sixty-thousand-year-old Neanderthal remains were found to have been buried with a "bouquet" of medicinal plants containing yarrow, ephedra, mallow, groundsel, and centaury.

The written record of medicinal plant usage in modern humans dates back approximately five thousand years in Sumerian and Babylonian texts. Here in North America native peoples had a vast knowledge of medicinal flora, many species of which are the most popular herbs used today. Sassafrass root, for example, became the first major commodity crop exported from the United States back to England to be used in a variety of tonics and extracts. Colonists also brought several species of herbs to the New World, many of which quickly naturalized here and became the first nonnative weeds.

Jumping ahead to the early twentieth century, our grandmothers cut, steamed, and ate dandelion greens in the spring of the year because, in addition to their being really tasty, Grandma knew they were good for her. She had not read the research reports demonstrating that bitter sesquiterpene lactones in dandelion leaves stimulate bile secretions to aid in the digestive process. She knew this simply because eating dandelion leaf made her body feel really good. She used these plants because they were readily available, free for the picking, and remarkably effective. This was not uncommon. Most people of her generation had a connection to the plants in their backyards, woods, and meadows and utilized them for food and medicine, not only because these plants were free or easily obtained but because they were effective.

Then along came the technological revolution, giving rise to contemporary medicine and what many refer to as the "silver bullet" approach. This approach treated the patient with synthetic drugs or surgery to alleviate symptoms associated with disease or injury first and foremost, while addressing the root cause of the ailment secondarily, if at all. In the United States this type of medicine became the mainstream, and "healing" became the dominion of health-care providers. The word "drug" itself comes from the Old French *droge-vate*, literally translated as "plant medicine stored in a barrel," but things certainly changed as that old word took on new meaning. Many people like our grandmothers and others in subsequent generations eventually became more disconnected from the plants and their local medicines and tonics, as convenience trumped quality.

Fortunately, there is transformation afoot, in large part as a result of the insightful and hard work of the herbal community. While "silver bullet" methodologies still predominate in most of our hospitals and medical clinics, we are increasingly empowered to take a more preventive approach to our own health care and to address the root cause of disease or injury. Phytotherapies (plant-based therapies) have been instrumental in this "new" understanding of healing. Many refer to this paradigm shift as "the herbal renaissance." Herbal healing, in the west forced underground or neglected for periods ranging from decades to centuries, in many cultures is resurfacing. Thanks to benefits provided by modern technological advances and newfound appreciation, in many ways herbal healing is now becoming better and more effective than ever. We have an amazing and invaluable cache of information and resources that began long ago as spoken word from village healers and shamans, evolved through written texts, and has now entered the electronic data-sharing realm via the World Wide Web. We have come full circle, returning to the wisdom of our elders while building on the research, knowledge, and insights provided by the scientific community, with botanicals front and center.

Education has also played an essential role in this herbal revolution. Clinical herbalists are studying at accredited colleges, universities, and other educational organizations at an incredible rate. They bring this ancient yet newfound knowledge back to their families and communities. Physicians, veterinarians,

**Figure 1-4.** Words that helped define and shape the Western herbal movement.

and other health-care practitioners are recognizing the efficacy of botanicals and are far more open to the integration of medicinal herbs in their treatment protocols. After all, almost 25 percent of modern pharmaceutical drugs are now or have been derived from plant compounds.[2] This integrative approach is the key to bridging the gap between old-world and new-world medicine and is a vital component to improving and maintaining the health of our planet and its inhabitants.

With the resurgence of herbal medicine, this is an exciting time for those of us who use, gather, or grow healing herbs. People need these plants. They want them live, dried, and fresh—for the medicine they make, the gardens they grow, and the classes they teach. As a result there is a growing need for medicinal herb farms, and there are many ways herb growers can participate in and become an integral part of this green movement.

Although there is great demand for medicinal herbs and amazing opportunities exist for farmers, there are still hurdles facing this emerging industry. In 2013 the

**Figure 1-5.** ZWHF crew hand harvesting calendula blossoms.

**Figure 1-6.** Colorful sacks display medicinal herbs in an herb shop in Cairo, Egypt. Photograph © 2014 Steven Foster

United States imported 70 percent of the raw botanicals used in the manufacture of $5.6 billion dollars' worth of herbal supplements. In the same year we exported 30 percent of our commodity crops, such as soy and corn.[3] If farmers in the United States have surplus crops to export and there is a growing demand for herbs, why aren't more farmers growing medicinal plants? Why are we continuing to produce record surpluses of low-value crops such as corn and soy?

In part because old habits die hard. If we ask the Nebraskan commodity farmer to convert his thousand acres of corn and soybeans over to such high-value crops as dandelion and chamomile, he will likely fall off his combine laughing. That combine isn't set up to handle crops like chamomile, and the gigantic grain bins he uses to dry crops aren't set up for dehydrating botanicals. Even if he is motivated to adapt his equipment to deal with these specialty crops, he'd have to be willing to forfeit his U.S. Department of Agriculture (USDA) subsidy checks when he stops producing these commodities—crops for which at times he actually gets paid by the U.S. government to *not* harvest when the market is flooded. It's a hard sell, to say the least. He is as addicted to corn as is the rest of the country, through no fault of his own. He is trying to feed his family and put a little aside for recreation and retirement in a challenging economy, just like the rest of us.

Now what if we told that same farmer who is grossing $1,100[4] per acre on his corn/soybean rotation that he could gross $20,000 to $40,000 per acre growing dandelion root and leaf? He could add another ten species of crops to diversify his offering and cover-crop half of his land as part of a rotation to begin restoring the organic matter depleted from years of poor farming practices. He could hire a consultant specializing in botanical production, retrofit that combine to harvest chamomile and other botanicals with it. He could convert his grain bins to herb dehydrators and contact some of the thousands of small and large herbal product manufacturers in the United States who are desperately searching for domestic sources of high-quality bulk herbs. Sounds simple, right? Well, if only it were that easy.

**Figure 1-7.** Dandelion is a beloved plant at ZWHF. Photograph courtesy of Bethany Bond

# Hurdles in the Field and Marketplace

Technological advances in agriculture over the past century in the United States have led to incredible gains in commodity crop production efficiencies on a per-acre basis. Almost all of this technology has gone into developing bigger and faster machinery to produce more food, fuel, fiber, and shelter, not medicinal plants. The world's leading producers of botanicals, which include China, India, and several eastern European countries, are way ahead of the United States in terms of equipment, techniques, and volume for botanicals production. The USDA doesn't even collect data regarding the production of medicinal herbs for commerce in the United States, so as far as they are concerned, it's almost as if commercial medicinal herb farmers don't exist here.

The simple fact is that the resurgence of herbalism in the United States is relatively new. On a national scale we are just catching up to what the rest of the world has been doing all along: integrating modern technological medicine into an already existing foundation of natural healing practices. Botanicals are the first choice in preventive health care in most countries outside the United States and Canada. These countries have developed specialized equipment and techniques to not only supply their own raw material needs but also produce enough surplus to export to countries like ours that are way behind the curve. Now not only are we buying their herbs, we are also buying and importing (most often at exorbitant costs) the specialized equipment they have developed to produce botanicals in an attempt to supply our own domestic herb needs.

As a nation we need to shift our attention away from a myopic focus on commodity crops and utilize our vast agricultural technologies, resources, and knowledge to diversify and include growing more medicinal plant crops on a commercial scale. Because large agribusinesses have been reluctant to do this, there is opportunity here for local growers on small farms around the country to meet this demand in an ecologically, financially sustainable and moral way.

Yet as farmers embrace the idea of growing medicinals, they are faced with another learning curve in the botanicals marketplace. Unlike the markets for soy, corn, and other commodities, the botanicals market can be extremely volatile and is often subject to the latest trends and research reports. Many in the industry refer to this as the "Dr. Oz (a cardiothoracic surgeon, author, and popular TV host who acts as a health guru to the masses) phenomenon." Here on our herb farm the phone rings daily with calls from customers inquiring about or placing orders for the bulk herbs we grow. Occasionally there will be a noticeable spike in demand for a particular and oftentimes relatively obscure species of herb. We and many others have noticed that there seems to be a correlation between these spikes and what Dr. Oz and others like him have talked about on their latest network television episode. While this can be great publicity, there is also inherent risk in it. In the early 1990s, for example, when a major television network news show produced a story touting the virtues of *Echinacea purpurea* and its potent immune-boosting activities, sales of herbal products containing echinacea went through the roof, as did the price of bulk echinacea root, leaf, and flower. Demand soared, and many farmers made their first attempt at diversifying their tobacco or vegetable farms by planting several acres of echinacea.

In the two to three years that it took for the perennial echinacea crop to mature, prices and optimism soared as more echinacea was planted. Then reality set in. After three years of waiting for the crop to mature, more media reports surfaced; only this time, instead of touting the benefits of this potent plant, new studies came out that attempted to debunk previous findings. Headlines such as "Echinacea leaves users in the cold" sent sales of supplements containing echinacea plummeting at the same time that many farmers were getting ready to harvest their new "cash crop." Suddenly the echinacea market was flooded, and the price plummeted. Many farmers were forced to plow their crops under to cut their losses. This was a tough initiation into growing medicinals for many farmers at the time and has had lingering effects on the uncertainty of the medicinals market.

Another challenge facing beginning herb farmers is the lack of educational resources available to guide them in establishing their crops and markets. There is an overwhelming database of information available to growers of conventional crops and such products as vegetables, milk, and grains, but when it comes to alternative crops such as medicinals, there is a serious deficit of easily accessible and accurate knowledge. The goal of this book is to help pave the way for a new breed of farm and farmer.

Like most other agricultural enterprises, medicinal herb farms vary widely in size, market, methods used, and products grown. There are some gigantic commercial herb farms in eastern Europe and India, for example, with more than a thousand acres in production, and then there are tiny herb farms all over the world farming an acre or less. Although medicinal herb farms are unique and relatively uncommon as far as farms go, especially in the United States, they are not all that different from many more "conventional" type agricultural operations.

To the casual observer Zack Woods Herb Farm (ZWHF) would likely appear much like that of a small, diverse vegetable farm, with greenhouses and row after row of various species of uniformly planted crops tended by field workers driving tractors and working the earth with their hands or with tools. If that observer were to look closely, though, she might notice that many of the crops grown on this farm resemble plants she would more likely plant as ornamentals in her flower beds rather than plants that would produce food to stock her pantry or feed her animals with.

The vast majority of species we grow on our herb farm are perennial herbaceous flowering plants, several of which are in fact often marketed and grown as ornamental plants. Some of the species of plants we grow could even be thought of as "crossover" plants, meaning they could be considered both food and medicine. Hippocrates was definitely on to something when he famously said, "Let food be thy medicine and medicine be thy food." Garlic is probably the most obvious crossover plant, but there are others, including burdock root, dandelion greens, stinging nettles, and many, many more, including most of the common culinary herbs, which serve dual purposes.

While there are some similarities between vegetable farms and medicinal herb farms, the differences are numerous. Whereas vegetable farms are primarily growing annual crops harvested fresh and brought to market rapidly before they spoil, medicinal herb farms are growing mostly perennial crops, some of which need three or more years of growth before they are harvested. While some herbs are marketed fresh, the majority of the herbs produced for the bulk medicinal herb market are dehydrated and processed before sale. There are many more differences between medicinal herb farms and other more common agricultural enterprises, which we will discuss in later chapters.

While small compared to many farms, sometimes our farm feels enormous to Melanie and me and our main crew of five employees as we try to manage its multiple components, oftentimes simultaneously. ZWHF is a good example of a diverse, small to midscale certified-organic medicinal herb farm. We currently own ten acres of land, and centrally located within this ten-acre tract of cropland and woodland are our home, irrigation pond, and various farm buildings (which we will discuss in chapter 4). In addition to the ten acres of land that we own, we also lease from two neighbors an additional twenty acres. Out of these thirty acres that we either own or lease, our total crop production has varied in size from one acre of crops in 1999 (the year we started

farming) to as many as ten acres of crops currently in production as of the writing of this book.

One may wonder why we don't just utilize our own ten acres of land for crop production instead of leasing additional land, since that is the scale we are currently working on. The answer to that question is threefold. The primary reason we lease more land than we need to plant our row crops on is that we rotate crops in various years from field to field depending on plant and soil needs. Second, we wild-harvest (also known as wildcrafting) various species growing wild on different areas of these biologically diverse tracts of land; finally, we feel it is important to have an extra cushion of opportunity in case we decide to expand our scale slightly. While the average ten acres of land we have planted from year to year is primarily utilized for row cropping, the remainder of the land is either fallow with natural sod cover or planted with soil-building cover crops in preparation for future plantings.

Though relatively rare as far as farmers go, the medicinal herb farmer is not unlike most other agricultural entrepreneurs in business today. We rise at dawn and work hard with the land, limited mainly by our physical stamina and daylight hours. We go to bed to recharge, and we rise again, often faced with new challenges, expectations, and rewards. The rewards aren't always monetary or even acknowledged by others. Sometimes they are as simple as kneeling down on an April morning while the last of the snow melts away and witnessing the unfurling of the first bloodroot flowers to open. Sometimes the rewards can be tremendous as discovering that the ginseng seeds that you planted in your secret hardwood sanctuary germinated and survived the first several years of life in their new home without being devoured by deer and rodents or stolen by poachers. Those ginseng plants will hopefully in time produce a crop of viable seed for you to plant to replace the mature roots you will harvest and market when they reach maturity.

The challenges with growing and marketing medicinal herbs commercially, on the other hand, can be numerous, even downright maddening at times, and demand that we utilize innovative methods to

**Figure 1-8.** Motherwort in full flower and thriving. Photograph courtesy of Bethany Bond

ensure financial as well as spiritual gains while work-ing in partnership with the nature that surrounds us.

## Diversity: The Benefits of Polyculture and Perennial Crops

Despite numerous challenges, there are many ways growers can find success working with healing plants. When we look at the way these plants grow naturally, unassisted by humans, we see incredible diversity in the landscape. In the high meadow on the hill above our farm, for example, the land has been fallow for years. The dairy farm that previously occupied the land is but a skeletal remnant of its former self, the farmer having sold the land for a development that has fortunately failed to materialize. The old pasture is being reclaimed by the so-called "pioneer species,"

the plants and trees that are the first to occupy the niche left open after the cows came home one last time. The brambles were first, their tenacious thorns establishing a natural barbed-wire fence to protect the new residents of this piece of earth. Growing in and among the woody brambles are a dozen or more species of herbaceous plants, fungi, grasses, ferns, and legumes. The trees are starting to stand up proud in the meadow. Now we see what an amazing job the birds, deer, and other winged and four-legged creatures have done in "seeding" this meadow. Wild apple, hawthorn, pine, pin cherry, quaking aspen trees, and other newcomers are establishing the foundation for what will eventually become first an arboreal softwood forest. When these short-lived softwoods die off and their decaying bodies

**Figure 1-9.** Diverse woodland beds of goldenseal, wild ginger, bloodroot, and mayapple. Photograph courtesy of Bethany Bond

contribute to the humus layer, the hardwoods, such as sugar maple, American beech, and yellow birch will dominate and eventually grow to a magnificent climax forest. Underneath all of this, we see the thin layer of humus just starting to form. Next, we come into the topsoil, which is just starting to regain the delicate biological balance it had before it was disturbed. It is playing host to a thriving community of fungi, bacteria, protozoa, and invertebrates. The soil is thin up here on this hardscrabble Vermont hillside, but it was much thicker before humans first broke ground and will, we hope, return again someday to its former glory, but only if the housing development doesn't materialize.

The soil is relatively thin in this field, not only because of the harsh location but also because the dairy farmer, with all good intentions for making a living on this hillside, removed an incredibly diverse woodland, rich with fungally dominant soils, and replaced it with one single species: corn. This monocrop of corn probably yielded a bountiful harvest in the initial years, but eventually, no matter how much manure he spread on the land, he noticed his yields declining from weed, disease, and insect pressure. Then he turned to chemicals to assist him. The chemicals worked well at first by fertilizing the soil, killing insects and weeds, and reducing diseases. But eventually, they too failed to return the land to its high-yielding glory days.

The farmer then realized that the land was tired and needed a rotation, so he planted red clover and timothy grass to graze the cows on and cut for silage.

He may or may not have known that red clover is a legume that takes nitrogen out of the air and helps turn it into free fertilizer via a symbiotic relationship in its roots between bacteria, nitrogen, and oxygen. He probably did notice, however, after a year or two of growing along with seasonal manure application, that the mixed forage crop he planted to replace the corn was healthy, productive, and free of pest and disease problems. This was a start in the right direction.

Diversity is one of the most important factors in a healthy ecosystem. At ZWHF we grow more than fifty different species of medicinal herbs. Although we plant most of these species separately to ease in cultivation and harvest, there is still an incredible amount of diversity within small tracts of land. We plant beds with multiple rows of plants, and within these beds, when possible, we plant several different species side by side to attempt to maintain the ecological balance and diversity that we found here on this land before we farmed it.

We also employ permaculture methods whenever possible by growing primarily perennial crops. A vast majority of the medicinal herb species commonly grown and used today are perennials. This gives herb farmers a great benefit in comparison to many other commercial crops such as vegetables and grains that need to be replanted each year. Here on our farm we generally get three to five seasons or more of growth and harvest from these perennial crops before the plants' vigor wanes, the weed pressure builds, and the plants "show us" that they are getting tired. After we till the old beds in, we often replace them with "green manures" or cover crops consisting of a nitrogen-fixing legume combined with a biomass-producing annual. We also apply compost and mineral powders at planting time and as needed to feed our soil. This semipermanent system reduces labor and materials costs dramatically and allows the soil to maintain a healthy, static balance rather than being tilled every year. This is polyculture, not monoculture. We are simply attempting to imitate nature on a smaller scale, and in so doing, we are maintaining the balance of a healthy and diverse ecosystem while simultaneously maximizing profits.

## Lower Pest and Disease Susceptibility

If you ask most farmers to name the biggest challenges in growing their crops profitably, they are almost certain to list pest and disease issues. There are reasons these challenges are so common on farms and in gardens. Instead of hurling wrathful curses and chemicals at these culprits, perhaps we should step back and take a look in the proverbial mirror. Virtually all species of food, fiber, and fuel plants that are grown and used today came to us from plant breeding. These plants have been crossbred by humans for thousands of years to produce new varieties with desirable traits and characteristics. For example, a flavorful and disease-resistant but relatively low-yielding tomato is crossed with a bland-tasting, high-yielding tomato, with the end goal being to improve upon the flavor, yield, and disease resistance of a single variety. Other species of plants are genetically modified by adding specific genes into a plant in an attempt to "improve" it.

Plant breeding has certainly held an important role in improving plants to clothe, fuel, feed, and shelter us, but it has come at a cost. Whether it be through classical breeding or genetic modification, many of the varieties of plants that have emerged from this technology have suffered from weaknesses such as pest- and disease-damage susceptibility and possibly even declining nutritional value. A study published in the *Journal of the American College of Nutrition* in 2004, entitled "Changes in USDA Food Composition Data for 43 Garden Crops, 1950 to 1999," compared nutritional analyses of vegetables done in 1950 and in 1999, and found substantial decreases in six of thirteen nutrients measured, including 6 percent of protein and 38 percent of riboflavin. Reductions in calcium, phosphorus, iron, and ascorbic acid were also found. The study, conducted at the Biochemical Institute of the University of Texas at Austin, concluded in summary: "We suggest that any real declines are generally most easily explained by changes in cultivated varieties between 1950 and 1999, in which there may be trade-offs between yield and nutrient content."[5]

High pest and disease susceptibility are often the bane of the commodity and veggie farmer but fortunately rarely challenge the medicinal herb farmer who employs polycultural methods. Over 90 percent of the plants commonly used for medicinal purposes are cultivated, domesticated versions of wild plants. These domesticated plants are still relatively indistinguishable from their wild counterparts. However, oftentimes, through what is known as selection, these species of plants have been "improved" upon. Selection can be natural or unnatural. An excellent example of natural selection is the Tibetan Snow Lotus. "The height of this plant at flowering has nearly halved over the past century as a result of the flowers being picked for use in traditional Tibetan medicine."[6] This change, although cased by humans, was not purposeful and is therefore referred to as natural selection. A good example of unnatural selection (also known as selective breeding) is found in the popular medicinal herb German chamomile. Seed growers established new varieties of this herb, carefully selected to produce seeds that grow plants at a uniform height to allow for ease in mechanized harvesting. Although these new varieties of chamomile are considered an improvement because of the standardized growth, they are still virtually the same species as the original wild chamomile.

Fortunately for medicinal herb growers, almost all of the plants we cultivate have been altered little, if at all, from their wild ancestors, and as a result the incidence of pests and disease problems is relatively low. These plants have evolved over tens of thousands of years to be naturally resistant to these threats. In fact, many of the medicinals we grow, such as yarrow (*Achillea millefolium*), valerian (*Valeriana officinalis*), and angelica (*Angelica officinalis*) are known as "insectaries," plants that attract beneficial predatory insects that parasitize and prey on malevolent insects. Thus, they are incredibly effective at maintaining the balance of beneficial and potentially harmful insects in the landscape, farm, and garden.

We see very little disease or injurious insect pressure on our farm; therefore, we don't need to rely on pesticides or fungicides. These treatments can be costly, not only to the farmer's bottom line but also to

**Figure 1-10.** Close-up of angelica blossoms. This is an excellent insectary.

the farmer's health, as well as the health of beneficial pollinating and predatory insects and the surrounding environment. The reasons for lower insect and disease pressure on polycultural herb farms are relatively simple: diversification, thoughtful planting of insectaries, and growing of "wild" plant species. This all adds up to a healthier, more profitable enterprise. It allows us to focus on growing healthy plants while maintaining diversity in our landscape. It is a win-win situation for the humans, the plants, and the insects and other creatures that dwell on this land and its air and water.

## Dehydration Reduces Perishability

Perishability is another challenge that many "conventional" farmers face. Whether it be milk, vegetables, flowers, fruit, meat, or even some grains, it often comes down to a race against time to get the food into customers' kitchens before the bacteria and fungi come in and spoil the party.

**Figure 1-11.** Dried herbs ready for shipment.

A majority of the bulk medicinal herbs produced and utilized for the manufacture of herbal products and teas are dehydrated. This process requires that shortly after harvest the fresh herbs be brought to a drying facility, where they spend a period of time having their moisture removed. Directly from there, they are processed and packaged. Once packaged, as long as they are stored in cool, dark, airtight containers, they can be warehoused for a year or more, as is the case with leaf crops and blossoms. Root crops can be stored much longer in general, often three years or more. After this period of time a very gradual decline in quality takes place due to oxidation. This lengthy shelf life gives the herb grower ample time to make the sale and deliver the product. It gives the herbal product manufacturer time to make the herbal product without having to refrigerate or freeze the bulk herbs while in transit and storage. For the herb grower this lengthy shelf life helps extend inventories for year-round retail sales.

Another benefit to dehydration is the amount of water weight that is removed from the product, saving on physical labor, shipping costs, and sore backs.

## The Savvy Consumer

The typical herbal product consumer is relatively well-educated and knowledgeable about uses for specific herbs and is most likely more concerned about high quality than about affordability when it comes to purchasing bulk herbs or herbal products. In general, these educated consumers want their products to be free from harmful chemicals, certified organic when available, and sourced as locally as possible. They want to have a relationship with the company or farm they are supporting. Herbal product retailers are often very knowledgeable about the products they sell and play an important role in guiding consumer choices. "The natural channel has always been more educational and missionary in its sales approach," according to Mark Blumenthal, founder and executive director of the American Botanical Council (ABC). "A lot of retailers have talked about this, that they are acting like nutritionists for their customers."[7] People are becoming more aware of and concerned about what they are putting into their bodies. For years we trusted our government to ensure the safety of our food and drugs through such agencies as the Food and Drug Administration (FDA) and the United States Department of Agriculture (USDA). We also trusted pharmaceutical companies to produce medicines for us that were well researched, well manufactured, well tested, and proven safe. All this has changed in a relatively short period of time. We see the recalls and hear the horror stories, and heck, if we haven't fallen victim to the negative side effects of these products ourselves, we definitely know people who have become ill or worse from putting something in their bodies that they thought had been tested and proven to be safe. As far as pharmaceutical drugs go, that can also be a roll of the proverbial dice. When the list of possible side effects sounds worse than the original symptoms, you have to ask yourself if it is worth the risk.

For years we have trusted the herbal products industry to provide us with safe herbal products. We never really had much reason to doubt that what was stated on the product label was accurate and that what was in the bottle was safe and effective. This too changed dramatically as examples of adulteration of ingredients used in herbal products have come to light.

In response to product adulteration and many other concerns, in 1994 the federal government initiated the Dietary Supplement Health and Education Act, or DSHEA. This act defines dietary supplements (including herbal products) as food, not drugs. This saves manufacturers from having to weather the FDA's drug approval process that can cost many millions of dollars for a single product approval. However, this process has been challenging for small herbal product manufacturers, who are facing the high costs associated with becoming compliant. DSHEA mandates that supplement manufacturers follow Good Manufacturing Practices (GMPs), as established by the FDA. Another requirement is that product labels be "truthful and not misleading" and state the name and quantity of all active ingredients.

While many companies were doing this long before DSHEA and GMPs, not all were, and that was often problematic. Some in the industry were more focused on profit than quality and safety. Consumers often lacked confidence that the products they were buying were produced with high-quality botanicals and were efficacious. The DSHEA, and specifically the GMPs, rapidly changed the way herbal product manufacturers do business. This may actually benefit the herb farmer in two ways. First, it could help to open possible markets that were not there before. Companies will need to look for growers that can meet these new standards. The new regulations place strict guidelines on producers to test the purity, identity, and safety of ingredients contained in their products and to provide accurate documentation of these tests. "With GMPs in full force several companies are realizing that the supply chain has shrunk dramatically due to ingredients that cannot pass the identity and quality GMP requirements," says Roy

Upton, president, American Herbal Pharmacopoeia (AHP), Scotts Valley, California. "In other words, ingredients that used to readily pass manufacturer specifications are now failing when proper identity and quality tests are applied."[8]

While these new regulations are anything but a panacea for the herbal product manufacturer, they may benefit the herb farmer in a second way by helping to restore consumer confidence in the herbal marketplace; consumers will increasingly trust that they are getting what they pay for and that it will be potentially beneficial to their health, not harmful. This can help to pave the way for success in an industry where forming relationships and trust with customers generally pays dividends through brand loyalty.

## Profitability and Markets

As with most crops grown for the marketplace, the relationship between supply and demand is one of the primary factors in determining profitability. We have seen what the consequences can be when the latest, greatest herb hits the market. Demand soars, and growers jump on the bandwagon to try to get in on the action. Planting large acreages according to market hype can be a recipe for disaster because of the high probability of the market's being flooded by other like-minded zealots. Growers who attempt to "read the tea leaves" and plant large acreages purely according to speculation also do so at a significant risk.

The market for raw bulk botanicals is now, and probably always will be, volatile. When people ask us how we can afford to do business in such a volatile marketplace, we tell them, "We try to grow for sales instead of speculation, and we diversify." At ZWHF we grow over fifty different species of medicinals, instead of only focusing on the ones that may seem, at first glance, to be the most popular. Yes, there is certainly speculation involved in guessing how much of what to grow, but making "educated" guesses has been the key to our success. Almost every seed we plant in the spring of a given year is planted according to what we are already in contract to provide on the wholesale end or is based on careful analysis of projected sales on the retail end. Every year we can count on at least one crop failure or deal breaker, but with so many species in the rotation we hedge our bets against the inevitables and unforeseens.

## Mass versus Niche Market

We like to refer to the medicinals market in terms of the *mass market* and the *niche market*. The mass market is the larger-scale side of the industry, which consists primarily of high-volume bulk herb and

**Figure 1-12.** Hannah bagging herbs for mail orders.
Photograph courtesy of Kate Clearlight.

herbal product retailers. This is where we deal in the wholesale realm by growing larger quantities of bulk herbs for larger companies, which we usually sell at a discounted price based on volume. There is almost no speculation in growing for wholesale accounts because almost all of these sales are contracted for at the beginning of a given growing season. The mass market makes up the bulk of our gross sales, and the higher sales volume helps to offset lower profit margins. It is crucial that we keep our costs of production to a minimum in this market because of the thinner margins between profit and loss.

The niche market side of our operation is where the profit margins are generally higher but volume is generally lower. Here is where we have to make educated guesses as to how much of which species to grow. These are primarily mail-order, telephone, Internet, and other direct-retail sales to individuals of dried and fresh herbs and value-added products such as herbal teas and live plants. This is where we endeavor to increase sales volume and to achieve higher profit margins. A vast majority of the niche market customers we sell to live within a hundred miles of our farm. This is beneficial on so many levels to us, our customers, and the earth and its natural resources.

Fortunately the medicinal herb industry is currently experiencing strong growth. In 2010 the botanical and natural ingredient export trade reached approximately $33 billion, according to the Market News Service (MNS) "Medicinal Plant and Extracts" report, published in the MNS December 2011 bulletin.[9] By 2015 the international herb supplement and remedies market is expected to reach $93 billion, according to a report by San Jose, California–based Global Industry Analysts, Inc.[10]

In 2011/2012 dollar sales increased approximately 8 percent for herbal formulas, 9 percent for teas, and 13 percent for herbal singles and flower essences, according to SPINSscan data, for the 52 weeks ending March 17, 2012. The ten top-selling products addressed inflammation, prostate health, immunity, adrenal support, detox, stress relief, sleep, and libido.[11]

Of the burgeoning health-minded U.S. population, 23 percent use botanicals, according to a recent survey commissioned by the Council for Responsible Nutrition (CRN), Washington, D.C. These statistics and others show that the medicinal herb market, as a whole, is strong. The "must-have herb of the month" may change, but overall there is a solid foundation of demand for these products here in the United States on which to build a medicinal farming industry. We would advise growers new to the herbal industry to establish solid markets before committing to large plantings. They should look toward growing herbs with long-term favorability in the herbal marketplace while striving to diversify and innovate with new markets, new species of interest, and value-added products. So while we simultaneously discuss the risks of market volatility along with the promise of steady growth in the herb industry, let's address some important considerations medicinal herb farmers should take into account before venturing into the great unknown.

**Figure 2-1.** A glorious field of *Echinacea purpurea* at ZWHF.

# CHAPTER 2
# Size and Scale Considerations

One of the most important considerations when starting a medicinal herb farm or incorporating herbs into an existing agricultural enterprise is how much acreage to plant, and how much land to buy or lease. Although there is no "one-size-fits-all" answer to these questions there are many factors to consider when deciding on scale. Bigger is not necessarily better in farming. We have seen the effects gigantic factory farms have had on the environment and the smaller businesses around them. Sure, some statistics demonstrate that large corporate farms are on the average more profitable than small farms, but that profitability often comes at a cost to the environment and sometimes even the health of its consumers.

The industrial agriculture model is set up to produce enormous volumes of low-cost goods with high profit margins, with generally little regard given to the quality of the end product. Economies of scale tend to favor this mass production model because most of the production costs can be spread out through massive volumes, thereby decreasing the cost to produce each given unit. Large farms can also benefit from agricultural subsidies and greater access to capital and mass distribution channels. Industrialized agriculture has basically taken the notion that farming is a noble and spiritual pursuit and turned it upside down in order to shake profits from it at any cost. Instead of working in harmony with the natural environment and its inhabitants and increasing diversity and beauty in the landscape, factory farms tend to work counter to natural processes. By monocropping huge swaths of land, spreading toxic chemicals, utilizing questionable labor practices, and producing cheap, low-quality products, these operations have given the many faces of farming a somewhat sinister side.

Small farms, on the other hand, have many advantages in the marketplace because of their ability to capitalize on higher-quality products, to develop unique or "niche" markets, to promote healthy and sustainable practices, and to build and maintain relationships with customers. Small farms succeed by focusing on increasing value rather than volume and are often more profitable when analyzed on a per-acre basis because of their ability to diversify and maximize the potential of every inch of ground they farm. Small-farm owners tend to operate their businesses as an integral part of the natural environment they inhabit by working with rather than against natural processes. They also tend to view profitability in terms of such factors as quality of life and contributing to the greater good rather than just viewing it in terms of mere dollars and cents.

**Figure 2-2. Combine harvesting wheat.** Photograph courtesy of Lars Plougmann

As Vandana Shiva said in *Earth Democracy*: "Living economies are based on working for sustenance. They put human beings and nature at the center. In living economies, economics and ecology are not in conflict, they are mutually supportive."

When deciding on how much acreage to plant, one of the most important considerations is to decide how you are going to enter the marketplace and in what capacity. Many people considering growing medicinals for the first time ask which comes first, the market or the product? This is one of the toughest questions to address because when it comes down to it the real answer is you can't really have one without the other. To establish markets, growers need to have a product for potential buyers to sample because it is highly unlikely that a buyer will purchase or contract for bulk herbs sight unseen. On the other hand, we wouldn't recommend growers spend a lot of time and money to plant large amounts of crops that may not sell or may take so long to sell that the quality deteriorates.

One recommended solution to this dilemma is to start out small, spend little money, grow several species in small plots, build a small drying facility, learn how to produce a high-quality product, then shop it around. When markets are secured, grow more than you think you will need because the odds favor the house when it comes to gambling on yield projections. You may as well throw a lot of what you have heard or read about potential yields into the compost pile because there are so many variables in the mix with weather, fertility, and maturation of perennial crops, as well as many other potentially complicating factors. Each farm is unique, so yield values are very subjective and are best used as a starting point for your own trials and data collection.

The answer to the question, "How big should my medicinal farm be?" should begin to emerge when you consider what your primary market(s) will be and how much you want to manage in terms of crops, land, employees, equipment, bookkeeping, and so on. Sorting out all of those considerations without having experienced the actual management realities can be daunting, so again we advise that new farmers start out small with perhaps an acre or two in production yet have the potential to expand as you wish and as markets are developed.

At ZWHF we entered our first growing season in 1999 in a purely experimental manner. Melanie and I had a fairly good sense of which species of herbs were in demand from our experience owning Sage Mountain Herb Products, but when we sat down and made a list of the herbs we knew we could sell, we felt a bit overwhelmed. There were over one hundred species of herbs on our "good sales potential" list. The thought of growing and managing that many species, especially during our first year, seemed like a stretch, so we began paring the list down. Species that were considered questionable for growing in northern Vermont's short growing season (USDA hardiness map zone 4b) were the first to be eliminated from the list. Others that we deemed "risky" because they required multiple (more than three) years of tending before a first harvest could be obtained (*Astragalus membranaceus*, for example) were also removed from consideration for the first year. A few species such as American ginseng that were not only risky to grow but also required more than five years to mature along with very selective growing conditions were also scratched off the initial list to be added in subsequent years. After much careful thought and deliberation, we finally settled on a list of sixty species that we would attempt to propagate, grow, process, and market. We decided to plant all sixty species in single rows of various lengths on one acre.

In hindsight that list should have been pared down significantly that first year because while some species germinated and grew well, even in our poor soils, others took so much time and effort just to get them to survive (and some didn't survive at all) that it took valuable time away from tending to the plants that had good potential. We were so naive on so many levels.

Our first rays of hope that first year came when the *Calendula officinalis* started blooming profusely. We were so excited when our first harvests started coming in, filling our baskets with bright orange blossoms. It wasn't until we weighed the dried blossoms that one of the cold, hard realities of herb production started coming to light. Most plants, like people, are composed mainly of water in terms of weight, so when you remove that water from plants, what's left is usually only about 20 to 30 percent on average of the original fresh weight. What seemed like a bounty to us in volume at first glance turned out to be meager in terms of dried weight after processing. "I guess we have to plant more next year" became a common refrain during the early years.

The impetus for us to keep farming in spite of the challenges came primarily from the feedback we were receiving from the customers who purchased the herbs we produced and secondly from our intense desire to see our dreams through to fruition. People who were used to buying low-quality herbs imported from foreign countries were really excited to be able to purchase locally grown, high-quality, certified-organic herbs grown, tended, and harvested by hand from people on a farm they could identify with. Suddenly one of our biggest challenges was producing enough herbs to keep up with the demand; this challenge still remains after fifteen years in business.

Luckily, during the first five years we both had jobs off the farm, so we weren't dependent solely on farm income for survival. We weren't making a profit

farming yet. However, we gradually began to realize that we had something really special going and that if we worked hard enough and smart enough, we could make a go at this career full time without having to work off the farm. In year five we finally started seeing some return on our investment and realized a small profit at the end of that year. This was the year that I stopped working off-farm and became a full-time farmer. Melanie continued her work as a middle school educator before finally leaving that profession to become a full-time farmer in 2013.

We finally realized our dreams of becoming full-time herb farmers by starting out small and expanding our farm gradually. We decided early on to avoid borrowing money and going into debt to expand our business rapidly. Instead, we invested any profits we made back into the business while making capital improvements and expanding slowly only within our means. Starting out small worked well for us and many others like us, but some farmers have certainly had success with starting out on a larger scale and entering into the profession going full speed ahead.

For those wanting to jump in deep right off the bat and start growing full time on a larger scale, go

**Figure 2-3.** Small trial plots of herbs during our first year on the farm.

for it, but be prepared by establishing a good dry storage facility for the surplus herbs you produce and have some good marketing strategies planned for the off-season to sell off existing inventories and establish new contracts.

Perhaps you already own or have use of an existing farm and want to diversify your offerings. That is a great way to start. Our friends and mentors Andrea and Matthias Reisen of Healing Spirits Herb Farm in Avoca, New York, started out dairy farming and realized there was more and easier money to be made than by milking cows. They sold their cows and used the funds to retrofit their barns and equipment to produce medicinals. They had no experience growing medicinals, but like us, they had a passion, and once they realized they could make a living doing what they loved, they never looked back. They are in their late sixties now and still happily and profitably farming herbs to this day and say that was the best move they ever made. They started out small the first year, using a push lawn mower and walk-behind rototiller to establish planting beds on approximately one acre, and have evolved from there a little at a time to their current size of approximately ten acres in production. They were our mentors early on and taught us a lot of what led to our success with farming herbs.

If some of this sounds overwhelming or scary it shouldn't be. Like almost all new business ventures, there is a period of time (generally three to five years) when you can expect to spend more than you earn. Take solace in the fact that if you work hard to establish a quality product and a good reputation, in due time you will be justly rewarded for your patience and efforts. Growers would be wise to start out small, yet have more potential land available to increase scale as sales warrant.

## Leasing versus Owning Land

Let's face it: Buying enough land to make a go at commercial farming can be a major investment. Even the process of searching for and finding the right piece of property can become a job in itself. I

can't even begin to count the number of times that I have heard prospective farmers say, "I would love to own my own farm, but I just can't afford to buy land." Gone are the days when old farms with quaint farmhouses, big barns, and overgrown pastures sold for bargain basement prices. Yes, affordable land can certainly be found, but in general you get what you pay for. Here in Vermont and elsewhere, wooded, hilly, ledgy, or swampy land can easily be found for under $1,000 an acre. Wooded land can be cleared, but the costs of clearing it well enough to plant field crops can be prohibitively expensive. One benefit to wooded land is that agroforestry crops can be grown in the woodland (we will discuss this further in chapter 5). Swampy land can be drained but often at exorbitant costs to the farmer and with a potentially negative impact on the downstream ecology. Hilly land can be farmed but not easily and is prone to soil erosion unless it is terraced well. Ledgy or extremely stony land usually doesn't offer enough soil to grow crops on and can be extremely hard on equipment and personnel.

Owning land can be incredibly rewarding (most of the time), but for those who can't afford to buy land, there are options. We own our farm, which is great, but more than half of the land we farm we lease from our neighbors. Leasing land is a viable option that should be considered by anyone eager to make the leap to owning her own farm business while saving up to purchase her own farmland. Many landowners with level, cleared, tillable land would rather have a farmer use their fields for agriculture rather than having to pay to have it mowed to keep it open.

A typical arrangement for agricultural land leases in rural areas usually doesn't involve much money; rather it is often the exchange of use of the land by the landowner for keeping the land well maintained and open by the farmer. Here in Vermont as well as in many other states there are tax incentives for landowners to keep their land undeveloped. To take advantage of these programs, the land often must be in "current use," which requires an agricultural, timber management, or agroforestry plan to be established by a qualified consultant. This is a huge

## The Pros and Cons of Owning versus Leasing Land

*Owning Land: Pros*

- The sense of ownership is invaluable.
- It is one of the best investments in today's economy.
- It is almost guaranteed to increase in value if well maintained.
- You are free to do what you wish with your land (for the most part).
- When you finish paying the mortgage it's yours.
- You can leave it as an inheritance to family members, thereby ensuring its legacy.

*Owning Land: Cons*

- It is expensive to buy.
- It is expensive and laborious to maintain.
- Property taxes can be expensive.
- Return on investment can be a very lengthy process.
- If you decide you don't want to own it anymore, it can take a very long time to resell.

*Leasing Land: Pros*

- Leasing land is generally much more affordable short-term than owning land.
- You don't have to pay property taxes.
- Most landowners who have surplus land that they aren't using benefit from farmers using their land (a mutually beneficial arrangement).
- If you learn that farming isn't your thing you can discontinue the lease and move on.

*Leasing Land: Cons*

- Any money you invest in the leased land isn't benefiting you long term.
- The landowner has final say in what goes on.
- Owner/lessee relationships can go sour fast.
- Lots of variables can lead to uncertainty (e.g., higher risk factor).

tax incentive for landowners to lease their land to farmers and is mutually beneficial to both parties. As in any financial agreement, it is best to have a clear written contract detailing the lease arrangement.

# Hiring Employees versus Doing It Yourself

Most people contemplating owning a business start with a fairly good idea about what the scale of their business will be, how much time they want to spend working, and how many employees they would like to have, if any. We sure had a pretty good idea about all this when we started our business, but when we look back, in retrospect our original plan has been altered significantly. Our plan in 1999 when we started farming was to keep the farm on our own ten acres and not get any larger than that. We knew we would be working incredibly long hours every day during the growing season, but we dreamed about having the winters off to ski. We had hired our friend Brian to work for us, and he was the only employee we thought we were going to have.

That was fifteen years ago, and a lot has changed since then. We now lease two additional tracts of land from neighbors, and while we do get to ski in the winter, we usually only have time to sneak in a couple of hours in the morning before work. We now have one full-time year-round, two full-time seasonal, and several part-time seasonal employees, including Brian. All this changed from our original plan in our attempt to meet the demand for our products while increasing the size of the operation in a sustainable and manageable way. We are still increasing our production and are now faced with the question of when to stop expanding. The answer to that question will likely come when we sense that the quality of our product and our quality of life could be compromised; then we hope it will be obvious to us that we have reached our limit.

**Figure 2-4.** 2013 ZWHF farm crew.

**Figure 2-5.** Jeff leading a tour for students from the Vermont Center for Integrative Herbalism. Photograph courtesy of Larken Bunce

Small farms can definitely be run without employees for those not interested in managing other people. We know of several very successful one- or two-owner farms with no employees who have found their niche and are quite comfortable working independent of outside assistance. As far as the scale of farming acreage per person, a common guideline is that one person can generally manage up to two acres of row-crop-type herb production independently. Anything larger than that, unless ultraefficient weed control and harvesting is utilized, could potentially become an unmanageable situation. With herb farming the limited window of opportunity to harvest herbs at their peak potency while having favorable weather can be extremely challenging. Sometimes getting behind on harvests because of inclement weather, as well as the time it takes to dehydrate crops, can have a domino effect on several other aspects of the operation. It would be wise for independent growers to consider having backup part-time help from friends, family, or temporary employees to avoid getting caught in a perpetual delay and have crop health, product quality, and sanity suffer as a result.

In addition to our regular employees, we have had excellent experiences with hosting college interns and farm apprentices each year since we started farming. These options, along with the WWOF (Willing Workers on Organic Farms) program, are excellent alternatives for small farm owners to consider, be sure to understand state and federal labor laws. Our intern program at ZWHF has been mutually beneficial for us and for the interns, who are required to fulfill college credit requirements by working for a set number of hours in a "hands-on" environment. College internships are customized to the individuals based on their credit requirements and time availability. Our apprentice program involves the exchange of knowledge and credit toward plants and herbs for time working on the farm and is also customized according to individual as well as farm needs. We interview several interns and apprentices each year and choose the individuals that we feel best fit our requirements and needs.

## Herb Pharm: A Large-Scale Farm and Product Maker Committed to High Quality

*Founded in 1979 by our friends Sara Katz and Ed Smith, Herb Pharm is one of the United States' leading producers of high-quality botanical extracts. In 1993 Sara and Ed purchased a lovely eighty-five-acre farm called the "Pharm Farm" in the foothills of the Siskiyou Mountains in southern Oregon with a goal of growing botanicals for use in the company's herbal products while restoring fertility to the land, which is once again becoming abundant with a diverse assortment of herbs. Native medicinal herbs preexisting on the land include trillium, lomatium, Oregon grape, spikenard, wild ginger, elder, and others. Through replanting of native shrubs and trees they have worked to restore the riparian zone around two creeks to reduce erosion and help renew salmon habitat in the watershed, along with creating new shade beds for establishing native medicinal woods-grown herbs.*

*The Pharm Farm currently produces approximately twenty to thirty tons of botanicals per year from sixty acres of certified organic cropland. This volume equals approximately 60 percent of the raw material used in the manufacture of Herb Pharm products. Herb Pharm has created infrastructure that allows them to produce large volumes of herbs while maintaining quality and caretaking the land. They also use both mechanization and manual labor to achieve these goals.*

Matt Dybala (farm manager) and
Michael Gillette (director of marketing) write:
Working steadily over the last five years to wean ourselves from harvesting plants from the wild for our herbal products, we are currently growing about eighty different species of crops to use in making our herbal extracts. The development and maintenance of this farm requires an immense amount of work. Besides a dedicated staff of four to five specialized workers and ten field workers, we also offer a work-study program called the HerbaCulture Intern Program from March through June taught by the Herb Pharm staff, herbalists, and farmers of southern Oregon. Well over a thousand students have been through the program, many going on to become full-time organic farmers, registered herbalists, naturopathic physicians, and various other roles in the holistic health and nutrition field.

While many aspects of our work are mechanized (cultivation between rows, harvest of root material, sickle-type blades used to harvest long-cut stem herbs that are thick in diameter, and grain combines for large seed quantities), there is some work that is best done by hand. We use hand harvesting for flowers, low-growing aerial harvests, and crops with a high percentage of weed competition. We chop roots by hand to clean soil debris from the crown and remove dried leaves and flowers from stems by hand rubbing them over screens. Hand-processed herbs consistently result in a higher percentage of clean whole material and retain quality. Specifically in our operation, small quantities and quality results mean we will continue to process by hand. In our farming experience, large quantities of machine-harvested material complicate our final product achievement of consistent quality results. Mechanized farm equipment is not yet capable of human judgment when material selection is critical to your ingredients and meeting high-quality specifications.

Having our own farm ensures consistency and quality from seed to shelf. This includes procuring organic seed, planting and cultivating in appropriate soil conditions, harvesting at the optimal time of season and day, then carefully drying and storing for future use or direct delivery of the fresh herbs to our production facility for immediate extraction.

**Figure 3-1.** The view from our office window. The work in the office and the field build on one another. The farm's success is dependent on both.

# CHAPTER 3

# Thinking Like a Business Manager

Many farmers, Melanie and I included, get into farming because in general, we prefer physical work over static work. We like to be outside interacting with the natural world while actively engaging our minds and bodies. We have chosen this profession over the typical 9-to-5 desk job with the predictable paycheck because most of us can't bear the thought of having to sit physically still indoors for that long day after day. There does come a time, however, that even the farmer has to face his or her "inner demons" by coming inside and sitting down at the desk or computer. It is an unavoidable aspect of doing business and one that can help to make or break the bottom line. The most productive and innovative grower in the world is extremely limited if he or she can't understand and manage the paperwork side of the business. Sure, we can hire someone to do most of the paperwork for us but at the very least, we need to have a thorough understanding of what our business looks like from the inside out rather than just from the outside in.

## The All-Important Business Plan

We had avoided establishing a formal business plan at ZWHF for thirteen years, which was thirteen years too many. In the beginning, as we contemplated buying a farm and growing herbs, numerous people had recommended that we write a business plan. Sure, it sounded like a good idea, but when it came down to really engaging in the process, we were so focused on establishing the fields, growing the plants, hiring staff, and developing and refining the equipment and infrastructure that it kept getting put off. Every year, generally in the fall, as we sat down to review the previous season's numbers we would say, "Let's do the business plan this year." This got to be a yearly mantra for us without the goal ever materializing. Each year we repeated the same old refrain: "Well, we had a pretty darn good year, and we kept good records, so let's focus our time and energy on increasing production rather than crunching numbers."

In the fall of 2010 we purchased a book titled *The Organic Farmer's Business Handbook* (Chelsea Green, 2009), by Richard "Wiz" Wiswall. We had visited Wiz's farm several times through the years and were impressed by his knowledge and methods. He seemed willing to take the time to share his knowledge and offer encouragement. In fact, we began to model a lot of what we were doing on our farm based in part on the methods and equipment we had seen him use on his farm. He had been incorporating a few species of medicinals into his vegetable production and had developed some really innovative techniques.

When his book came out we eagerly dove into its pages, excited to have access to more insights from his successful formula. When we read his book, one thing that he emphasized really stood out. He said, "Farm for profit, not production." Those five simple words really held a mirror up to our self-limiting view of focusing on increasing production above all else. Sometimes it's easy to get so caught up in a concept or pattern that it becomes hard to think outside the box until someone else opens the box to let some light in.

Farming for profit rather than production, as Wiz had recommended, gave us a new paradigm to work with. We realized it was time to establish a formal business plan, which would force us to do an in-depth examination and analysis of the strengths and weaknesses of our farm and to make a "road map" to help lead us into the future. We applied for assistance through a federally funded program called the Vermont Farm Viability Enhancement Program, got accepted into the program, and received grant funds to hire a consultant to help us establish a business plan and increase the future viability of our farm. When we got the list of consultants who were available to assist farmers, we were excited to see Richard Wiswall's name on the list of consultants, so we asked to have him become our advisor. Our request was granted, and for two growing seasons, Wiz came to our farm once a month, helped guide us in the process of writing our business plan, and offered us a wealth of invaluable suggestions on how we could improve upon everything from organizing our office space to revamping the plant and row spacing in our beds to utilize our land more efficiently.

Writing a business plan was a lot of work. We had compiled thirteen years of good data but hadn't really done much with the data, so this was our opportunity to put reams of good information to use. It is important to say that in all of the investments in equipment, infrastructure, and time we have made on the farm, writing a business plan has probably yielded the greatest return dollar for dollar. We gained tremendous insights on aspects of our business that immediately improved both our production and profitability.

Writing a business plan need not be a laborious, tedious, lengthy process. Farm owners can customize it to their unique situation. The business plan can and should be an evolving exercise that gets updated from time to time as the business grows and changes. There are several resources for business plan guides and formats. I highly recommend the format that Richard Wiswall presents in *The Organic Farmer's Business Handbook*.

# Managing the Paper Trail

We often joke about the paper trail and data management system being the "not-so-sexy side of farming." Everyone who comes to the farm wants to dig into the soil, pick calendula blossoms, plant seeds, or drive the tractor, but to date we haven't had any visitors who want to belly up to the computer and look at spreadsheets. The outside work and time spent with the plants is a huge draw for folks. We get it. That's why we got into farming, too. However, as more farmers analyze their profitability, conversations are shifting to include a deeper discussion about what the numbers tell us. Not only will we be talking soil science and medicinal uses of plants, we will also be talking yields, cost of production (COP), and crop budgets. The paper trail is crucial to establishing a "road map" and making accurate assessments of what to grow, how much to grow, what is profitable, and where advancements can be made in efficiencies. At the end of the day good data management is what will help you make important decisions, possibly/probably increase your profitability, and keep you doing what you love doing—farming!

What data should herb farmers keep? Good question. It's not easy to find a lot of data on herb farming, especially when compared with vegetable and dairy farming. For most herb farmers the data will need to be generated from your own work because there are very few resources available, and every farm is unique. In the beginning years we were hard pressed to find information about herb cultivation for our region, realistic yield projections, or useful planning guides. So we planted our first fields and set out to learn through hands-on experience based on trial and error. We were really gung ho, and to be perfectly honest, we kept so much data at first that it wasn't entirely helpful. In fact, tracking every little piece of data drove us both crazy, and in the end we didn't know what to do with it all.

But as the years rolled by we began to notice patterns and started to see what was helpful to track and what we could let go of. After years of data keeping, then through our work with Wiz, it became apparent that one of our primary issues in the beginning was that we weren't clear on the questions we were trying to answer. Once we had clear questions the data we collected sorted itself out, and the answers became clearer. Then the issue became keeping accurate data and finding timely ways to analyze and use it for decision making.

The key questions we ask fall into these categories: production, sales, and profitability. The data that we keep for production and sales are later utilized to answer the profitability questions. As your farm grows and you get a foundation of information, you may ask different questions and need to collect different data sets. For example, some data we collected in the early years we no longer need to gather because we "get it" (e.g., dry-to-fresh ratios for each species). We collect some of the more variable data every year to monitor our profitability and cost of production (e.g., inputs and labor cost). Other questions we are asking for the first time and are developing ways to gather the data (e.g., should we develop another tea blend? If so, what kind? Do value-added products significantly help our sales?).

Table 3.1 is a sampling of questions that we've asked over the years and ones that would be useful to consider as you begin outlining your data management system.

Once you've established your guiding questions, we recommend setting up a data management system that is simple to put into operation and one that you can commit to. Our data system utilizes the strengths of our team and works for the scale of our farm. We designate responsible employees to help collect data, schedule frequent check-ins, and meet together to review results. The data we have found helpful to collect focus on five main areas, which we'll cover in the following sections:

1. Seeding and Vegetative Propagation (Cuttings) Data
2. Cultivation Data and Crop Budgets
3. Harvest and Drying Data
4. Field Maps
5. Labor, Sales, and Expenses

## Seeding and Vegetative Propagation Data

Seeding and vegetative propagation data is kept on a calendar in the greenhouse. Each season we have two to three people involved with seeding and

**Table 3-1.** Guiding Questions

| Production Questions | Sales Questions | Profitability Questions |
|---|---|---|
| How much land does each crop utilize? | What herbs do people/businesses want and in what quantities? | How much does it cost to produce each crop? |
| What are the inputs needed to grow each crop (soil amendments, seeds, water, etc.)? | How much of each herb do I produce in a year? | How much does it cost to produce each value-added product? |
| How much time do I spend planting, weeding, harvesting, drying, and processing postharvest? | How much of each herb do I sell in a year? Do I have shortages or surpluses? | What are the "overhead costs" on my farm? |
| What are the crop yields for a given field in a season? | How do my sales and/or herb availability fluctuate throughout the season? | What are my biggest expenses for each crop, and are there ways to lower my cost of production (COP)? |
| How much fresh plant material is needed to get one pound dried? | When do harvests occur? What are the yields? | What efficiencies (cultivation techniques, machinery, reallocation of resources) can I incorporate to lower cost of production or overhead? |
| Are there any special considerations that I need to be aware of when growing/harvesting/drying a specific crop? | What are my top selling herbs or value-added products? | Are my wholesale and retail prices set correctly to reflect COP and profitability margins? |
| | Where are my markets: wholesale, retail, fresh, dried? | |
| | Who are my top customers? | |

| Species Seed | Source/Year | Tray Type & Quantity* | Time | Person Seeding |
|---|---|---|---|---|
| Tulsi (Ocimum tenuiflorum, syn. O. sanctum) | Horizon Seeds 2013 | 3 × 128, 12 × 72 | 30 min | Hannah |

\* 3 trays of 128 cells and 12 trays of 72 cells

**Figure 3-2.** Sample seeding and vegetative propagation data entry.

propagating cuttings. They record their data on the calendar as they work. We record the species, seed/cutting source and year, quantity of seeding/cuttings, time it took to seed/make cuttings, and the name of the person doing the work in case questions arise. When seeding we also like to know the size, type, and number of flats used for each species so we can track how much greenhouse space and potting soil it requires and the amount of row feet it will plant.

A calendar is a simple way to track what type of work is being done and when it occurs. This information helps us map out germination times and plan transplanting. If you use a biodynamic calendar, you can also see how your work aligns with planetary rhythms and cycles. Our biodynamic calendar hangs right above the seeding bench so it doesn't get misplaced and we don't forget to enter our logs—which can be easy to forget on a busy day. To make year-to-year comparisons, we just flip open the calendars.

## Cultivation Data and Crop Budgets

With over fifty different species of crops growing at any given time on our farm, it can easily become overwhelming if we try to keep production and harvest data on every species. This is what happened to us in the early years. We tried to collect too much data, on too many species, and it became untenable, so we changed our practice. We find it more helpful to *closely* monitor ten crops a year—to really dig into what is happening with those crops, then rotate our attention to other crops in subsequent seasons. How do we choose the ten crops? It varies.

We rotate different crops through our data system based on what information we need and the questions we are trying to answer. We always monitor crops that are new to us to establish cost of

production data in order to set price points. We also use data to capture insights on how these plants like to grow and to see if we can get them to thrive and produce high-quality plant medicine on our farm. In addition, we frequently monitor our top sellers to see if we have our price points set correctly or if we need to make adjustments either in price or in production techniques. At other times we monitor crops that seem to be requiring extra care or labor to see if there are ways we can work "smarter, not harder." Any time we want to buy a piece of equipment or make a considerable change in the growing or harvesting practices for a crop, we monitor and record it. This helps us assess the efficacy of the purchase or practice at the end of the season.

Unlike the seeding and vegetative propagation data system, there are a lot of moving parts in the production and harvest data collection. First, we spread the work around. We all keep track of the hours we work and the basic tasks being done. This information is kept in a column on the employee time sheets. This helps us track, in general, what's happening with each crop on any given day. Second, we delegate primary data collecting on the designated ten species to one reliable employee. That person has a small field notebook and a timepiece that he carries into the fields. It is his responsibility to record the work or inputs being done on the designated species. For example, this past year we tracked data on nettle (*Urtica dioica*). Whenever people worked in the nettle beds it got recorded. In general, what gets tracked are labor rates for specific tasks and inputs of materials. We also record tractor work done on designated crops. We keep another small notebook on the tractor, and the tractor driver records the work being done and the implements used. *It is important*

## ZACK WOODS HERBS

Hyde Park, VT
Today's date: 6/28/2014
Worksheet adapted from Richard Wiswall

**Projected budget for:**

| | |
|---|---|
| | Nettles |
| | 1st harvest of 1st year plants |
| Size of planting: | 1400 bed feet |
| Field name: | Howard |
| Notes: | plants at 12" in row, 2 rows/bed<br>Labor rate: $13/hour<br>Tractor rate: $7/hour |

### Production Labor

| | Time | Labor | Machinery | Product | |
|---|---|---|---|---|---|
| Spread amendments | 30m | $6.50 | $3.50 | $28.00 | half a spreader load of composted chix manure, or I yard. $700 for 50 yards |
| Disk | 30m | $6.50 | $3.50 | | |
| Till | 60m | $13.00 | $7.00 | | |
| Seeding | 4 h | $52.00 | | | 2800 bare root cuttings in fall moved directly from one field to another |
| Transplanting | 1.5h | $19.50 | $10.50 | | |
| Flaming | | | | | |
| Reemay | | | | | |
| Weeding | | | | | |
| Hoeing | 30m | $6.50 | | | |
| Irrigation | | | | | |
| Side-dressing | | | | | |
| Mulch | | | | | |
| Tractor cultivation | 60m | $13.00 | $7.00 | | 3 passes w sweeps in May |
| Spraying | | | | | |
| Weed trimming | | | | | |
| Cutting & racking | 12h | $156.00 | | | |
| Garble 1st & 2nd | 9h | $117.00 | | | |
| **Total production expenses** | | **$390.00** | **$31.50** | **$28.00** | **equals $449.50** |

### Overhead Expenses

factored in on the end of year budget when all harvest data in

**Total of production and overhead expenses** **$449.50**

### Sales

Yield: 100 lbs dried in first cutting will get 1-2 more harvests of comparable size
Price: $24/lb dried retail

| | |
|---|---|
| Sales | $2,400.00 |
| Minus expenses | $449.50 |
| **Net profit** | **$1,950.50** |

**Figure 3-3.** Sample Crop Budget: Nettle.

*to record all of this data as a rate: the work done over time and distance* (e.g., forty minutes of hand weeding for two hundred row feet). This will help you later when analyzing and making projections.

## Sample Entries for Nettle

5/1/13  Top-dress—Bed 1\*—2 people × 10 min—
        1 tractor bucket composted chicken manure

5/10/13  Hoeing—Bed 2\*—2 people × 1h 40 min

5/30/13  Harvest & processing (1st harvest)—
        Beds 1 & 2—2 people × 56 min

6/2/13  Tractor cultivation—Bed 2—5 min\*\*

As mentioned before, all of these entries are kept in small notebooks that travel into the field as we work. Later we review the data, then enter it into our crop budgets on Excel spreadsheets. The template for these budgets was designed by Richard Wiswall and can be found in his book. They are elegant, both in their ease of use and for the powerful data that result.

This sample crop budget for nettle is the data recorded for one harvest of one bed. For nettle we have one-, two-, and three-year-old plants in different beds in different fields. Therefore, we need to do multiple budgets per crop per harvest. At the end of the season we compile all these individual budgets together for a final synthesis. These different budgets can be useful throughout the season to see how things are shaping up and for end-of-year analysis. Progress monitoring can be really useful if unexpected issues arise and you have to make decisions on how to move forward with a crop. We had this happen in 2013 with ashwagandha (*Withania*

---

\* We label each bed on a field map and know the dimensions of the bed and field. Later we can plug in these dimensions to get our rates. We find it is easier for folks to remember the bed numbers than the exact lengths of the beds because they tend to vary from year to year.

\*\* For tractor work we have a standard hourly rate "billed" to the crop for both the tractor use and driver's time. We calculate this standard rate using the formula: total cost of tractor, divided by its years of useful life, plus annual repairs and annual fuel expenses, divided by the number of hours used per year.

*somnifera*). The spring and early summer were extremely cool, and it rained pretty much every day for a month and a half. By mid-June the ashwagandha looked really poor, and the weeds, which thrived in the murky weather, had dwarfed the ashwagandha plants. We stood on the edge of a quarter-acre field, hoes in hand because it was far too wet for tractor cultivating and thought, "Are we really going to hand weed this? Is it even worth it?"

Instead of jumping right to work, we went back to the office and pulled out our crop budgets. We knew how much money and time we had invested in seeding and transplanting the ashwagandha. We could predict from previous budgets how much time it would take to hoe the field (based on past labor rates). Soon we were calculating the labor costs and weighing out all the variables. Machine cultivation (which is cheaper) was out of the question because it

**Figure 3-4.** All inputs on a designated crop are captured in the data tracking system from seeding to harvest. Here Beth is harvesting nettles. Photograph courtesy of Bethany Bond

was too wet to drive the tractors in the field. It was either hand weed or lose the crop. We knew approximately what a quarter acre of ashwagandha would yield in a good year (not a rainy one), and we knew the price we could sell it for. In the end it wasn't worth trying to save it. Based on the numbers the best thing to do was to plow it under when the field finally dried out, and as heartbreaking as it was, that is what we did.

This example shows an important paradigm shift that occurred on our farm. In our early days of farming when situations like this happened, we didn't have any data to help guide us, and as a result, we hand-weeded fields we probably should have let go. We worked for production, not profitability. In the beginning years we would have continued to grow and process the ashwagandha so we could sell it, not knowing if we were losing money in the process. On the other hand, sometimes you have to hang on to a lost crop—you may be under contract to a company and have orders to fill. You may be starting out, and it's your labor on the line, not a paid employee's, so you just do it. Sometimes, despite the numbers, it's just too agonizing to lose the crop and you do it because you love the plants and the intrinsic rewards outweigh the bottom line. In the end you do what works for you and your farm. Having the data simply helps you explore all the options, look for possible solutions, and at the very least, helps you make informed decisions.

## Harvest and Drying Data

Harvest and drying data are easy to track. In our dry storage, where we warehouse all our dried herbs postharvest, we have a three-ring binder that contains harvest and drying logs. These logs record the

**Figure 3-5.** Ashwagandha thriving in full sun.

species, field location, date of harvest, amount of fresh harvest, date the harvest is done drying, and amount of dried material yielded from the harvest. We also have a place to record comments and observations and assign each harvest a lot number. Lot numbering is important because it is required through our organic certification, and customers who are dealing with Good Manufacturing Practices (GMPs) guidelines will need this. It allows people to track where herbs are coming from and monitor quality.

Note the differences in the two harvest and drying logs. The first log was for elder flower and for illustrative purposes; it shows only one entry from the log, although there were other entries throughout the season. You'll note that the entry for elder was completely filled out, whereas the calendula entries were

not. We used this level of detailed record keeping for elder because in 2013 elder was one of the crops we wanted to monitor closely. We were thinking of dramatically increasing production of this herb because of increased demands for the blossoms and berries. We needed more information about yields and a better understanding of how much wildcrafted material can be sustainably harvested from our farm and how much we should expect to plant. This log helps us answer important questions and tells us many things:

- Elder is in bloom in early July in our region of Vermont.
- We did the wild harvest, so we know approximately how long it should take one person to harvest a particular amount. (This is helpful when employees

**Figure 3-6.** Melanie emptying the drying shed.

do the harvest because it gives them a labor rate to shoot for.) We also know the amount of blossoms we can yield from the elders we wild-harvest on our land and how much we will need to cultivate or wild-harvest in the future to meet demands.

- This record also gives us an understanding of the moisture content of elder blossom so we know how much fresh material is needed to yield a pound of dried elder blossom. This will be used in planning production and setting prices.

The drying and harvesting log for calendula is very different from that for elder. First, you can see we didn't record the *harvest date* or *fresh weight*. At this point in our farming career, we don't need that particular data. We know that once the blooms start in late June, we are harvesting every two to three days until late September/early October. We don't find it helpful to record each harvest date or fresh weight. We sell almost all of our calendula dried, and we know (from past years' data) how much dried calendula is yielded per row foot. We plant enough calendula to meet our projected sales and don't monitor the fresh-to-dried ratios; we can look that up from past seasons' data. The useful information we glean from the current calendula log is this:

- When harvest starts and how long it continues
- The yields of early harvests compared to yields later in the season

**Table 3-2.** Sample Harvest and Drying Log: Elder

| 2013 Harvest and Drying Log | | | Species: Elder flower (*Sambucus nigra*) | | | |
|---|---|---|---|---|---|---|
| Date Harvested | Lot Number | Fresh Weight | Date Dried | Dried Weight | Field Location & Bed Size | Comments |
| 7/06/13 | 19113 | 23 pounds | 7/09/13 | 4.4 | | Wildcrafted 2 people × 2 hours |

**Table 3-3.** Sample Harvest and Drying Log: Calendula

| 2013 Harvest and Drying Log | | | Species: Calendula flower (*Calendula officinalis*) | | | |
|---|---|---|---|---|---|---|
| Date Harvested | Lot Number | Fresh Weight | Date Dried | Dried Weight | Field Location & Bed Size | Comments |
| | 18813 | | 7/06/13 | 2.6 | | Early harvest, smaller blossoms |
| | 19113 | | 7/09/13 | 2.1 | | |
| | 19613 | | 7/14/13 | 6.01 | | Had to be reracked; not dried. Took one more day |
| | 19813 | | 7/16/13 | 8.75 | | |
| | 19913 | | 7/17/13 | 7.25 | | Rainy, drying slowly |
| | 20113 | | 7/19/13 | 9.25 | | |
| | 20513 | | 7/23/13 | 6.25 | | |
| | 20713 | | 7/25/13 | 10 | | |
| | 21013 | | 7/28/13 | 10.1 | | |
| | 21313 | | 7/31/13 | 9 | | |
| | 21613 | | 8/03/13 | 14.5 | | Blossoms starting to peak, from 2 harvests |

Note: These entries are a sample from the season. The log and the harvest continue until the beginning of October.

- When the plants are producing their biggest yields
- Because calendula can be challenging to dry, the comments about drying conditions or noting the need to rerack can be essential to monitoring for quality control.

Other information we often keep on the harvest and drying logs are the quantities of fresh orders that go out to specific companies and special adjustments to drying conditions, if we try something new or out of the ordinary for a crop. For example, we dry comfrey leaf differently than other leaf crops because comfrey leaf has a higher water content and is prone to browning. Specificity of drying conditions is often noted so we can capture successes and replicate them in the future.

## Field Maps

Field maps don't need to be fancy and can be simply drawn on regular or graph paper, delineating field dimensions, crop species, and plant and row spacing per bed. However, with the technology available today it's simple to download Google Earth or Google Earth Pro, which allows you to use satellite photos and measurement tools to measure the acreage of your farm or length of beds. It's an easy, accurate, and cool way to see what's going on from a bird's-eye view. Field maps will be useful in planning and keeping track of crop rotations year to year. They will also be required for organic certification.

## Labor, Sales, and Expenses

We use a simple bookkeeping system called QuickBooks Pro to record our payroll, sales, and expenses. We definitely do not endorse one bookkeeping program over another but have found that QuickBooks has what we need for the scale that we are at and is relatively user friendly. Employees record their labor hours on their time sheets, and using QuickBooks we can easily calculate payroll, including state and federal withholdings and liabilities, and print paychecks. Some farmers choose to get payroll and tax tables to calculate this themselves, which is great, but we choose to utilize our time in other ways and

would rather purchase a program such as Quick-Books (which includes a yearly fee) to do this for us.

This bookkeeping program and others like it help to capture the money going in and out of the business. In such bookkeeping systems you are able to set up accounts, create product lists for each species, and track expenses. Whatever program you choose to purchase or create, you should make sure that you can mine and sort the data in a way that helps you answer your management questions. At the very least you should be able to tease out specific data on customers, sales (both summative and species-specific data), product pricing, and types of expenses. This will allow you to see where you are making money (or losing it), which herbs are doing well in the marketplace, and where you are incurring costs. This data will be essential to answering your profitability questions and should be taken into consideration when developing your bookkeeping systems.

One other key feature of QuickBooks that we like is the ability to generate graphs easily and sort data for year-to-year financial analysis. It also has a simple accounts receivable program that keeps track of outstanding invoices and payments due. For me this is essential because we are hyperfocused during the season on growing plants and tend to be in the office less. Therefore we need a program that helps us monitor payment collections and deposits. It really doesn't matter what system you use, as long as you have a system that works for you—meaning that the system keeps track of the data you need and helps you keep some semblance of sanity during the growing season.

# Hiring and Managing Employees

Employees are often viewed by many business owners as necessary liabilities. "Can't live with 'em but can't live without 'em" is a typical business owner's refrain. Here on our farm we view our employees as assets. Yes, there are certainly challenging aspects to managing employees, but by hiring the right people, giving them clear expectations, fostering mutual

**Figure 3-7.** Some members of the 2014 farm crew wild harvesting Japanese knotweed. Photograph courtesy of Bethany Bond

respect and open communication, and paying fair wages, employees are definitely worth the investment of time and money.

Most potential employees looking for work on a small farm are doing so because they really like the variety of tasks that farm work provides and want to be outdoors performing physical labor. These employees aren't your typical fast-food-burger-slinging teenagers biding time in a minimum wage job to score some extra spending money. They are more likely to be mature, responsible adults looking for longer-term work in a pleasant working environment. These are generalizations, of course, and there are likely to be some workers who just don't fit the bill, but by carefully screening potential employees, we can usually avoid having to part ways prematurely.

Hiring employees is something we do in the off-season so we have plenty of time to interview and screen several candidates before we make a decision. When looking for help, we often place classified ads in local newspapers and through farming websites and newsletters. We generally receive dozens of applications each year, some responding to our ads and some just random calls or e-mails from people wondering if we are hiring. After we review the applications and select some good candidates, we begin the interview process. We schedule one-hour interviews with each candidate after we have called to check their references. Along with relevant experience, a good employment record, good reference checks, and a good interview, the most important factor guiding our decision-making process is our intuition. It is generally fairly easy to select the right person for the job based on the impression he or she makes in person, and we have been fortunate to have had a dependable, hardworking, and talented farm crew every year.

Although some small farms work independently, without outside help, we have found that having employees has been an invaluable investment for our business. The first year on our farm we had one part-time employee who would come and help Melanie and me when we needed an extra set of hands. We quickly noticed that the amount of work that three of us could do together in one hour was much more efficient than what one person could do in three hours or two people could do in an hour and a half. That one part-time position quickly turned into more of a full-time one, and in subsequent years we added more employees as we expanded.

There is definitely a synergistic effect to having more people working simultaneously on a given task. There is also often a direct correlation between the efficiency of a crew of farm workers and the organizational skills of the manager or crew leader directing and monitoring their work. Having employees stand around idly waiting while the crew leader tries to plan the day's work list and delegate tasks would be a waste of our money and their time. Therefore we always try to plan ahead to avoid inefficiencies with our labor force. After a typical workday is done, we will make a plan for the following day's tasks. We will also have a contingency plan in the event that something unexpectedly alters the original plan. For example, if we are digging roots on a Monday morning and the root digger malfunctions, we need to have a backup task so that no one is standing around waiting while we repair the equipment. Melanie or I usually meet with our crew leader in the morning and give her the work assignments along with the backup plan for the day so we can keep everyone moving efficiently even when things don't go as planned.

Farming is seasonal work, which can be challenging to some employees because the farming season generally runs from April through October, but we have the good fortune of being located near a ski resort town where jobs run from December to March, and many of our employees work there during the off-season. Fortunately for us, small farms in Vermont (and in many other states) that meet certain gross income requirements are exempt from paying unemployment compensation during the off-season because of the seasonality of our businesses. We do provide workers' compensation insurance so that if our employees are injured here while on the job they are covered. This insurance is affordable and is wise to have to protect our crew and our business. In some states it is mandatory, even for seasonal workers.

Once we find good help, we do our best to keep them by paying them well and showing our appreciation for their hard work. Our starting pay with new employees is prorated for a trial period of one month. After that month we sit down with the individual and discuss whether we feel we are a good fit for each other. If we decide we want to hire the employee permanently and the employee wants to work for us, we give him a raise reflecting our standard starting wage. For each subsequent year that the employee returns, provided both parties feel good about the arrangement, we offer another pay raise. We make sure to honor and thank our employees for their help and invite them to share any suggestions they may have as to how we may improve our practices. If Melanie and I are the only ones trying to envision ways to do things easier and better, that would be pretty self-limiting so we encourage our employees to come up with innovative ideas and methods—which they often do.

We also offer our employees discounts on herbs and live plants, help pay for their attendance at herb conferences and workshops, and let them glean surplus veggies and herbs from our gardens. We have found that when employees feel appreciated and feel that they are an important asset to the success of our business, they seem to exhibit more of a sense of ownership. Employee morale seems to correlate directly to productivity, and a happy workforce is generally a more productive and dependable workforce.

# CHAPTER 4

# Facilities Layout and Design

Have you ever sat in your home or workplace while looking at the way the rooms are laid out and thought to yourself, "I could have done a much better job designing this place"? I have certainly spent many hours mentally rearranging the layout of our home. Melanie and I didn't design and build our house; if we had, we would have made the living room larger and the front porch smaller. We also would have added on another bathroom. These additions will happen in time but will be much more costly than they would have been had the person who built our home taken the time to plan the layout more thoughtfully and efficiently. In this chapter we will discuss the importance of thoughtful planning in regards to the layout and design of your farm and facilties.

Fortunately, when we bought our farm the existing farmhouse was located roughly in the middle of the ten acres of land that came with it. For us, having our home located within our business certainly has its pros and cons. Being able to step outside your front door and be at work can be a blessing, but it can also be a curse when you are trying to have some downtime but your "to-do" list is literally staring you in the face. Ultimately, Melanie and I wouldn't have it any other way, but for some people being able to leave work at the end of the day and *really leave* work is a better choice.

The central location where our home was built had plenty of open space around it to build our greenhouse, drying shed, barn, and other facilities. Since we started small and built up our infrastructure gradually, we had plenty of time to evaluate thoroughly how we would locate the individual components efficiently. One of the most important

considerations when planning the layout of your own operation is the efficient flow of material, equipment, and personnel. The difference between locating buildings fifty feet apart and putting them one hundred feet apart may not seem that vast in the short term, but when you factor in the amount of steps that additional fifty feet will add up to over the course of several years, especially when carrying bulky or heavy materials, that difference can become enormous over the long term.

Now here you are with a relatively clean slate. You have just purchased or leased your farmland and you are ready to design the layout of facilities you will need, such as the greenhouse, a drying shed, a processing room, dry storage, and possibly a retail store or farm stand. We will refer to all of these facilities (including greenhouses) as "buildings" to keep it simple. You basically have a great opportunity to get it right the first time; otherwise, you could be faced with spending time and money rebuilding or moving buildings that really shouldn't need to be moved. For those with farms that have buildings already in place, let's hope the layout of your already existing buildings was well planned in advance so that you are already one or several steps ahead. If not, you may have to get creative, which is not always a bad thing.

## Vehicle Access

Since the buildings will likely be the center of activity for the farm, there are many good reasons to try to locate them near the center or hub of the farm as a whole, including the fields where the crops will be grown. This centralized location isn't always possible

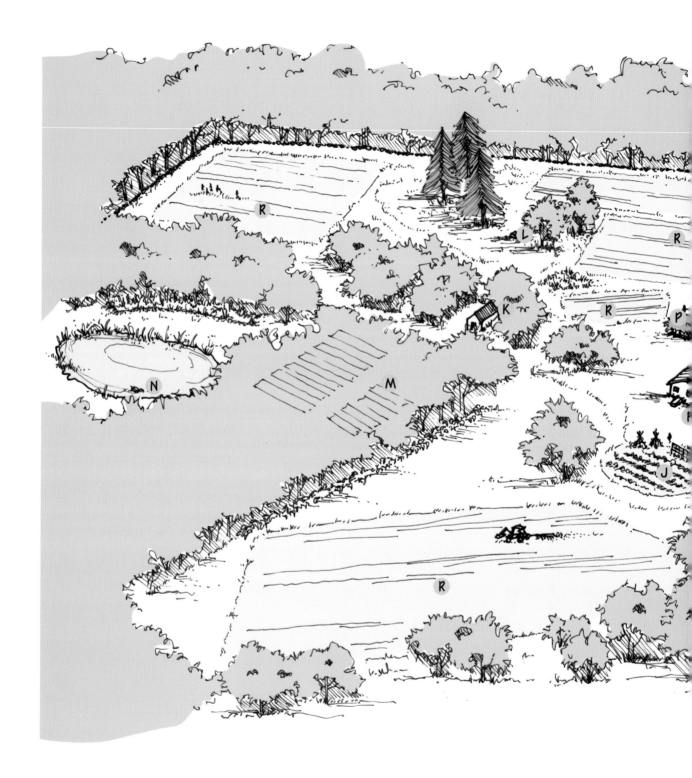

**Figure 4-1.** Artist's rendition of ZWHF. Illustration by Cornelius Murphy

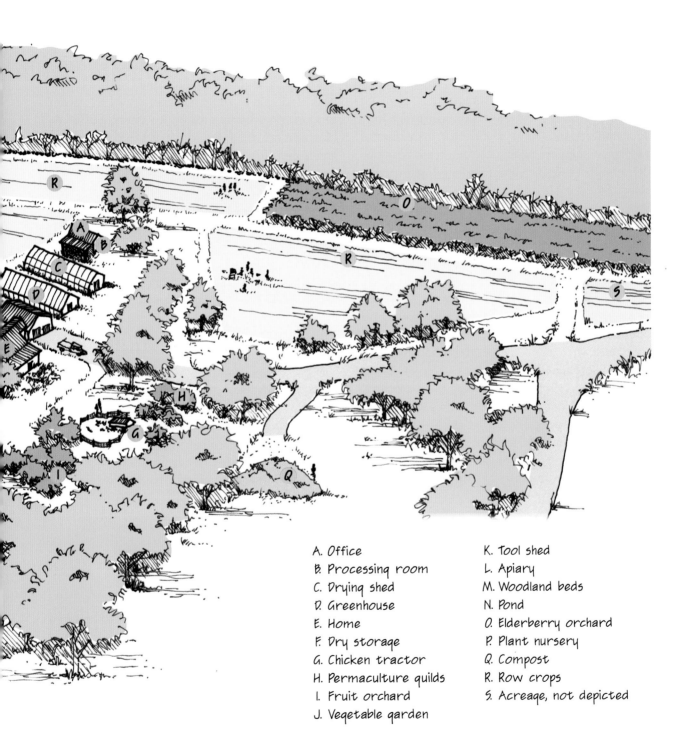

A. Office
B. Processing room
C. Drying shed
D. Greenhouse
E. Home
F. Dry storage
G. Chicken tractor
H. Permaculture guilds
I. Fruit orchard
J. Vegetable garden
K. Tool shed
L. Apiary
M. Woodland beds
N. Pond
O. Elderberry orchard
P. Plant nursery
Q. Compost
R. Row crops
S. Acreage, not depicted

given the variability in land borders and features, so the next important consideration would be to locate the buildings as close to a road or driveway as possible, along with having easy access to any necessary electrical and plumbing utilities. If you are able to design the driveway, you will want to allow plenty of room for large delivery vehicles to maneuver. Don't wait until it's too late to realize that the pallets of cardboard shipping containers you ordered can't be delivered because the sixty-foot-long tractor-trailer can't turn around in your driveway. You would also be wise to have the driveway located near the dry storage area so that when the shipping truck comes to pick up the boxes of herbs you are sending to your customers the truck driver can back up to the building instead of having you or the driver carry boxes to the truck.

Another important consideration with all facilities in cooler climates is snow removal. Snow sliding off buildings and greenhouses accumulates rapidly and can freeze solid. Without the ability to move this snow, you could find yourself limiting access to buildings or even possibly compromising their structural integrity. Greenhouses are particularly susceptible to snow and ice damage, and most greenhouse manufacturers recommend a ten-foot spacing at the very minimum between greenhouses or between greenhouses and other structures built in snowy climates.

## Dry Storage

The dry storage facility can be as simple or elaborate as you wish, but it absolutely needs to be waterproof and rodentproof. Ideally it would also be cool, dark, and able to be hermetically sealed to attempt to maintain product quality. We utilize half of the basement below our house, which has a cement floor and is spray foam insulated from wall to ceiling. This dense foam insulation keeps the room cool, dark, waterproof, and impermeable by rodents and insects. There is also a dehumidifier in the room to maintain humidity levels below 50 percent. Herbs store very well inside airtight bags in packing tape–sealed cardboard boxes or gasketed drums in this environment, and accessing the room while moving large boxes and drums is relatively easy through bulkhead doors.

## Processing Room

The processing room is where herbs are processed from their dry state either by milling them with machinery or by rubbing the dried plant material over a stainless steel mesh sieve by hand to remove stems (also called garbling). We are currently working with an engineer to design a device to partially automate the garbling process, but for now we do it

**Figure 4-2.** Removing snow from between greenhouses.

**Figure 4-3.** Dried herbs being processed by rubbing them over stainless steel mesh to remove stems.

solely by hand. The processing facility can be a simple open-shed-type structure to keep equipment and personnel dry in inclement weather, or it can be as elaborate as needed. We recommend planning ahead for the potential addition of larger milling equipment and increased "throughput" for those considering the possibility of future increases in production.

You should consult with a qualified electrician to assess and establish the potential electrical requirements of current and future equipment here. Milling and processing herbs produces an enormous amount of fine airborne particulate matter; therefore, it is advisable to install dust removal systems and/or the ability to open doors and windows and direct fresh airflow via fans. Laminar flow hoods, which can remove airborne particles from a small work area, are another option for dust control. After the herbs are processed, they are weighed, packaged, labeled, and moved into dry storage ready for sale. Ideally the processing room will be located as close to the dry storage as possible to allow easy transfer of material.

## Drying Shed

The drying shed could be considered the pièce de résistance of the medicinal herb farm. This is where we remove literally tons of water from fresh plants in a given season through the process of dehydration. The way in which water is removed from the plants has enormous bearing on the potency, quality, and value of the finished product, so it is imperative that the design and construction of this facility be well executed. We will detail drying shed construction in chapter 14, but for now, let's talk location.

Since the drying shed is where most of the crops grown on an herb farm are processed, ideally it should be located as close to the crops as possible. This isn't always practical, given the vagaries of land layout, so at the very least it needs to be located within a short (ten- to twenty-minute) drive from the farthest field. The longer herbs sit in piles in transport after harvest, the faster anaerobic and aerobic bacteria are able to multiply and degrade the quality of the herbs or, in the worst-case scenario, contaminate them to the point that they would not pass the microbiological testing required by many buyers. One of our land leases is a ten-minute drive from our drying shed, and during hot weather we can utilize large screens stacked in the back of a truck during harvest to keep piles of freshly harvested herbs cooler and well aerated during harvest and in transit.

The drying shed should have ample electrical capacity for running fans, lights, and heating equipment and be built with a waterproof foundation or

**Figure 4-4.** A look inside the drying shed. Photograph courtesy of Bethany Bond

**Figure 4-5.** Past drying sheds.

vapor barrier on the ground, with curtain drains installed around its perimeter to direct rainwater away. Not only should the drying shed be located in close proximity to crops, it should also be as close as possible to the processing facility and the dry storage, since material will flow from the field into the drying shed, next into the processing room, and finally into the dry storage. Having the ability to drive vehicles loaded with herbs right up to the drying shed entrance doors is also highly recommended.

## Greenhouses

The farm's greenhouses don't necessarily need to be located in close proximity to the dry storage, processing room, or drying shed because they are utilized mainly to start plants and not necessarily for the processing and handling of dried product. An exception to this could be for the beginning or low-budget

herb farmer, who may wish to utilize the greenhouse for starting plants in the spring, then convert this same structure into the drying shed as harvest time nears. Our first greenhouse was a multipurpose facility that served as greenhouse in the spring, drying shed in the summer and fall, and storage shed in the winter and filled this role remarkably well until we needed to upgrade to handle our increased volume.

One of the most important considerations in selecting a site for the greenhouse is exposure to the sun. Since the purpose of the greenhouse is to maximize the sun's potential by allowing UV light to reach plants and to capture heat through solar gain, we have to select a location where there is full solar exposure without the shade from trees or other objects limiting its potential. I can remember a friend who was working on our farm last summer eagerly planning her own herb farm. She shared with me what a difficult time she and her partner were having

**Figure 4-6.** Our drying shed in the early spring.

deciding where to locate the greenhouse and told me that the greenhouse manufacturer stated that the greenhouse must be oriented from east to west, but she didn't have a good site near her main buildings that would allow for that. She was considering cutting down trees in a different location to orient her greenhouse according to the manufacturer's directions.

After listening to her describe her dilemma I quickly pointed out to her the orientation of our greenhouse, which runs lengthwise from north to south, and mentioned to her that never once in fifteen years of growing plants in that greenhouse did we feel that this north-south orientation noticeably limited the growth of plants inside it. Sure, there is definitely a slight increase in solar potential from having the east-west orientation, but it is fairly minimal. We would recommend, however, that greenhouses oriented north-south be equipped with a clear gable end on the south end of the house if possible to maximize

**Figure 4-7.** Herb plugs growing in the greenhouse. Photograph courtesy of Bethany Bond

the lower angle of the sun during the cooler months. The south gable end on our greenhouse is covered with clear greenhouse poly, and when the sun is low it still shines directly into the house, offering plenty of solar gain even during the cooler months when the sun is low in the sky.

Another important consideration for those utilizing roll-up ventilation sides on their greenhouses is the direction of the prevailing wind. For example, our greenhouse and drying shed are both oriented north-south and are set up well to receive ventilation from our prevailing westerly winds.

## On-Site Retail

For those of you possibly interested in doing some direct-retail marketing from your home or farm you may want to consider a present or future building site for this when planning your farm layout. A roadside stand–type of setup is one option to consider for selling bulk herbs or value-added products such as live plants, tea blends, and herb extracts. Here at ZWHF we are located in a very rural area without much vehicular traffic, but we do have several customers who like to come to the farm to shop or to pick up their orders. For this we built a small timber-frame barn that serves as our farm office and is a place where we can set up small displays for retail sales during farm tours and for people stopping by to shop.

**Figure 4-8.** People shopping after a farm tour.

Ideally, a retail location would be located close to a driveway or parking area and have utilities such as plumbing and electricity for lighting and restrooms.

Another consideration if you are planning on hiring employees is a designated employee area. This need not be an elaborate room or building and could basically just be a place for employees to have lunch breaks; leave rain gear, gloves, and other personal items; and sign in and out via time sheets or a punch clock.

Planning which facilities to build, where to locate them, and especially what size they should be can definitely seem overwhelming. In fact, each year we ask ourselves whether we need to increase the size of our greenhouse, drying shed, and barn, and while the answer to these questions is more often than not "probably," sometimes we just make do with what we have for another year or two until the answer becomes a definite yes. We can usually add another layer of racks to the drying shed or utilize space more efficiently in the barn and greenhouse if need be for a year or two before committing to expensive capital improvements.

Melanie and I both trend toward the fiscally conservative side, so unless there is a specific reason we need something to be bigger and better, we try not to overextend ourselves. What we do, instead of building things proactively much larger than they need to be, is to build them the size we foresee they will likely need to be within the realistic short-term three to five years or so ahead. We also plan and build our facilities so they can easily be expanded without having to do a major rebuild. Greenhouses—and drying sheds constructed using greenhouse frames—are very simple and inexpensive to add on to.

Buildings can also easily be designed to have simple additions made without the need for wasteful demolitions. Again, starting small and growing incrementally within one's means has been part of our formula for success and one that we recommend to others new to herb farming. In the next chapter we discuss similar considerations around the land itself. Many of these same concepts regarding size and scale apply to both the facilities that we use to produce and process our crops in and the land where we grow them.

# CHAPTER 5

# Field and Crop Considerations and Planning

I still vividly remember the first week that Melanie and I moved onto our new farm in the spring of 1999. To say we were enthusiastic would be a gross understatement; in fact, we were ecstatic to be finally realizing our dream of owning our own farm. I had spent some time hand-digging test holes to try to determine soil types and drainage characteristics prior to purchasing the land, but those test holes barely scratched the surface of what was underneath the ten acres of overgrown pasture surrounding the farmhouse. We were certainly naive back then. Our thoughts had been occupied by pastoral visions of long sunny days harvesting bountiful crops and not at all with the pitfalls we could potentially encounter while bringing those visions to reality.

It wasn't until we got ready to start prepping fields for planting that reality started sinking in. Suddenly I became very nervous about what I would find when plowshare met soil for the first time. Would there be large outcroppings of ledge, lurking just under the sod? Would the land be so stony as to render it difficult to work with? Would the water table come up so high as to cause plants to drown in prolonged rainy spells? Would the soil be so acidic or infertile that we would have to spend a fortune applying mineral powders and composts just to get the plants to grow well? Would the sloping portions of the land be prone to erosion in torrential downpours? Would the well that supplied the house run dry while we were trying to irrigate withering plants in spells of drought?

Luckily, most of these concerns had favorable outcomes, but not all of them did. There was definitely some shallow ledge, but none of it was in the areas we tilled and planted. Some areas of the land were stony, but we removed most of the stones and are now just having to perform maintenance stone picking at the beginning or end of each season. The water table was definitely high during rainy spells on some of the lower portions of our land, but we expected that, and the higher ground was well drained. The soil was acidic, but luckily it had decent fertility and good organic matter. The sloping portions of the land were definitely prone to erosion, but we mitigated that by orienting the beds across the slope, leaving sod paths with drainage swales in between them and keeping the soils planted with crops or cover crops to help anchor the soil in place. Our well had plenty of capacity for the house and vegetable and flower gardens, but it didn't have

**Figure 5-1.** Google Earth view of the farm.

**Figure 5-2.** A glimpse back in time: our first year on the farm, circa 1999.

enough extra to irrigate the row crops. Fortunately, there was an excellent pond site on the premises, and we quickly turned a swampy spring into a great swimming, fishing, and irrigation pond.

We had definitely put ourselves in a risky position purchasing land that hadn't been thoroughly evaluated from an agricultural standpoint, but the bottom line was that this was the most affordable piece of land we could find at the time that seemed to offer the greatest potential for serving our needs. We don't regret this purchase at all, but in hindsight we were very fortunate that things worked out the way they did. Perhaps Demeter, the Greek goddess of agriculture, was smiling down on us at the time, or maybe we just got lucky.

## Thoroughly Evaluate Your Land before You Commit

For prospective herb farm owners and operators we highly recommend thoroughly evaluating any potential purchase or lease for all the reasons stated above. It is definitely worth the trouble to hire a mini-excavator or backhoe to perform a thorough survey of the soils. It can also be helpful to glean as much information as possible from neighbors and current and former landowners. Most people wouldn't purchase a home without performing a thorough evaluation of the building's structural integrity, and it would be wise to take the same cautious approach when it comes to buying land. Soil maps can be a good starting point for determining soil types, but they can at times be somewhat inaccurate because of local geographical and geological variations. Soil testing is a good way to start to get a baseline for pH, organic matter, and fertility levels, but they too can be somewhat variable in their accuracy unless performed methodically and thoroughly.

Another important consideration when purchasing or leasing land is the history of agricultural inputs previously applied to the land. Becoming certified organic is beneficial on so many levels and is highly recommended, both from an ecological and a marketing perspective. If the land you are interested in has had any inputs that are prohibited for organic certification applied to it in the previous three years, there could be a delay in the process of becoming certified organic. If getting certified isn't important to you, it would still be prudent to determine what if any chemicals have been applied. Herbicide residues, if present, could still be concentrated enough to damage plant tissues even several months after their application, and dangerous pesticides could linger for years in the soil and potentially be transferred into plants and beyond.

## Proximity to Processing

Since we have already discussed how important it is to get freshly harvested herbs into the drying shed before anaerobic conditions occur, locating crops as close to the drying shed as possible will help to reduce time in transit for freshly harvested plants. Designing efficient paths and roadways for vehicular access within and around planting fields should be considered the foundational framework of field layouts. Maximizing yields by efficient use of available acreage for planting is great, but without vehicular access to get these plants to the processing facility, labor costs for hauling materials by hand can add up fast. Oftentimes we will have one person tasked with driving the truck loaded with herbs to the drying shed and back and two people in the drying shed placing the fresh plant material on drying racks

**Figure 5-3. Soil profile.** Photograph courtesy of Lynn Betts, USDA Natural Resources Conservation Service

**Figure 5-4.** Melanie working quickly to unload nettle and get it into the drying shed. Photograph courtesy of Bethany Bond

while the rest of the crew harvests. This can all occur simultaneously during harvest time so that there is no downtime with harvesting, transporting, or racking. This is another example of how we have gained efficiencies by hiring employees.

# Proximity to Water

It's a commonly known fact that people and plants cannot live without water. And while yes, rains will eventually materialize at some point during times of drought to nourish thirsty plants (at least they will where we farm in Vermont) and people can carry water-filled vessels with water to hydrate themselves, the farmer who relies solely on Mother Nature to fulfill irrigation requirements does so at significant risk.

Here at ZWHF we spent the first four years farming without an irrigation system. We simply couldn't afford the costs of installing a new well or digging a pond, and we had no other water source in close proximity to our fields. We were extremely lucky during those four years because we usually had either too much rain or just enough to get by on. During dry spells the rains would eventually come just in time to prevent big losses. However, we knew that it was only a matter of time before our luck ran out. We also knew that during those dry spells we could have gotten higher yields from healthier plants if we had been able to get more water to them. We finally saved up enough money to dig a pond and invest in a solar-powered irrigation system after going without it for four years and realized almost immediate return on that investment.

Medicinal herbs vary greatly in their water needs. While some species such as calamus (*Acorus calamus*) thrive in prolonged swampy conditions, others such as hoodia (*Hoodia gordonii*) can thrive for years without any rainfall or irrigation. Most medicinals fall somewhat in the middle, needing an average of approximately one inch of rain or the equivalent in irrigation water per week during their active growth period. Some species of herbs, as we will discuss in chapter 9, actually become more potent when drought stressed.

**Figure 5-5.** Our irrigation and swimming pond.

# Efficient Bed Layout

The term "bed" is used to describe the portion of a field where plants are grown in rows. These beds may be raised beds formed either with hand tools or tractor implements or they may be flat beds at ground level. Rows are the lines of plants within the beds. Paths (often called *middles*) are the space between each bed where people walk and tractors and other vehicles drive while straddling the beds. Beds vary in width and shape from farm to farm according to the type of crop grown, the amount of space available, and the type of equipment used. Each farm typically uses one standard bed width that is primarily determined by the size of the tractors and implements that will be used in them. Standardization of bed width and row spacing is highly recommended and allows for planting, cultivating, and harvesting implements to be adjusted uniformly to avoid spending time readjusting them when working in different sized beds.

Determining the size of planting beds while maximizing available acreage can be a complicated task given so many variables, including how much to plant, how to orient fields and beds, what type of plant spacing to use in rows and between rows, what species to intercrop, how to get irrigation water to the crops, how to control weeds, how to harvest, and on and on. We have found that using aerial photographs of our land accessible on the Internet via Google Earth or through the local agricultural Cooperative Extension Service gives us a much better perspective to measure and lay out our fields before establishing them than by trying to do it from ground level.

After we have determined our bed configurations, we decide which crops will be planted at ground level in flat beds and which crops will be planted in raised beds. This is determined by several variables, including crop type and field drainage characteristics. The width of our beds is determined by the wheel spacing of the tractors we use for cultivating and harvesting. We work with either two or three rows per bed at a fourteen-inch row spacing (between rows) in raised beds that are approximately forty-six inches wide and six feet from the center of one bed to the center of

**Figure 5-6.** An example of bed efficiency: herbs planted three rows to a bed.

**Figure 5-7.** Raised-bed and row-spacing dimensions.

another. This six-foot-on-center bed width is a fairly standard size for most farms with tractors having tires spaced sixty inches on center. It also allows us to drive pickup trucks down the beds during harvest to pick up harvested material. Beds that are planted with large plants, such as stinging nettles, utilize only the two outer rows, spaced twenty-eight inches apart. Beds that are planted with smaller plants, such as chamomile, use all three rows spaced fourteen inches apart. Spacing of plants within the rows varies according to the plant's growing habits and harvesting methods. Bed sizes and row spacing can easily be adapted to narrower beds for those farming with walk-behind or compact tractors, or wider beds for those using larger equipment and acreage.

The benefits of raised beds are that they are extremely uniform in size and shape (which assists us with mechanical cultivation), they are well aerated

and well drained to promote deep root penetration, and they form a nice clean and level seedbed for direct seeding or transplanting. They also assist us with hand-harvesting blossoms by slightly reducing the distance we have to bend our bodies to harvest. Raised beds also ease mechanical root harvesting by lessening the depth that the root digger has to penetrate to dig roots. Raised beds can also be beneficial for land that tends to stay wet for prolonged spells from seasonally high water tables and rainy spells because the plants are slightly elevated above the sodden soils. This helps the plants' roots receive more oxygen and assists in the prevention of fungal diseases.

Ground-level beds, or "flat beds," as they are commonly referred to, have the advantage of not requiring as much soil preparation. We plant most of our leaf crops in flat beds, which allows for ease in harvesting with equipment such as sickle bar cutters.

We form raised beds with a raised bed former/mulch layer combination unit. This implement attaches to the tractor's three-point hitch and is pulled along through the field, where it compresses the loose soil into a rectangular-shaped raised bed approximately six inches high and forty-six inches wide. The mulch layer unit attached to this implement can cover the raised bed with a layer of plastic or biodegradable mulch, which we will use with some crops for weed control and to minimize the soil residue that can splash on plant leaves. Please note: Biodegradable mulch is currently under review for approval for certified-organic use.

One of our goals in planning bed layout is to try to maximize the potential of every square foot of open ground. In our early years farming we planted single rows of crops spaced thirty-six inches apart, which left an excess of unused space between rows.

**Figure 5-8.** Jeff forming raised beds.

This unused space allowed for more open ground for weeds to multiply and it also exposed more of our precious soils to the erosive elements of wind and rain. It also meant that we were only really using about a third of the actual acreage we were tilling. It was inefficient, to say the least. Now that we are planting with the two- and three-row beds, there is a much smaller niche for weeds to occupy; there is much less soil exposed to the elements. The herbs we plant, by the time they reach maturity, are literally filling a majority of the fields we till. We will discuss how we prepare fields prior to planting in chapter 7.

# Intercropping and Permaculture with Medicinal Crops

*Permaculture is a philosophy of working with, rather than against nature; of protracted and thoughtful observation rather than protracted and thoughtless labor; and of looking at plants and animals in all their functions, rather than treating any area as a single product system.*

— BILL MOLLISON[1]

Most medicinal herbs used today are plants that humans have taken from the wild and domesticated for our own uses. What many herbalists and plant lovers who have worked intimately with these plants for years often observe is that these wild plants don't always like to be "told" when to germinate or where and how to grow. They seem to prefer growing on their own terms—where, when, and how they want to—as they have for eons without our help. Yes, we can usually tame them somewhat and get them to abide growing in our controlled garden environments. But what happens when we encourage and assist them to grow in a more natural way, in wild settings as they would without our intervention? Our experience has demonstrated that they usually thrive in these "wild simulated" settings much more than they do in linear rows planted within hundreds or thousands of their own kind. They prefer to grow in diverse communities, as they do when we see them growing in the wild.

The challenge to the commercial herb farmer is that growing them in wild-simulated environments can be counterproductive to commercial production techniques that benefit greatly from economies of scale. When we harvest peppermint, for example, we are literally mowing down a large field of peppermint by hand with field knives or with a sickle bar cutter without having to spend excessive time avoiding plants that we don't want to harvest. In wild settings with lots of diversity, harvesting can be much more laborious and time consuming because there are often lots of other plants growing in and among each other. When we plant a field of peppermint, we attach the transplanter to the tractor and can plant an acre easily in a couple of hours. When we plant polycultures on a commercial scale, we aren't able to use these efficient methods and equipment because of the variation in size and growing habit of multiple species, as well as the nonlinear spacing we use to mimic more natural growing environments. Planting this way becomes much more labor intensive. However, in spite of these challenges, there are ways that we can profitably and sustainably grow medicinals in wild-simulated settings using common permaculture practices.

**Figure 5-9.** Wild plant communities: nettle, valerian, and angelica naturalized in a wild-simulated bed. Photograph courtesy of Bethany Bond

The word permaculture was coined to describe "permanent agriculture" and later adapted to become defined as "permanent culture" to include the social relationship among humans, plants, insects, birds, fungi, bacteria, minerals, elements, and so on. The philosophy of permaculture involves observing and imitating patterns in nature where individual species benefit from complex relationships with other species sharing the same habitat. The synergistic effect provided by diverse communities of trees, plants, fungi, bacteria, insects, birds, animals, and humans interacting within constructed natural habitats is thought to contribute to a sustainable, regenerative resource system in which nutrients are recycled to provide fertility for food, fuel, fiber, fodder, and "farmaceuticals." The harvests of *individual* species within these systems are often lower yielding than if they were grown in monoculture-type settings where fertility, water, sunlight, and weed control are more intensively managed. However, when viewed as a *whole*, these polycultural communities can be very high yielding, self-sustaining, and relatively maintenance-free. Permaculture can often be viewed as more productive in terms of a whole system being much greater than the sum of its parts.

Permaculture designers refer to the forest's vertical structure as having between five and eight layers or horizons that function both independently and as an integral component of a larger ecosystem. Following are examples of layers with species of plants, trees, and fungi recommended for permaculture guilds for the medicinal herb farm, along with a brief description of their medicinal uses.

## Layer One: Overstory or Tall Tree Canopy

This layer dominates the forest and utilizes the highest percentage of nutrients, sunlight, and water of all the layers in the ecosystem. The trees within this layer generally prefer growing in full sun at the apex of the forest canopy.

## Layer Two: Understory Tree Layer

This layer of trees is generally composed of lower-growing trees that are adapted to growing in full sun or partial shade, depending on the species.

## Layer Three: Shrub Layer

This layer supports lower-growing trees and multistemmed shrubs that will grow in full sun or partial shade.

## Layer Four: Herb, Fungi, and Ground Layer

This layer is composed mainly of herbaceous plants and fungi of variable sizes adaptable to growing in partial shade.

## Layer Five: Vine Layer

This layer is composed of vertically growing vines that utilize the other four layers in the system for support. These vines are generally shade tolerant.

One of the biggest challenges we have faced here on our farm has been attempting to grow medicinals commercially in a polycultural setting while keeping costs of production to a minimum. If we were only growing these plants for our own use on a small scale, it would make sense and be relatively easy to produce our plant medicine from wild-simulated settings. However, we are trying to make a living by growing them commercially, and almost everything we do has to be done as efficiently as possible in order to keep our costs of production down. That said, every year we attempt to incorporate permaculture designs into our production system because we feel it is important and we know that success can be gained by utilizing the right approach.

Some of the plants we grow using permaculture methods actually perform much better and are more profitable when grown in wild-simulated settings. An example of this is in our woodland beds where we grow black cohosh, goldenseal, ginseng, bloodroot, Solomon's seal, and Siberian ginseng. These plants prefer to grow in a diverse guild in the understory of deciduous trees, like they do in the wild rather than growing in monoculture in rows under a shade house or in an open bed. Once planted, many of these woodland natives multiply and spread on their own via lateral runners and by self-seeding. We also collect seeds from mature plants and propagate them in the greenhouse to fill in any spots left open. By

**Table 5-1.** Overstory Layer

| Species | Use |
| --- | --- |
| Black cherry (*Prunus serotina*) | Sustainably harvested limb wood provides wild cherry bark, which is used in cough syrups for its expectorant, antispasmodic qualities. |
| Black locust (*Robinia pseudoacacia*) | Fixes nitrogen, providing fertility to surrounding soil. Source of extremely rot-resistant wood for fence posts, stakes, and other construction needs. Rich source of nectar for honeybees and other pollinators. |
| Black walnut (*Juglans nigra*) | Valuable timber species. Delicious edible nuts. Walnut hulls are a strong vermifuge (antiparasitic), antiseptic, and antidiarrheal. *Cultivation consideration:* Juglone produced in the tree's roots can be toxic to some plants and trees (such as fruit trees) growing in close proximity. |
| Ginkgo (*Ginkgo biloba*) | Ginkgo leaves are prized for their antioxidant and circulatory benefits. Known to help improve memory and cognitive function. |
| Poplar (*Populus* spp.) | Buds are the source of the medicinal aromatic resin, balm of Gilead. Also a source of propolis from honeybees. |
| Sassafrass (*Sassafrass albidum*) | Roots are utilized for their tonifying, antiviral, and aromatic properties. Used to promote liver health. |
| Slippery elm (*Ulmus rubra*) | Sustainably harvested limb wood provides slippery elm bark, which has excellent demulcent properties. Useful for soothing irritations of mucous membrane linings. |
| Sugar maple (*Acer saccharum*) | A mineral-rich natural sweetener, maple syrup is obtained from boiling sap in the spring. Calcium-rich leaves help to fertilize understory woodland medicinals, such as ginseng and goldenseal. |

**Figure 5-10.** Ginkgo tree. Photograph courtesy of Bethany Bond

**Table 5-2.** Understory Tree Layer

| Species | Use |
| --- | --- |
| Fringe tree (*Chionanthus virginicus*) | Fringe tree bark is traditionally used as a liver and gallbladder tonic. |
| Hawthorn tree (*Crataegus* spp.) | Hawthorn flower and berries are incredible cardiac tonics. They lower blood pressure and cholesterol levels and help alleviate blockages in arteries. |
| Linden tree (*Tilia* spp.) | Linden (also known as basswood) flowers are an excellent cardiac tonic, helping to lower blood pressure and calm the nervous system. Leaves are edible. Also a prime nectar source for bees. |
| Moringa tree (*Moringa oleifera*) | Moringa leaves and seedpods are extremely nutritious: high in vitamins and minerals, especially calcium. This tree is often called the "Tree of Life" for its ability to grow in drought-stricken regions and provide nourishment when food is scarce. The seeds are also used to purify water for drinking. Not cold hardy in the north. |
| White willow tree (*Salix alba*) | Bark contains salicin, which has analgesic and antiinflammatory properties. One of the original sources of aspirin. |

**Figure 5-11.** Hawthorn tree in berry.

**Table 5-3.** Shrub Layer

| Species | Use |
|---|---|
| Aralia (Spikenard) (*Aralia racemosa*) | Roots, leaves, and berries are used to treat respiratory infections. A strong expectorant that helps clear deep infection and excessive mucus. |
| Chaste tree (*Vitex agnus-castus*) | The berries help regulate hormones and balance the production of estrogen and progesterone. Used as a tonic for treating PMS, irregular menstruation, and menopausal problems. |
| Cramp bark (*Viburnum opulus*) | The bark is used as a strong uterine nervine with antispasmodic and anti-inflammatory properties. |
| Elderberry (*Sambucus nigra, S. canadensis*) | Blossoms and fruit have strong antioxidant and immune-enhancing properties and are excellent for preventing and treating colds, flu, and respiratory infections. |
| Oregon grape (*Mahonia aquifolium*) | Containing berberine, the root is highly effective at fighting infection. Antiviral and anti-inflammatory. Used internally and topically. |
| Red raspberry (*Rubus* spp.) | Leaves are used to tone and strengthen the reproductive system in both men and women. An excellent source of iron, niacin, and other minerals. |
| Roses (*Rosa* spp.) | Flowers and hips are astringent, nutritive tonics full of vitamin C. Delicious and uplifting to the spirits. Used in teas, glycerites, and jams. |
| Siberian ginseng (*Eleutherococcus senticosus*) | The roots are an effective yet gentle adaptogen, helping the body deal with stress, exhaustion, and illness. Enhances the body's immune response and liver function. |
| Witch hazel (*Hamamelis virginiana*) | The leaves and bark are used to make astringent skin care preparations. |

**Figure 5-12. Rose.** Photograph courtesy of Bethany Bond

**Table 5-4.** Ground Layer

| Species | Use |
| --- | --- |
| Alfalfa (*Medicago sativa*) | Popular nitrogen-fixing forage crop, beloved by bees. Leaves and flowers are highly nutritive and contain chlorophyll, protein, and vitamins A and C. Used to help lower cholesterol and help regulate and balance estrogen levels. |
| Angelica (*Angelica archangelica*) | Roots are rich in essential oils and are used for their bitter and warming properties. Aids in digestion, clears liver stagnation, and is good for treating flus, colds, and bronchitis. Also good for the female reproductive system. |
| Black cohosh (*Cimicifuga racemosa,* syn. *Actaea racemosa*) | Roots are used to aid in hormone regulation and help ease muscle spasms and tension. Indicated specifically for use with menopausal symptoms and delayed menstruation and to aid in delivery, often given in the last week of pregnancy. |
| Bloodroot (*Sanguinaria canadensis*) | Roots used in very small doses internally for bronchial infections, topically used to treat skin growths and as an antibacterial mouth rinse to prevent plaque. |
| Blue vervain (*Verbena hastata*) | The leaves and flowers are used to strengthen the nervous system and to help relieve stress and tension. |
| Boneset (*Eupatorium perfoliatum*) | Aerial parts are highly beneficial for treating flus and bringing down fever. It also works as an expectorant and clears excess mucus from the respiratory system. |
| Burdock (*Arctium lappa*) | Roots can be used as a food source and are highly nutritive. An excellent liver tonic and bitter. Used in treating a multitude of skin ailments, including acne and psoriasis. Considered invasive by some. |
| Comfrey (*Symphytum officinale*) | Leaves and roots are highly demulcent. Used topically to promote cell regeneration and to heal damaged joints and ligaments. Soothes irritated and inflamed tissue. Commonly used internally for ulcers and bronchial inflammation. Considered invasive by some. |
| Dandelion (*Taraxacum officinale*) | Leaves, roots, and flowers are all edible and highly nutritive, containing calcium, magnesium, and iron. They also are a great source of vitamins A and C. Bitter properties make dandelion a prized digestive aid and liver tonic. Considered invasive by some. |
| Elecampane (*Inula helenium*) | Roots are used to treat deep and persistent lung infections. Antimicrobial, the roots help to fight infection and expel excess mucus. |
| Fungi | There are dozens of species of edible and medicinal fungi that are well suited for cultivation in a permaculture system. |
| Garlic (*Allium* spp.) | The bulbs help to stimulate the immune system, fight infection, and warm the body. Used to lower cholesterol and blood pressure. Excellent for the respiratory system and for internal parasites. |
| Goldenseal (*Hydrastis canadensis*) | The roots contain berberine and hydrastine and are highly effective at treating infection. Used topically and internally for its astringent, antibiotic, and antimicrobial properties. |
| Horseradish (*Armoracia rusticana*) | Pungent and warming, the roots are antimicrobial and excellent for clearing congestion in the sinus cavity and treating the common cold. Also a wonderful digestive aid. |
| Lemon balm (*Melissa officinalis*) | Leaves and flowers are full of volatile oils and are used to calm the nervous system, aid in digestion, and promote sleep. They also contain antiviral and antiseptic properties and can be used topically and internally. |
| Marshmallow (*Althaea officinalis*) | Roots and leaves are demulcent. They sooth irritations to mucous membranes, are indicated for treating respiratory infections, and are excellent tonics for the kidneys and urinary system. |
| Meadowsweet (*Filipendula ulmaria*) | Aerial parts are rich with salicylic acid. They help to reduce fevers and inflammation. Specifically used to treat the digestive system by soothing ulcerations and neutralizing excess acidity. |

| | |
|---|---|
| Motherwort (*Leonurus cardiaca*) | The leaves are used to strengthen and tone the heart and blood vessels. Also indicated for easing menopausal symptoms. Considered invasive by some. |
| Nettle, stinging (*Urtica dioica*) | Leaves are used as a nourishing tonic, full of vitamins, minerals, and chlorophyll. One of the most nutritive herbs. The roots are used to reduce inflammation of the prostate. All parts are used to prevent the symptoms of seasonal allergies and to strengthen the urinary system and liver. Considered invasive by some. |
| Peppermint (*Mentha piperita*) | Aerial parts are used to enliven the spirit and sooth the nervous system. Also contains volatile oils that have a carminative affect, helping to aid in digestion and relieve gas. Considered invasive by some. |
| Red clover (*Trifolium pratense*) | Blossoms and leaves are wonderful at detoxifying the blood and are full of calcium and iron. Also used to strengthen the respiratory system. |
| Skullcap (*Scutellaria* spp.) | Leaves are a strong nervine and help relieve stress, insomnia, headaches, and nervous tension. A powerful yet safe tonic to use over long periods of time to strengthen the nervous system. |
| Valerian (*Valeriana officinalis*) | Roots are used to help relax muscle spasms, ease tension, and alleviate insomnia and headaches. One of the most commonly used nervines. *Note:* For some people valerian is contraindicated and can act as a stimulant. When first using the plant, start with small doses. |
| Yarrow (*Achillea millefolium*) | Leaves and flowers are used to stanch blood flow and have antiseptic properties, making yarrow extremely useful for treating wounds. Also used to promote sweating and reduce fevers. |
| Yellow dock (*Rumex crispus*) | Roots are extremely bitter and are used for liver health and as a digestive aid. Helps promote proper hormone levels and can ease constipation. Considered invasive by some. |

**Figure 5-13.** Bloodroot.

**Table 5-5.** Vine Layer

| Species | Use |
| --- | --- |
| He Shou Wu (*Polygonum multiflorum*) | Roots are used to strengthen the adrenals and build energy and vitality in the body. A wonderful long-term tonic. |
| Hops (*Humulus lupus*) | Flowering strobiles have a sedative effect on the nervous system and are used to treat insomnia and tension. The bitterness of hops makes it a fantastic digestive tonic. |
| Passionflower (*Passiflora incarnata*) | Excellent for calming the nervous system and encouraging restful sleep. Often used for treating panic attacks. |
| Schisandra (*Schisandra chinensis*) | Berries are used as a reproductive tonic to improve vitality and stamina. Also used as a highly effective adaptogenic tonic for strengthening the body's ability to manage stress. |
| Wild yam (*Dioscorea villosa*) | Roots are used to promote liver function and also help the body regulate hormone production. Used to treat infertility and to strengthen reproductive health in men and women. |

**Figure 5-14.** Hops in search of support. Photograph courtesy of Bethany Bond

## "Invasive" Medicinal Plants

Scientists currently estimate there are approximately 8.7 million species of living organisms on earth. Every one of those organisms (including humans) shares a few common traits with the others. We all metabolize, grow, adapt to change, respond to stimuli, and reproduce. Every single one of us is on a similar mission to survive and multiply, yet most of us carry out those processes in vastly different ways. We also all need a place to live. Melanie and I moved around our state of Vermont for a while until we found our home and moved in. Now our home and land supplies us with most of what we need in food, fuel, and shelter. What we did is typical of all 8.7 million or so living organisms: We found a niche to occupy and to sustain us, we inhabited it, and we multiplied. Humans are extremely mobile compared to most species. We occupy almost every inhabitable place on the face of the earth, and we keep growing, moving, consuming, adapting, and multiplying. We just keep performing natural biological and social processes, but at the same time we humans (well, some of us, that is) have decided that certain organisms (plants in this case) who perform the same natural biological functions that we all do should be deemed "invasive." The word "invasive" carries with it a lot of negative connotations (aggressive behavior, attack, encroachment are some that come to mind) and puts us on the defensive (and sometimes even aggressive attack) against these perceived threats.

Although plants have been doing what is natural for some 700 million years, humans who have only been around for a fraction of that time have recently decided that some plants should be deemed invasive plant species for doing what is totally natural. The great irony in this discussion comes from the fact that most, if not all, of the plants that we call invasive have been given that dubious distinction as a direct result of our actions. Not only have we created new niches that these eager plants occupy, we also assist them greatly in their migration to their new homes. So while we sit in our glass houses and throw stones, there is definitely reason for us to be concerned. The concern should mainly be centered around the fact that extremely toxic chemicals are being widely used in a (usually futile) attempt to control these so-called invasives. There is also concern that other species, especially native* plant species, are being displaced and even becoming threatened with extinction in some cases by these so-called invasives. While I, as well as many others, prefer to use terms such as "opportunistic" or "persistent" to describe these plants, the U.S. government and many private citizens have taken the term invasive and turned it into a household word.

"Invasive Plant," as classified by the USDA per executive order 13112, is defined as a plant that is:

1. Nonnative (or alien) to the ecosystem under consideration and
2. Whose introduction causes or is likely to cause economic or environmental harm or harm to human health.

The problem is that there is little to no scientific method for validating which species are truly causing "econonomic or environmental harm." There also seems to be no accurate method for determining exactly what qualifies as "native." While executive order 13112 officially lists hundreds of species of invasive plants, state by state, there are also many more species that have come to be referred to as invasive by the general public for the mere fact that they are persistent or hard to control once introduced into the landscape.

---

* The term "native" when referring to plants and their habitats is a fairly subjective term and can carry with it a certain degree of controversy. Since plants have been dispersed far and wide both naturally and anthropogenically since the dawn of humans, it can be challenging to determine exactly which species qualify as "native" and which ones were "introduced." Here we are referring to species that existed in their native environments in North America before European colonization.

Many species of medicinal plants that are grown or wild-harvested and marketed today are considered invasive plants. Following is a list of some of the more common "invasive" medicinal plants, all of which have fair to excellent marketing potential.

Barberry
Blackberry
Burdock
Chamomile
Chickweed
Coltsfoot
Comfrey
Dandelion
English ivy
Goldenrod
Japanese knotweed
Milk thistle
Mint
Mullein
Plantain
Purple loosestrife
Saint John's wort
Sea buckthorn
Self-heal
Stinging nettle
Sweet Annie
Wild rose
Wormwood
Yarrow
Yellow dock

harvesting mature rhizomes and leaving the immature plants to grow on, this system becomes almost completely self-supporting. Leaves from mature deciduous trees in the forest canopy fall and decay, feeding the rich humus layer, supporting the plants growing within it, and weeds are kept at bay by the lack of sunlight reaching the forest floor. The cost of production for these wild-simulated woodland beds is very low and aside from applying a little bit of supplemental calcium every couple of years when soil tests show deficiencies and removing a few saplings and any large branches falling from trees above, they require very minimal inputs once established.

We are also experimenting with growing a combination of black locust and hawthorn trees with schisandra vines, black cohosh plants, and medicinal mushrooms. In this system the black locust trees in the canopy layer will fix nitrogen into the soil, which fertilizes the plants growing around them. The wood from their limbs will provide a sustainable harvest of decay-resistant hardwood stakes and posts. Hawthorn trees in the understory tree layer will provide a yearly harvest of blossoms and berries. Schisandra vines growing up the black locust trees will benefit from the shade and produce berries to

sell. Medicinal mushrooms will recycle nutrients and help encourage fungal dominance in the soil and can be eaten or marketed for medicine making. Black cohosh, growing in the understory or herb layer, will benefit from the shade of the overstory and provide a yearly sustainable harvest of mature rhizomes for sale, while producing seed and immature rhizomes to propagate more plants. This system has recently been established and is in the development phase as of the writing of this book, so we have yet to determine how efficient it becomes. Our hope is to design and implement many guilds like these in order to capitalize on nature's brilliant model.

## Bottomland or Upland?

When we view the land around us in terms of population density, in general the closer you get to bodies of water such as rivers, lakes, and oceans, the more densely populated these areas tend to be. There are obvious reasons for this. Rivers provide transportation methods for commerce and electricity in the form of hydroelectric power. They also support fish and other aquatic life for food and recreation and provide drinking and irrigation water. The valleys and lowlands surrounding these bodies of water have acted as catch basins for the minerals and organic material deposited there for eons by gravity and the erosive forces of nature.

Over the course of history seasonal floods have deposited rich river bottom silt onto the fertile floodplains. It was obvious to Native Americans and early settlers that the soil in these lowlands tended to be richer than in the surrounding hills, so they preferred these fertile valleys for growing their food and raising livestock. Today we see that most of the agricultural operations still dotting the landscape primarily exist in these valleys. Farmers in the floodplain generally accept the trade-off that although floodwaters can wreak havoc on their crops, the benefit of richer soils in these fertile plains can tilt the balance in their favor.

However, with climate change comes increased unpredictability of the occurrence of extreme weather events. Here in Vermont, for example, as of the writing of this book in 2014, the state has experienced two of the worst flooding events in recorded history in the last five years alone. More and more farmers are heading to the hills, where the soil may not be as rich, level, and stone-free as in the bottomlands but, instead, where the rivers and streams can't wipe away their livelihoods in a moment's time.

## Full Sun

While all plants need sunlight to grow, some need more than others. Plants that are categorized as needing full sun generally require at least five to six hours of direct sunlight per day and preferably more. Most of the medicinal plants grown and used in the world today prefer growing in full sun. Bear in mind that there are local geographical influences that affect the amount of solar radiation plants can tolerate and that full sun in Vermont doesn't necessarily equate to full sun in Texas.

For example, here in the northeastern United States we can grow stinging nettles in the full sun, and they thrive in this environment. However, if one were to plant stinging nettles in the full sun in a place like Texas, they would probably wither in the intense heat

**Figure 5-15.** Bottomland.

**Figure 5-16.** Nettle growing in full sun.

## Considerations for Growing Herbs in a Changing Climate

*There are no passengers on spaceship earth. We are all crew.*
— BUCKMINSTER FULLER

Climate change is a reality. Whether it is anthropogenic, part of a natural cycle, or a combination of both may be open to debate, but the undeniable fact is that no matter where you live on the earth, the air and sea surface temperatures around you are rising at a rate faster than at any other time in recorded history. Some farmers we have heard from view global warming as somewhat of an advantage in terms of potentially longer growing seasons. However, most of us see it as a very serious and immediate threat to the well-being and livelihood of the current and future generations of people, plants, animals, and other organisms that call this planet our home.

There is no denying the fact that our weather is changing dramatically and unpredictably. As stewards of the land farmers are faced with taking a proactive approach to ensuring that our soils remain intact and healthy so they may continue to serve their critical role as a foundation of life. As growers of trees, plants, and fungi that serve as powerful agents capable of preventing or treating illness and injury we are responsible for ensuring that the species of plants we are in partnership with continue to thrive, not only in the cultivated setting on our farms but most importantly in the wild in their natural habitats. We are also responsible for ensuring that our actions as farmers and business owners are well thought out and executed in terms of contributing to the health of our planet and its inhabitants while mitigating ecologically harmful impacts. Humans are a resilient species, well adapted to countering adversity, but we have yet to experience the types of changes that are predicted to come if we continue on the path we have been treading.

Here are a few simple things medicinal herb farmers can do to contribute not only to our own well-being and longevity but also to the greater good during these changing times.

1. **Think and act positively even when it seems hard to do so.** The power of intention is incredibly strong, and the first step toward effecting positive change is believing that it can happen. The next step is to act on that belief system.
2. **Feed and protect the soil.** Soil is considered the largest carbon-storage system on the planet and should be viewed as a renewable resource capable of helping to mitigate elevated atmospheric carbon levels. Bare soils can rapidly oxidize, releasing gaseous carbon into the atmosphere, and are prone to erosion, which can carry organic carbon into our waterways. Feeding our soil through remineralization and the addition of beneficial microbes produced by decomposing plant residues can increase its ability to sequester carbon and to produce high-yielding, healthy, nutrient-dense plants. Minimizing tillage practices and keeping bare land cover cropped helps protect mycorrhizal fungi, which are a link between plants' roots, soil nutrients, and water. These fungi produce glomalin, an organic gluelike substance that plays a crucial role in soil carbon sequestration.
3. **Grow more plants, both in quantity and variety.** This is simple. From an environmental perspective, plants, especially perennials (which most medicinals are), draw carbon out of the air via photosynthesis and use it to fuel their growth. Surplus carbon not needed by the plants is transferred to the soil via fungal hyphae, where it can be sequestered. From a humanitarian perspective our role as growers of plant medicine cannot be overstated in its importance for helping to heal the planet and its inhabitants. It is also our responsibility to protect threatened species of medicinal plants by expanding their populations, both in the wild and in the managed setting. Plants provide critical habitat and food for pollinators and countless other organisms that contribute to the biodiversity within our

landscape. Closer to home, growing plants contributes to our own food and medicine security.

4. **Proactively adapt land for greater resiliency.** As the climate continues to change we are likely to see increased volatility in our weather patterns, as we have in the last two to three decades. With this increased volatility comes the likelihood that we will experience weather events that exceed the scope and severity of events we have previously experienced. While we as humans can likely adapt to increasingly volatile weather, thanks to our resourcefulness and mobility, our local ecosystems can't get up and run away and are more likely to suffer from the consequences of acute weather events. Instead of waiting until the next flooding event rips a gully through our fields, carrying away our precious topsoil, we should create swales and mounds and plant vegetation to help distribute water more uniformly. Instead of waiting until the drought causes plant stress and undermines our food, water, and medicine security, we should locate springs and other water sources and build ponds and other containment systems to capture and store water for drinking and irrigation. We should also assume that frosts will come earlier or later than average, winters may be colder or milder than average, and the whole concept of assessing an average based on past weather events will be a less effective predictor when it comes to future events.

5. **Mitigate potentially harmful impacts through our business practices.** As small to midsize organic farm owners it is easy for us to point our fingers at industrialized chemical-based agriculture and scoff at its apparent disregard for its negative impact on the environment. We all have an impact, some greater than others, obviously, but we can take measures to ensure that our practices are well planned and executed to minimize their environmental impact. Nutrient management is one of the most obvious and important practices we should implement. Soil testing gives us a road map for nutrient management by providing us with information regarding what our soils need and may not need. After we apply soil amendments based on soil testing recommendations it is up to us to ensure that the nutrients already present in the soil as well as those we add will remain as stable as possible. Remineralization and maintaining good pH levels and ample organic matter percentages are a good foundation for nutrient stabilization. Plants and cover crops that utilize and help store soluble nutrients make up the cornerstone of our "nutrient bank accounts," while surface and groundwater management help to serve as "overdraw protection." There are many more ways we can lessen our impact by making thoughtful choices with how we run our businesses.

at that latitude. Given some shade and rich, moist soil from an irrigation source, however, nettles could probably adapt to growing in the harsh arid conditions of places like Texas. Most commercial medicinal herb crops grown in the full sun are perennials that are grown in open fields either in densely planted blocks or in row-crop-style plantings designed for mechanical planting, cultivation, and harvest.

Not only is some sunlight necessary for all plants to photosynthesize and grow, it can also play a significant role in the development of secondary plant metabolites that are the primary source of the bioactive compounds that give plants their medicinal activities. Studies have shown that abiotic stresses, such as those caused by increased ultraviolet radiation, can have a positive effect on the production of secondary metabolites such as essential oils, antioxidants, and many other valuable and important bioactive compounds that give plants their medicinal activity.

## Farming on the Edge

When we walk or drive through the countryside near where we live, we see many examples of species of herbs that we grow on our farm growing naturally in the wild. We can often gain valuable insights into growing healthy plants by observing where these wild plants like to grow naturally. For example, skullcap (*Scutellaria lateriflora*) is an herb that we and many other herb growers have often found challenging to grow consistently well. Skullcap is a great example of a plant that seems to prefer very specific growing conditions. We have learned by observing where wild skullcap grows that it appears to prefer growing on the edge. By "the edge" we mean the margins, where one habitat borders another. In scientific terms this zone is referred to as the "ecotone."

The primary ecotone where we find many of our native and naturalized medicinal herbs, including skullcap, growing wild is where woodland meets grassland. Here the undisturbed soils tend to be fungally dominant, which many species of medicinals seem to prefer.

When we started growing skullcap, we read that the plant prefers growing in full sun to partial shade. We planted it in full sun, thinking that it would thrive in that environment, especially considering "full sun" is a very relative term here in Vermont, given the amount of cloudy, rainy days we have here. The first year we grew it was an extremely rainy and cloudy year, and it did very well growing out in the open. It overwintered well and emerged in the spring, but that next season was hot and sunny and unusually dry, and we didn't have an irrigation system set up. The skullcap withered in these arid conditions, became infected with powdery mildew, and was a total loss. We had seen very few isolated pockets of skullcap growing in the wild in Vermont, and there seemed to be a very specific type of growing environment that it had adapted to. It seemed to like growing on the edge of woodlands, where it had plenty of sunlight

**Figure 5-17.** An ecotone, where the edge of the field meets the forest. Photograph courtesy of Todd Lynch, Ecotropy LLC

for the early half of the day and plenty of shade for the later, hotter part of the day. It also did not seem to like having to compete with other plants; in fact, it often grew where a river or stream had previously flooded and left pockets of moist, fertile soil in its wake when the floodwaters receded. The skullcap was able to fill that niche before the meadow grasses and sedges could come in to outcompete it.

In subsequent years on our farm we have tried to mimic this pattern by planting it in moist, fertile soils on the western side of our fields up close to the woodlands, which provide protection from the afternoon sun. This approach to growing skullcap "on the edge" has proven to be much more successful than growing it in the full sun.

## Growing in the Shade

Growing woodland plants presents many unique rewards and challenges. There isn't much work that we do here on our farm that is as enjoyable as planting seeds or young rootlets in the dappled sunlight of an early spring morning while songbirds joyfully announce their return from southern migrations. On the hottest days in mid-August we often seek shelter from the blazing afternoon sun by venturing into the woodland beds to gather seed, weed saplings, or remove tree branches blown down from the canopy above. The woodlands are a sanctuary for the farmers here as well as for the plants we grow, and the benefits of growing crops in this environment transcend the financial realm.

### Woodland Medicinals

Some of the most effective and sought-after medicinal plants in use today are native to North America's forests. Unfortunately the effectiveness of such plants as ginseng and goldenseal has contributed to the fact that they and many of their woodland-dwelling counterparts are faced with the threat of extinction unless we work to replant wild populations and cultivate them in a sustainable manner rather than harvesting them from the wild. There are tremendous opportunities for growers to not only help protect

**Figure 5-18.** American ginseng and goldenseal growing together in the shade of a mature sugar maple. Photograph courtesy of Bethany Bond

the wild medicinals we use and cherish by growing cultivated alternatives but to do so in an enjoyable, profitable, and sustainable way.

Shade-loving woodland plants grow best in areas that provide shelter from excessive amounts of the sun's heat and ultraviolet radiation. There are many ways to provide the ideal growing conditions in which to nurture the sensitive nature of these important medicinals. The preferred method of growing woodland plants is simple: Grow them in the woodland in fungally dominant soils where they prefer to grow. But not everyone has a woodland available to grow plants in. Fortunately there are alternatives to growing in the woods, including erecting artificial shade structures and even utilizing the shade from other existing objects, such as buildings or other trees or plants.

Woodland growing is classified using two methodologies, "woods grown" or "wild simulated." Woods grown describes the method of cultivating planting beds in a woodland setting. Wild simulated describes the method of simply planting seeds or young rootlets in natural undisturbed habitats without cultivating the soils in those habitats. The ideal conditions for both of these methods are well-drained, humus-rich, fungally dominant soils in stands of mature deciduous hardwood trees where dappled sunlight reaches

## Indicator Species for Determining Good Ginseng Habitat

- Arrowwood, *Viburnum dentatum*
- Basswood, *Tilia americana*
- Beech, *Fagus grandifloia*
- Birches, *Betula* spp.
- Black cherry, *Prunus serotina*
- Bloodroot, *Sanguinaria canadensis*
- Blue cohosh, *Caulophyllum thalictroides*
- Bunchberry, *Cornus canadensis*
- Christmas fern, *Polystichum acrostichoides*
- Hickories, *Carya* spp.
- Jack-in-the-pulpit, *Arisaema triphyllum*
- Jewelweed, *Impatiens capensis*
- Large flowered trillium, *Trillium grandiflorum*
- Maidenhair fern, *Adiantum pedatum*
- May apple, *Podophyllum peltatum*
- Rattlesnake fern, *Botrychium virginianum*
- Red baneberry, *Actaea rubra*
- Red maple, *Acer rubrum*
- Red oak, *Quercus rubra*
- Red osier dogwood, *Cornus sericea*
- Spicebush, *Lindera benzoin*
- Sugar maple, *Acer saccharum*
- White ash, *Fraxinus americana*
- White baneberry, *Actaea alba*
- Wild sarsaparilla, *Aralia nudicaulis*
- White oak, *Quercus alba*

**Figure 5-19.** Woodland beds of goldenseal emerging in the spring. Photograph courtesy of Bethany Bond

from the soils around their roots. This fact could necessitate supplemental irrigation during dry spells. Conifers also tend to grow in less fertile, more acidic soils than deciduous hardwoods, so it may be helpful to amend the soil with organic fertilizers and/or calcium when needed, as recommended by soil tests and specific plant requirements. Conifers are also very dense-growing trees, so it may be necessary to do some thinning to allow for more sunlight penetration, which the shade-loving plants need in small amounts to photosynthesize. We discuss preparing woodland beds in part two in the ginseng profile.

## Shade Houses

Shade houses are an alternative to the woodland culture of shade-loving medicinals. In fact, most of the commercial ginseng and goldenseal produced today is grown in shade houses. In the past these shade houses were constructed of wooden lathe strips fastened to removable frames, set on large posts sunk into the ground and supported by guy wires. These wooden frames were costly to build and maintain and needed to be removed and stored during winters to avoid damage from snow loading. Total costs of these lathe-type shade houses range from $20,000 to $40,000 per acre, which can be a risky investment,

the forest floor. There are many species of wild plants that are considered good "indicator species" for ideal woodland medicinal herb habitat.

There are ways growers can make do with less than ideal woodland situations, such as planting under smaller trees and even conifers, but these environments require a more managed approach than the hardwood forest, which can be almost entirely self-sufficient (aside from preparation, planting, and harvesting). Conifers, for example, are primarily shallowly rooted and tend to pull a lot of moisture

**Figure 5-20.** In 2003 we grew goldenseal in this shade house. The increased weed pressure and maintenance costs caused us to reevaluate, and after two to three years we switched to growing in wild-simulated, woodland beds.

considering the fact that a marketable crop may not be harvested for four to five years and may only yield $30,000 worth of crops. Today woven polypropylene shade fabric suspended by wire grids strung between posts has replaced most of the wooden lathe-type shade houses. However, these shade cloth–covered structures are still laborious and costly to maintain.

There are inherent and substantial drawbacks to using shade houses. In addition to the high material, building, and maintenance costs, crops grown in shade houses are often monocropped and are extremely susceptible to disease. As a result, some nonorganic farmers have turned to using toxic and costly fungicides. Soils used for shade house production tend to be bacterially dominant, which can be a challenging medium for growing plants such as ginseng that prefer fungally dominant woodland soils. Also, roots harvested from these monocrop

**Table 5-6.** Cultivation Methods for Ginseng: Comparison of three growing methods with approximated costs, yields, and profits[2]

| Method | Artificial Shade | Wood-Cultivated | Wild-Simulated |
|---|---|---|---|
| Time to first harvest | 3–4 years | 6–8 years | 7–12 years |
| Seeds planted per ½ acre ($85/lb) | 50 lbs | 24 lbs | 12.5 lbs |
| Total labor per ½ acre ($10/hr) | $1,500 | $1,950 | $825 |
| Tools, pest control, fertilizer, and other expenses | $14,250 | $2,595 | $590 |
| Total cost per ½ acre | $33,500 | $24,135 | $9,690 |
| Root yield per ½ acre | 1,125 lbs | 300 lbs | 80 lbs |
| Root price per dry lb. | $12/lb | $100/lb | $450/lb |
| Gross income per ½ acre | $13,500 | $30,000 | $36,000 |
| Net profit per ½ acre | -($20,000) | $5,865 | $26,310 |

Note: Figures will vary based on individual circumstances.

shade houses (namely ginseng) tend to fetch much lower prices on the market because they are viewed as being less potent than their wild or wild-simulated counterparts. Before investing in a shade house, one would be wise to explore the many detractors of this system and consider implementing a wild-simulated growing practice instead.

# Till versus No-Till

To till or not to till, that is the question many organic farmers ask themselves as they weigh the benefits of having a clean slate of bare soil into which to plant or seed versus the costs of potentially damaging soils while tilling. Modern no-till (also called zero tillage or direct drill) agriculture was developed during the dawning of the chemical age shortly after World War II. The weed killer 2,4-D was such an effective herbicide that farmers realized they could spray their fields with it instead of tilling to remove weeds in preparation for planting. They would then plant directly into the "clean" fields and apply more herbicide as needed throughout the growing season instead of cultivating the soil mechanically. Subsequent crops were often rotated and planted directly into the residues from previous crops. Soil conservation was a relatively obscure practice at the time, so farmers were practicing no-till mainly to reduce the costs associated with combating weeds but also to increase yields. In the 1970s the herbicide atrazine, and more recently glyphosate (Roundup) and many more, were added to the chemical arsenal, giving no-till farmers a broader spectrum of chemicals with which to battle weeds that were becoming increasingly resistant to herbicides.

While chemical farmers were utilizing no-till methods primarily to combat weeds and increase yields, others, especially in the growing organic movement, were viewing no-till methods not only in terms of cost reduction and increased yields but more importantly in terms of soil conservation practices. Tilling can upset the delicate biological balance that comprises healthy soil (we discuss this in more detail in chapter 7). It can also greatly increase the soil's susceptibility to erosion by gravity, wind, and water. Instead of using chemicals to prepare a weed-free seedbed, organic farmers have found ways to utilize mulches to smother weeds, then plant directly into the mulches, either by hand or with specialized machinery. During the growing season the mulches serve to keep weeds from germinating while also protecting the soils from erosion and biological disruption.

## Cover Crops

Both organic and conventional growers have had great success with growing cover crops, then utilizing specialized tractor-mounted roller/crimpers to knock the cover crops down and "crimp" the stems to halt growth without dislodging the cover crops' roots. Specialized tractor-mounted planters then slice into the cover crop residue with sharp steel coulters and plant seeds or transplants. The specialized equipment that currently exists for these methods is primarily geared for larger-scale agriculture, but there are small-scale alternatives currently being developed.

## Mulches

Another method of no-till agriculture well suited for small farms not wanting to invest in specialized equipment is to no-till plant directly into mulch. Clean, dye-free cardboard is a favorite mulch we have been experimenting with on our farm as well as on many others, as it smothers weeds by blocking light, allows sufficient oxygen and water to reach the rhizosphere, and slowly degrades over time, adding organic material to the soil. We use recycled boxes by placing them onto the planting beds either in the fall or early in the spring to smother weeds. We weight the edges down with soil and small stones where the cardboard is susceptible to being dislodged by the wind. We then cut or poke holes in the cardboard, dibble holes in the earth, and place our transplants or seeds into the holes. Sometimes a little hand weeding around the holes is necessary until the plants become well established. The cardboard often needs to be replenished in subsequent years when using it on long-term perennial plantings.

Other types of mulch, such as bark, untreated burlap, newspaper (printed using soy-based inks), straw, leaves, and hay are also good candidates for small-scale no-till methods.

As ideal as no-till agriculture may seem, it is not without its challenges. The primary drawback to mulching is that it can delay soil drying and warming up in the spring. Most mulches used for no-till agriculture are light colored and tend to reflect rather than absorb solar radiation. This can be especially challenging in northern climates where growing seasons are already short because cold temperatures slow plant growth. Evaporation of moisture is also slowed, which can be challenging in the spring or during prolonged rainy spells.

Another challenge growers have found is that if mulches are applied too thickly, they can upset the transfer of oxygen into and moisture out of soils, disrupting soil biological balances. The fine line between enough mulch to smother weeds and not smothering or drowning soil biota can be challenging to discern.

## Cost Benefits

The cost-saving benefits of no-till agriculture are obvious. No tilling means less labor, fuel, irrigation, and equipment costs, and crop yields can be much higher because of less weed competition and better soil health. Less obvious but perhaps even more important in the long term are the ecological benefits to no-till farming. Tilling releases carbon, both from the burning of fossil fuel–powered machinery and from the depletion of soil organic matter, whereas no-till can help to sequester carbon in the soil. Tilling can also destroy beneficial soil microbes and other organisms, whereas no-till methods allow soil biota to flourish. No-till can improve water infiltration into the soil, decrease the potential for erosion, and reduce irrigation costs by inhibiting evaporation. No-till farming methods should be considered, developed, and implemented by medicinal herb farmers growing long-term perennial crops to help increase profits and conserve our precious soils.

# Farming on Sloping versus Level Land

Wouldn't it be nice if we could all find the ideal combination of deep, fertile, well-drained loamy soils on level ground on which to farm? The reality of the situation is that most of the "perfect" agricultural land is either already being farmed or is slated for residential or commercial development. So where does that leave the aspiring farmer? It oftentimes leaves aspiring farmers with choices to make that may involve having to sacrifice one piece in the puzzle that could otherwise constitute an ideal farm site. Obviously this isn't a factor for everyone, as there are many regions where ideal cropland is more plentiful and affordable, but in general it is becoming increasingly scarce, less affordable, and more polluted by chemical farming methods or nearby industrial sources.

As to the choice of which piece of the ideal farm puzzle to consider sacrificing, it can involve choosing the least restrictive limiting factor, which often means having to farm sloping land. By sloping land I am referring to gently sloping land, not hilly land. Hilly land can certainly be farmed, but it is very challenging and requires thinking outside the parameters of conventional farming techniques. On hills, growers would have to implement either no-till or terraced systems to prevent soil erosion. If precautions are not taken, bare soil on hillsides will inevitably succumb to the forces of gravity and erosion, especially when assisted by torrential rains. Agroforestry and permaculture systems are often good candidates for hilly land, since woodland soils tend to be more stable because of tree roots and understory vegetation that help to stabilize the soil.

Farming sloping land is possible but requires an extremely cautious and thoughtful approach. Soils that took thousands of years' worth of decomposing minerals and biota to form can disappear downstream right before a farmer's eyes unless methods are utilized to ensure that the soil stays in the field where it belongs and not in the waterways. Much of the land we till and plant at ZWHF is gently sloping land, and even with our extremely conservative

**Figure 5-21.** One of our gently sloping fields.

**Figure 5-22.** Precision weeding with sweeps belly-mounted to a Farmall Cub.

approaches, we have experienced soil erosion first-hand. Fortunately for us, the erosion we have experienced has been minimal because of the soil conservation measures we adopted.

On the sloping portions of our land, instead of tilling large, open fields, we have tilled smaller (⅛ to ¼ acre) fields. These fields and the beds of plants within them are oriented perpendicular to the angle of the slope, similar to the way hillside terraces are formed. This perpendicular orientation means that when water flows down the slope, it encounters lots of small "dams," which force the water to move laterally, spreading out, percolating through the soil gently, and slowing its potentially destructive momentum. Between these small fields are eight-foot-wide sod paths. These wide paths act as secondary dams that help to stabilize the whole slope by catching, slowing, and diverting more water before it gets a chance to form gullies and erode large quantities of soil.

Unfortunately we still have to deal with occasional "sheet erosion," which occurs when torrential rains erode fine organic material from the soil surface and deposit it in the low points of the beds. To mitigate this we try to utilize cover crops and their residue where we can and fill as much of the beds as we can with plants to minimize the amount of unprotected soil. The riskiest time of year is in early spring, when soils are freshly tilled and plants have yet to establish good root systems to help stabilize the soil.

Another challenging aspect of farming on the slope comes with the use of mechanized equipment such as tractors. Tractors, especially when pulling implements, tend to "crab" or walk sideways downslope during inopportune times. This is particularly challenging when using precision cultivating implements to weed close to crops. The best way to address this situation is to plan ahead and bring some hand tools when tractor cultivating in case you have to replant any plants within the row that get uprooted from errant cultivating. There seems to be a direct correlation between the time the driver has to spend off the tractor seat fixing mistakes by hand and his or her improvement in skills performing precision cultivating.

# Irrigation Equipment and Methods

As we mentioned previously, plants need water to survive, and we can't always depend on Mother Nature to bring rain in times of need. There is a reason many farmers refer to their irrigation systems as their "irritation systems." The fact of the matter is that it's neither cheap nor easy to irrigate cropland, so these systems must be thoughtfully planned, installed, and maintained to avoid irritation with your irrigation.

There are many, many options when deciding on irrigation methods and equipment, but the first thing you need to consider is where the water for irrigation is going to come from. Whether it be a rainwater catchment system, spring, well, stream, river, pond, reservoir, or lake, it has to be a dependable source of water that won't dry up during times when it is needed the most. It also should be free from harmful contaminants and relatively clean. Purifying filters can improve silty water thoroughly enough to prevent irrigation equipment from clogging. But they require maintenance, and the dirtier the water flowing through them, the more often they need to be maintained. Ideally, the water source is uphill from the fields so gravity can assist with getting the water to the plants more efficiently.

The second consideration when planning irrigation is how to get the water from the source to the plants. There are four primary irrigation methods to consider: drip irrigation, spray irrigation, surface or flood irrigation, and subsurface irrigation.

## Drip Irrigation

Drip irrigation is generally considered the most efficient method at over 90 percent efficiency, because water is slowly emitted, either on the surface of the soil or just below it, and allowed to slowly percolate into the root zone through capillary action via pores in the soil. There are several affordable drip irrigation options, including soaker hoses, T-tape, and emitter style. There are two types of emitters: pressure sensitive and pressure compensating. Pressure-sensitive

**Figure 5-23.** Drip irrigation on skullcap.

emitters deliver a higher flow at higher water pressures. Pressure-compensating emitters provide the same flow over a variable range of pressure.

At ZWHF we utilize T-tape, which is a turbulent flow pressure-compensating product, and we find it to be affordable, relatively easy to install and operate, and dependable. T-tape requires such low pressure and flow rates that we are able to irrigate some of our fields using photovoltaic panels with small 12-volt RV pumps to carry water from our pond and stream to the fields. An inline filter strains particulate matter from the stream water and we have to rinse the filter element, on average, once for every two days of steady irrigation use to keep it clean. For the pond we are installing a do-it-yourself sand filter that will lessen the amount of time we have to spend cleaning the strainer on the filter there. You can find the directions from the University of Vermont Extension Service Agency on building a great do-it-yourself sand filter at http://www.uvm .edu/~susagctr/resources/SandFilterHowTo.pdf. The beauty of the solar-powered irrigation system is that it carries drip irrigation water to the plants' roots when they need it most, when the sun is shining and plants are at their highest transpiration rate. Another benefit of drip irrigation is that the water doesn't soak the aerial portion of plants, which helps to limit the spread of fungal diseases.

## Spray Irrigation

Spray irrigation is another option, especially for those with plentiful water resources. The downside with spray irrigation is that it is only between 50 and 70 percent efficient because of the amount of water that is evaporated into the air during irrigation. It also requires much more water at higher pressures, so in general the pumps and piping required to deliver these requirements can be expensive to buy and operate. Spray irrigation can also lead to increased rates of fungal diseases from wetting of aerial plant parts. There are many types of spray devices to choose from when installing spray irrigation systems.

## Surface Irrigation

Surface irrigation, or flood irrigation as it is often called, involves distributing water across the surface of croplands, where it percolates into the soil. Although from a historical perspective this is the most common type of irrigation used, it is rarely employed on small farms today. Although surface irrigation can be efficient at delivering water, the downside is that it requires large volumes of water to be effective and can cause rapid leaching of nutrients through the soil caused by the volume of water distributed. It also requires canals and ditches to carry water to cropland, which can be laborious to install and maintain. Efficiency varies between 50 and 90 percent with surface irrigation, depending on how it is applied and the amount of water that is utilized by the plants, versus water that simply flows through the ground without affecting plants.

## Subsurface Irrigation

Subsurface irrigation is a less common method of irrigation that involves raising the water table in grounds below cropland to irrigate the plants within. These systems vary in their cost and effectiveness and are rarely employed because of the complexities involved with groundwater management.

# CHAPTER 6

# Tools of the Trade

Father John Culkin, friend and student of the great Marshall McLuhan, once wrote, "We shape our tools, and thereafter our tools shape us." When we apply this metaphor to modern agriculture we can see how the tools we have invented to increase efficiencies in our work have not only influenced our social identities as farmers but also greatly influenced the methods that we use to farm. Imagine a

**Figure 6-1.** An assortment of hand tools. Photograph courtesy of Bethany Bond

painting of an older man and a woman in colonial period clothing standing in front of a house looking solemn. If asked to define who they are or what they do for work, one would likely be hard-pressed to supply the correct answer. Now imagine a pitchfork in the gentleman's hand in the same picture, and the answer to the same question becomes almost universally, "They are farmers." To the casual observer there is nothing remarkable about a woman walking down the road, but when she drives down the road on a tractor she personifies the farmer.

When we consider agriculture in all of its myriad forms there are many universal methods and tools employed, regardless of the period in time or the crop or animals in production. Somewhere in use on most farms around the world are tools consisting of long wooden handles attached to curved steel blades. Whether these tools are hoes (first mentioned in ancient documents such as the Code of Hammurabi [ca. eighteenth century BC] and the Book of Isaiah [ca. eighth century BC]) being used to terrace rice paddies in Cambodia or shovels used to scoop grain into a feeding trough on a five-thousand-acre Utah cattle ranch, these tools have changed very little since their inception. In fact, these tools are not that much more technologically advanced than the first known tools used by humans (*Homo habilis*) 2.6 million years ago in Ethiopia to remove meat from animal hides: 2.6 million years have passed, and the cave man's sharpened stone blade now has a longer wooden handle and curved blade made of steel. Now that is what I call slow product development. Those tools haven't evolved much because they work fine the way they are. Tractors have followed a similar trajectory but in a fraction of the time. The first tractors weren't as colorful, shiny, and efficient as the new ones, but their basic form and function haven't really changed much at all.

When I use a hoe or shovel it becomes an extension of my arm. When I get on my tractor I am so familiar with its mechanisms and movements that I am barely conscious of what my body is doing to drive the thing. Our tools certainly have shaped us as farmers, so let's talk tools.

# Mechanization versus Hands On

There is something so elegant and simple about a well-made hand tool—the way its smooth wooden handle feels in our hands, the way the curve of its steel blade slices through the soil, the lack of obnoxious fumes and noise emitted while using it. All these reasons and more make hand tools indispensable. They are also very affordable, and when they break they can usually be repaired or replaced easily. Small farms can operate on hand tools alone; I have seen it done, and it makes me somewhat envious.

As much as I hate to admit it, our business is totally dependent on machinery, which is both good and bad—good in the sense that I have been in love with tractors ever since I could crawl, and now I get to live out my boyhood fantasy of owning and operating them. I love tractors and their implements and all the other machines that make what we do so much easier . . . when they are running smoothly, that is. When they break down or run poorly, on the other hand, there's no love. In general (for better or worse) we farmers have a tendency to anthropomorphize our tractors. We often refer to them as "she," as in, "How's she holdin' up pulling that four-bottom plow?" I can't tell you how many people have seen our little Farmall Cub cultivating tractor and told us how "cute" it is. The tractors even have names. We call the larger one "Big Blue," and our daughter Lily has dubbed the Cub "Little Red." As with most relationships, ours with our tractors certainly has its ups and downs.

## The Negatives

The negative aspects of being dependent on machinery are numerous, the first being that the one thing you can always depend on with farm machines is that they will eventually malfunction in one way or another. These malfunctions often occur during the most inopportune times imaginable. Have you ever noticed that farmers tend to be more mechanically adept than the average person? They don't all start out this way, but most of them sure do end up this way because if they don't, they go either bankrupt or insane from having to pay repair bills. It took us several years of paying people to fix our equipment before we smartened up, bought a welder, and learned through trial and lots of error how to fix the equipment ourselves.

Another negative aspect of machines is that they are expensive to purchase and operate. Usually these expenses are justified by the savings in time and labor by using them, but they still require capital to purchase and maintain. Probably the worst aspect of operating machinery is the environmental impact they have. As much as we try to use "environmentally friendly" biodiesel blended fuels and lubricants to keep them running, they still guzzle petroleum and belch carbon, and they definitely consumed and polluted natural resources during their manufacture and shipment.

## Draft Animals

In spite of the challenges that come with owning, operating, and maintaining machines, the fact of the matter is that they are definitely worth the hassle. We couldn't be doing what we are doing on the scale that we are doing it without machines. Some people have asked us why we haven't considered using draft animals to farm. My answer is that I have an enormous amount of respect for anyone using draft animals to farm, but I just simply do not have the patience, skills, time, or desire to adopt those techniques, nor do I think, even if I became the highest skilled draft animal farmer in the world, that I could even come close to matching the productivity we have achieved with the help of the internal combustion engine.

Believe me when I say that I am morally challenged by the amount of carbon this farm emits from its machinery. However, when I consult with my moral compass for guidance, our contributions to stewarding the land, the healing benefits of the products that we provide, and the fact that we are producing locally grown herbs that are commonly sourced from thousands of miles away allow me to sleep better at night knowing that the resources we use are helping the greater good.

# Basic Hand Tools

We love our hand tools and prefer their simplicity and low-impact nature when it makes sense to use them instead of a machine. There are so many options when it comes to hand tools that it can be overwhelming at times deciding which one is the best for a given task. I am continually amazed when I open new tool catalogs and see the latest gadgets, some ingeniously designed and fabricated and others . . . well, not so much. If I had to keep track of every minute I have spent at the local farm and garden store trying to decide which of the spading forks or shovels would be strong enough to withstand digging four-year-old black cohosh rhizomes from the woodland beds or which pruning shear is going to feel comfortable in my hands and remain sharp while taking hundreds of hardwood elderberry cuttings, it would seem excessive.

We have all heard the saying, "Things just aren't made as well as they used to be." That statement is as true as the day is long. We have a shed full of hand tools, and the ones that have survived the test of time are the "old-school" tools that were handed down to us from previous generations. In general, the newer the tool is, the shorter its life span seems to be. Yes, there are some very well-made tools still being manufactured and sold, but they are becoming exceedingly hard to find and expensive. Many times an expensive tool is a well-made tool, so it is often worth spending some extra money for something well designed and manufactured. Following is a list of our favorite hand tools.

## Pruning Shears

We like the Swiss-made Felco pruning shears and probably use them more than any other hand tool on the farm. Felcos are expensive, but they are worth the price. Don't be fooled by the imitation Felcos that are less expensive because you definitely get what you pay for. Japanese Okatsune shears are also well made and worth the extra expense.

## Field Knives

Steel field knives come in a variety of designs, sizes, and shapes and are our preferred tool for hand-harvesting leaf crops. They are amazingly affordable, durable, and easy to sharpen and maintain. We get ours from the Johnny's seed catalog or Peaceful Valley Farm Supply.

## Scythes

Scythes are some of the coolest and most elegantly designed hand tools ever invented. I have met some scythe fanatics, and some of them are as twisted as their favorite tool's handle. Scything is considered by some to be akin to a form of meditation. I can definitely think of higher-quality spiritual pursuits than swinging a scythe, but there is certainly a graceful, rhythmic feeling involved with scything grasses or herbs. This tool works well for harvesting many types of crops, as long as their stems aren't woody, at which point scythes become ineffective. The older ones seem to be the best, and there seem to be plenty of them around old barns and sheds. Scythe Supply company from Perry, Maine, is an excellent source as well.

## Hoes

Hoes have come a long way since the short-handled gooseneck hoe your grandmother probably used to hoe her beans. Nowadays hoes are ergonomically designed, with longer handles, smaller lightweight blades, and better angles to enable the user to weed in a more efficient, upright manner. We use a variety of long-handled hoes made famous by Eliot Coleman. They can be purchased through Johnny's seed catalog. Some people like stirrup hoes, but we prefer the thin-bladed "collinear hoe," as well as the trapezoidal hoe for its lightweight design and simplicity.

## Garden Rakes

Garden rakes or "steel rakes," as they are sometimes called, are great for leveling soil by hand in beds, raking small amounts of seed in, and any general grading work. We like the larger landscape rakes for hand bed prep and grading.

## Spading Forks

Spading forks, also called digging forks, are ubiquitous on farms and in gardens around the world. They are the preferred tool for hand-digging root crops, since their tines have the ability to probe down into the soil and pry roots out without cutting their lateral branches off. They also tend to remove less soil from the hole than shovels, easing the strain caused by lifting heavy soil. These are the tools that we break most often when digging roots by hand, and I have yet to find a modern spading fork that is anywhere near as good as any of the antique forks we have in our tool shed.

## Broadforks

Broadforks are like spading forks on steroids and are an excellent tool for providing deep tillage by hand. They operate with two long handles attached to a row of long steel tines that are inserted into the ground and pried backwards to break up compacted soil. Some people use them to dig roots as well. Johnny's seed company and Peaceful Valley Farm Supply sell them.

## Hay Forks

We have found that hay forks are great for moving large piles of freshly harvested herbs. For example, when we bring a truckload of stinging nettles to the drying shed, these long-handled forks make it easier to unload the truck onto tarps or carts, which we drag or roll into the drying shed. These tools are also useful for applying mulch to beds and paths.

## Shovels and Spades

Shovels and spades are great for moving materials such as soil, sand, grain, and snow; turning compost; edging gardens; and digging holes. There are hundreds of variations to choose from, and it definitely pays to buy the highest-quality versions you can find.

## Sharpening Stones

Sharpening stones are the tool that no farm should be without because they keep all of the above tools sharp. Dull tools are slow tools. We like to use a double-sided whetstone with a coarse side and a fine side. Apply a little saliva, water, or mineral oil for lubrication, and these stones do an awesome job of keeping our tools sharp. Grinding wheels on a bench or angle grinder hone the bigger tools and blades.

## Earthway Seeder

The simple push-type Earthway vegetable seeders are very affordable and effective for direct seeding small acreages. They can even be ganged together with crossbars to seed multiple rows at a time. Several of the standard vegetable plates can be used for medicinal herb seeds with a little experimentation or tinkering. For a small fee the manufacturers can also make custom seed plates if you send them the seed you wish to use.

## Broadcast Seeder

We use an Earthway broadcast seeder for all of our cover crop seeding. This device consists of a nylon bag with a shoulder strap to hold seed, attached to a hand-crank-operated spinner that distributes seed as the operator walks along the field. Broadcast seeding large acreages this way is impractical, but we have used our seeder to sow up to an acre of cover crops at a time.

## Backpack Flame Weeder

A gas-grill sized propane tank mounted on a backpack frame connected to a hand-operated torch makes for an excellent tool for killing small weeds. We bought our torch from Harbor Freight Tool Company with an adjustable pilot flame and lever-operated torch. This tool works great for creating stale seedbeds and flaming small weeds before they become established. We often use this for "touch-ups" after we have prepared stale seedbeds using the tine weeder and for weeding small weeds growing close to sturdy plants. Although the sturdy plants may get a little wilted, they are large enough to withstand the flaming without being seriously injured. Tiny weeds, on the other hand, get obliterated by way of the heat bursting their water-filled cells.

## Draw Knife

Draw knives consist of a single steel knife blade, often slightly curved, with a handle on each end of the blade. We use this tool to peel medicinal barks off branches by grasping the two handles and drawing the knife down the branch at a shallow angle to peel the bark

just under the inner cambium layer. Draw knives are elegant tools that must be kept sharp to work well.

# Tractors

There are a countless number of options when it comes to buying tractors, and an entire book could be written on tractor choices alone and probably already has been. Let's just discuss a few of the

**Figure 6-2.** Backpack flame weeder.

**Figure 6-3.** Draw knife.

**Figure 6-4.** Walk-behind tractor (rototiller). Photograph courtesy of Bethany Bond

common and basic options. I am not going to even mention pricing because there are far too many variables involved with discussing the value of tractors for me to mention here, and you can buy them used or new. Most tractors are equipped with a three-point hitch and a power take-off (PTO) that allows implements such as rototillers to be attached to the back of the tractor and powered by a gear-driven rotating PTO shaft attached to the tractor.

## Walk-Behind Tractor

Walk-behind tractors are a great choice for the beginning or small-scale farmer because of their affordability, versatility, relative ease of maintenance, and reduction in soil compaction from their lighter weight. Basic two-wheel tractor configurations come powered by gasoline or diesel engines and accept an extensive array of add-on implements that can be driven via PTO. We use a walk-behind Troy-Bilt rototiller on our farm to till our vegetable and flower beds and to cultivate weeds in pathways between rows of plants that are too tall for the tractor to cultivate.

**Figure 6-5.** Two-wheel-drive tractor. Photograph courtesy of Bethany Bond

### Two-Wheel-Drive Tractor

The mainstay of modern agriculture is the two-wheel-drive tractor. These machines helped pave the way for farmers to make the leap from horse-powered to horsepower and haven't really changed a whole lot since that time. Available in diesel, gasoline, electric, photovoltaic, and even propane-powered versions, two-wheel-drive tractors are best for relatively level land and can be outfitted with an enormous array of tow-behind or three-point-hitch implements. At ZWHF we use a 1950 gasoline-powered, four-cylinder, two-wheel-drive Farmall Cub tractor with belly-mounted basket weeders and sweeps for cultivating weeds.

### Four-Wheel-Drive Tractor

Also called "front wheel assist," the four-wheel-drive tractor is extremely versatile. The driver has the ability to engage the front wheels to improve traction and to help the machine perform better when using front-end loaders, pulling heavy loads, or using three-point-hitch implements. Four-wheel-drive tractors are recommended for hillier terrain because of their increased traction and stability. We use a 55-horsepower four-wheel-drive tractor with front-end loader on our farm for the bulk of our tractor work.

# Implements for Tillage

Tillage simply means preparing the soil for agricultural purposes. If you hear a farmer say, "I just tilled the back 40," it doesn't necessarily mean she used a rototiller to till it; it means that the soil in the back 40 was prepared at a depth adequate for planting using any one or more of a multitude of tillage implements, including hand tools such as broadforks. Tillage is often classified as "primary tillage" and "secondary tillage." Primary tillage is more of a rough tillage, where we open up the soil to begin the process of breaking up compaction, smothering weeds, and incorporating residues. Secondary tillage comes later and is more of a fine tillage, where we are doing a final preparation of the field for planting or seeding. Following are some of the most

**Figure 6-6.** Four-wheel-drive tractor.

common tractor-mounted tillage implements and a description of their uses.

## Plow

The old-fashioned moldboard plow or plough is the original tillage tool and represents one of the first modern advances in agriculture. The relatively simple design and operation of plows make them a versatile and dependable choice for primary tillage. Basically they can be described as a steel wedge pulled below ground through the soil on which a "moldboard" or wing is attached, which creates a furrow and overturns a continuous strip of sod. This overturning action inverts the sod and brings topsoil up with which to cover weeds. The primary benefit to moldboard plowing is that the equipment has very few moving parts, is pulled behind a tractor without using a PTO, is widely available and affordable, and does a great job at tilling virgin ground and incorporating dense sod. Some disadvantages are that good traction and horsepower are required, especially when pulling more than one plow unit or "bottom" at a time. These tools can also have a tendency to create a "plowpan," or compacted layer of soil, that can hinder root growth and water infiltration caused by the smearing action of the plowshare. We use a three-bottom plow on our farm to till new ground and incorporate dense cover crops.

**Figure 6-7.** Three-bottom plow.

## Rototiller

The oft-maligned rototiller is one of the most common secondary tillage implements used on small farms. Rototillers could be compared to guns in that they are both great tools that when used carelessly can cause tremendous harm. As with guns there are people who view rototillers as tools of destruction (which they absolutely can be when used improperly). It is amusing to see some people's reactions when they come to our farm and see the rototiller parked with all the other implements. They may say something like, "Ughhh, you actually use that thing?" "Yes, I do, but I use it wisely," I respond, "and it's a great tool." Perhaps I should start a new version of the NRA and call it the National Rototillers Association.

On our farm we use a rototiller that runs off the tractor PTO. We use this tool thoughtfully and sparingly to do final bed prep before we form raised beds. We can also use the rototiller to prepare our flat (ground level) beds and simultaneously mark the rows by attaching eyebolts onto the bottom of the tiller apron. These eyebolts do a great job delineating the rows, which guides us when direct seeding or transplanting. We find that one pass through a field with the rototiller helps to break up the large clods of sod left after doing primary tillage. It also works great at incorporating soil amendments such as compost and rock powders into the topsoil. Unfortunately rototillers only offer shallow tillage (four to six inches deep at most) and can create or exacerbate hardpan conditions because their rotating tines smear and compress the soil below. We avoid soil compaction by running the rototiller at low RPMs and relatively high ground speeds after we have shattered any hardpan with the chisel plow.

When rototillers are used too often or in wet conditions, the delicate biological and structural makeup of good soil can be destroyed by their pulverizing action. This can undo thousands of years of natural processes that it took to create good topsoil, and that damage can take many, many years to repair. The best approach with rototillers is to use them minimally and thoughtfully at higher ground speeds and lower PTO rpms than the manufacturers suggest. In

**Figure 6-8.** Rototiller.

general, the soil is too wet to rototill if you pick up a handful of your soil, squeeze it into a ball, and bounce it in your hand once or twice, and it doesn't crumble apart. Excessively dry soil is not good to rototill either, so if clouds of dust are kicking up behind the rototiller, that is another sign it may be time to park the machine. All of that said, however, if I could only afford one tractor implement with which to start a new farm, the rototiller would be it because of its versatility.

## Reciprocating Spader

Reciprocating spaders are well-designed implements that perform a very efficient job of deep primary tillage while causing minimal damage and compaction to the soil. Unfortunately, like most things that sound almost too good to be true, spaders have some drawbacks. These PTO-driven implements can be very expensive to purchase, tend to break easily, and can be difficult and expensive to repair. They like to be run at very slow ground speeds and can therefore be time consuming for tilling large acreages. They

also do not tolerate rocky soils well, and large stones are often the culprit with broken spaders.

For those with soils relatively free of large stones and a larger budget, a spader can be a great choice. They offer deep primary tillage, work well with low-horsepower tractors, and can be used with wetter

**Figure 6-9.** Reciprocating spader.

**Figure 6-10.** Disc harrows.

soil than for most other implements without deleterious effect. They operate by way of a reciprocating row of mini shovels that are inserted into the ground one at a time, incorporating residue and loosening the soil deeply without pulverizing and damaging its delicate structure.

## Disc Harrows

The disc harrow is the workhorse of tillage tools. No tillage tool gets as many hours of use on our farm as our set of disc harrows. This tool is extremely versatile in that it doesn't require a PTO to operate, is affordable, rarely needs repair (I'm knocking on wood as I write that), is relatively gentle on the soil, is easy to maintain and operate, and is our tool of choice for secondary tillage to incorporate soil amendments and crop residue and prepare fields for seeding and planting. Disc harrows are made up of rows of concave steel discs placed approximately twelve inches apart and set on axles in rows mounted at offset angles. When dragged through the soil behind a tractor, these discs slice into and gently mix the soil back and forth via the opposing angles they are set at.

The main disadvantages of disc harrows are that they do not provide deep tillage and can, like many other tillage implements, create or exacerbate hardpan conditions.

## Mini-Chisel Plows

The mini-chisel plow is our favorite tool for primary deep tillage. This tool consists of two or more rows of spring-trip-mounted solid steel shanks mounted twelve to eighteen inches apart on a toolbar that are pulled through the soil behind a tractor. The shanks can be used at depths of up to 12 inches and gently shatter hardpan to improve rooting depth and water infiltration without pulverizing the soil or burying rich topsoil. They are also excellent at pulling up stones in rocky soils so they can then be removed from the field. Like the moldboard plow and disc harrows, a PTO is not required to operate chisel plows, and they are relatively affordable, extremely durable, and easy to maintain.

We use the chisel plow as soon as the soil is dry enough to be worked in the spring to begin aerating the soil and to loosen it before performing secondary tillage with the disc harrows and rototiller. The main drawback with chisel plows is that they tend to clog in fields with high amounts of crop residue and therefore perform best on relatively bare ground or in fields where crop residues are more finely chopped. They also work best with tractors that have good traction and/or higher horsepower because of the resistance encountered when pulling them deeply through the soil.

## Field Cultivators

Like mini-chisel plows, field cultivators consist of two or more rows of steel shanks with sweeps mounted on a toolbar and spaced twelve to eighteen inches apart and pulled through the soil behind a tractor. The main difference between chisel plows and field cultivators is that the shanks on field cultivators are made of C- or S-shaped flexible spring steel tines that flex as the implement is pulled through the soil, whereas chisel plow shanks do not flex but instead have heavy-duty spring trips to keep them from breaking when encountering obstacles.

Field cultivators are used as a secondary tillage tool because of their shallow-running nature and are excellent in moderately low-residue situations for killing small weeds, incorporating soil amendments, and preparing fields for planting. These implements work well with lower-horsepower tractors and are affordable and easy to maintain.

## Bed Former/Mulch Layer

If you were to ask many vegetable or other row-crop-type farmers what single tractor implement purchase has provided the greatest return on investment, a majority of them would likely name the combination bed former/mulch layer at the top of their list. Cultivating weeds by hand or by machine is one of the biggest costs associated with farming, and this tool has dramatically decreased those costs on many farms. Herb farmers, ourselves included, are following suit and realizing the utility of these implements, not only from a weeding labor cost-reduction standpoint but equally or more importantly as a way to affordably and practically improve

**Figure 6-11.** Chisel plow.

**Figure 6-12.** Field cultivator.

**Figure 6-13.** Combination bed former and mulch layer.

hygienic conditions in botanicals that are susceptible to microbial contamination from soil deposition during heavy rains and winds.

Bed former/mulch layers are two tools in one. Although both components of this tool are available separately, the combination units allow for two procedures to occur simultaneously during one pass of the tractor. The bed former consists of a frame attached to a toolbar with various furrow shanks, wings, and shaping discs that loosen the soil and pull it into a sheet metal pan that compresses the soil and extrudes it into a uniformly shaped raised bed. Mounted to a frame behind the bed former is a set of rollers, press wheels, and hilling discs. A roll of poly or bio-based plastic mulch is pulled off the rollers as the machine travels down the row. Press wheels gently stretch the mulch over the raised bed while hilling discs cover the edge of the mulch with soil to keep it intact.

Most of these devices also come with attachments that allow for the simultaneous installation of drip tape under the mulch. The height and width of the bed as well as the size and type of mulch used are adjustable on most of these units, adding to their versatility. Some drawbacks in purchasing and using these machines are that they are relatively expensive to purchase, they require higher-horsepower tractors to pull, and they can be challenging to use on sloping land. They also require that thorough tillage be performed before their use to allow for adequate soil consistency for forming raised beds. Bed formers used independently of the mulch layer attachment are great for forming uniform raised beds without applying mulch. Mulch layers can also be used independently for mulching nonraised (flat) beds.

## Cultivation Implements

The term "cultivation" has many broad definitions. So broad, in fact, that you may hear a farmer say something like, "I have ten acres in cultivation." In that sense cultivating basically means growing crops. The word cultivation can also be used interchangeably with "tillage," meaning to work the soil to incorporate residues and weeds and prepare fields for planting.

Figure 6-14. Cultivating tractors: Brian cultivating garlic with the Farmall Cub.

Generally, though, cultivation refers to the act of using tools to remove or kill weeds growing within crops, and that is the context we will use here. Following are some of our favorite cultivation implements.

## Cultivating Tractors

Cultivating tractors are manufactured and used specifically for cultivating row crops. Most cultivating tractors are designed to provide the driver with optimum visibility to perform precision weeding without damaging the crops being cultivated. Visibility is enhanced by the design of these tractors either by having the driver's station offset to one side of the motor and drivetrain or having the motor and drivetrain mounted behind the driver. With a clearer line of sight to the crops below, these tractors can be outfitted with mid- or "belly-mounted" weeding implements that allow the driver to steer the tractor while remaining in a forward-facing position to precisely guide the weeding implements below. Some of the more popular cultivating tractors used today are old Farmall tractors such as Cubs and "As" equipped with offset seats, Allis Chalmer "Gs," Kubota L-245 Hs, and John Deere 900 HCs.

For years we cultivated our crops with rear-mounted cultivating implements attached to a three-point hitch behind our larger tractor. This method of cultivating was exceedingly painful and frustrating at times. Painful because the driver (me) occasionally had to look backwards over his shoulder to ensure that the cultivators weren't clogging with weeds or stones or ripping irrigation lines or plants out of the ground. Inevitably and occasionally, especially on our sloping land, I would get into a situation where I would lose the alignment required with precision cultivating and begin to wreak havoc on the crop I was weeding. This necessitated carrying a hand tool with me while cultivating to fix these mistakes.

After years of inefficient and frustrating cultivation, we finally purchased a Farmall Cub cultivating tractor outfitted with midmount cultivators, which has greatly improved the ease and quality of tractor cultivation here on our farm. The Cub is used exclusively for cultivating; this saves time because we don't have to switch implements on the larger tractor every time we want to cultivate. We still use the larger tractor to cultivate occcasionally, especially for larger weeds, but it is nice to have several options to choose from.

## Row-Crop Cultivators

Row-crop cultivators come in a wide variety of sizes and configurations and primarily consist of spring steel C- or S-shaped shanks or tines mounted with sweeps or knives attached to two or more rows of toolbars. There is a wide variety of other weeding tools that can also be mounted to these toolbars simultaneously. For row-crop cultivating, tools and shanks are adjustable and are spaced on the toolbar to enable the tractor and implement to straddle rows of crops while the tines cultivate as closely as possible to the crop. These can either be midmounted on a toolbar using hydraulic lifts, rear mounted via a three-point hitch, or towed behind a tractor using gauge wheels.

At ZWHF we use a two- and three-row cultivator mounted on the three-point hitch of our four-wheel-

**Figure 6-15.** Row-crop cultivator.

drive tractor using S tines and sweeps. We also mount sweeps to the midmount toolbars and behind the wheel tracks of our Farmall cultivating tractor.

The benefits of row-crop cultivators are that they are versatile, affordable, easy to maintain, and can be used with low-horsepower tractors. They also perform well with both large and small weeds, as well as large and small crop plants. The challenges are that sweeps can tend to hill soil up around plants (which can sometimes actually be beneficial for weed smothering). They are also not as precise as other cultivating implements.

### Basket Weeders

Basket weeders consist of sets of variously sized cylindrical rolling "baskets" made of quarter-inch spring wire that are spaced according to row dimensions and mounted on two rows of axles attached to a universal mounting frame. The two rows of baskets are connected via chain with a larger gear on the front axle and a small gear on the rear axle. The ground-driven action of the spinning baskets in front causes the rear-mounted baskets to spin faster, agitating the soil and dislodging small weeds. Baskets are sized and spaced according to plant size and row dimensions.

The benefits of basket weeders are that they can be run within one or two inches of plants in the row without damaging or hilling the plants with soil. The

drawbacks are that they don't handle crusted-over soils well, they can only be used for very small weeds, and they don't handle excessively rocky soils well. They are also only able to be used for cultivating plants up to eight to ten inches high because of limited clearance under the axles. They are relatively expensive new, so if you can find a used set in good shape, consider yourself lucky because they are in relatively high demand. We use a basket weeder belly mounted on our Farmall Cub for cultivating and find this tool to be our preferred implement for early-season cultivation while plants are small and weeds are eager to germinate and grow.

### Flex Tine Weeder

Flex tine weeders are composed of a toolbar with three or more parallel cylindrical mounting bars on which rows of light-gauge flexible steel tines are attached in a staggered, offset fashion. These tines are mounted at an adjustable angle on the toolbar, which is attached to a tractor's three-point hitch and pulled through the soil where the tines scratch the soil surface. The steeper the angle they are adjusted at, the more downward force they have to contact the soil and dislodge germinating weeds before they emerge. This weeding technique is referred to as "blind cultivation." This term refers to weeding the entire soil surface indiscriminately without trying to avoid the crop. This can be done at the "preemergent" stage

**Figure 6-16.** Basket weeder.

**Figure 6-17.** Flex tine weeder.

when the direct-seeded crop has not germinated, yet weed seeds have started to germinate, and also at the "postemergent" stage when a crop is well established enough to withstand the force of being run over by the spring tines.

Using a flex tine weeder requires a much different approach from that used with most cultivating tools because the weeds that are being eradicated are removed before they are barely able to be seen. This preemptive approach is very effective but must be performed regularly and thoughtfully. If you can see the weeds from the tractor seat, it is often too late to tine weed because the thin flexible tines are not aggressive enough to kill weeds past the first true leaf stage. These tools are relatively expensive yet durable, easy to maintain, and require very minimal horsepower and traction to pull.

The main advantages are that they can be used at fast ground speeds without having to worry about injuring crops and they are extremely effective at eradicating weeds if used correctly and timely. The challenge to using these tools lies in the timing factor. If you snooze, you lose. You basically only have a few days' window of opportunity to kill the newly germinated weed seeds in the "white-root" stage before they become too large for the tine weeder to kill, at which point it is time to pull out the "big guns" described above.

# Transplanting Implements

Transplanting is labor intensive when done by hand. There are a wide variety of mechanical devices for planting plugs into the ground, some simple and some more elaborate. All of them require manual assistance to feed plants into the device and to monitor its progress.

## Water Wheel

The water wheel transplanter is designed for planting into plastic or bioplastic mulch. This device is either attached to a tractor's three-point hitch or towed on wheels behind the tractor. It consists of a steel frame that supports an overhead water tank. Water or

Figure 6-18. Water wheel transplanter.

liquid fertilizer gravity-flows down through tubing to one or more steel wheels supported on a frame below and turning on axles driven by the wheels' contact with the ground as the tractor pulls the device along. The wheels have steel punches mounted onto them and are perforated at each punch. Water simultaneously flows through the holes while the punch perforates the plastic and makes a planting hole into the soil below. Two or more personnel riding on seats mounted to the rear of the device place plugs into the water-filled holes. The water irrigates the plant and pulls soil around the plug to initiate good root-to-soil contact.

Gauge wheels keep the transplanter centered over the planting bed and are adjustable to fit over raised or flat beds. Water wheel transplanters can also be used on bare ground without mulch. They are efficient and easy to maintain. They are also versatile, in that the punch wheels can be adjusted to accept various planting widths, numbers of rows, and plug sizes. The main challenge with water wheel transplanters is that they can be expensive to purchase. We use a Kennco water wheel transplanter adjustable for one or two rows and with two riders or more for planting into mulch.

## Mechanical Disc-Type Transplanter

Mechanical disc-type transplanters are used for transplanting into bare ground. Unlike water wheel transplanters, where plants are placed directly into holes made by the device, disc-type transplanters operate by feeding plants into the device, which then places the plants into a furrow. One or more riders place individual plugs between two flexible, thin-sheet-metal discs that gently squeeze the plants and rotate them at a predetermined spacing. As the plants reach the bottom of the discs, they are released into a furrow created by a steel furrower being pulled through the soil underneath the discs. Two packing wheels rotating alongside the discs firmly pack soil around the plants after they are planted into the furrow. The whole mechanism is ground driven as it is slowly pulled along attached to the tractor's three-point hitch.

Like water wheel transplanters, disc-type transplanters have the capability of watering the plants as they are placed into the ground, along with giving them a metered dose of granular fertilizer. The benefits of mechanical disc-type transplanters are that used models are relatively common and affordable because of the many thousands of them that were used on tobacco farms during that industry's heyday. They are efficient and versatile and do a great job planting when they are adjusted well.

**Figure 6-19.** Mechanical disc-type transplanter.

The disadvantages with this type of transplanter are that they can be a bit challenging to adjust to accommodate different row spacings and plant sizes, and they don't work particularly well in excessively rocky soil. We use a Powell disc transplanter for almost everything we plant into bare ground.

# Harvesting Implements

Harvesting crops is labor-intensive work. Automating any part of this process to improve efficiencies equates to increased profit margins. We have seen a very rapid return on investment for the money we have spent on harvesting implements.

## Chain Digger

Chain diggers, also known as potato harvesters, are made up of a steel wagonlike frame supported by two wheels and pulled behind a tractor. Within the frame lies a conveyor chain with long, spokelike links spanning the width of the bottom of the harvester. The implement is either ground driven (by being attached via a smaller chain to a drive wheel) or PTO driven, attached by a shaft to the tractor PTO. A heavy-duty steel plate or "nose" attached to the front of the harvester digs into the ground at an angle that determines the depth of harvest and is adjusted by a manual lever.

As the angled steel nose is forced down into the soil by the harvester's being pulled along through the bed, roots are pried out and lifted into the conveyor chain, which rotates along the length of the harvester. Underneath the chain are oblong-shaped gears that rotate and shake the chain, which in turn shakes excess soil off the root crop being dug. Roots are deposited on the ground behind the harvester as it is pulled along. Farm personnel then gather the roots and load them into bins to be transported to the farm for processing.

The benefits of this implement are that it saves an incredible amount of labor by our not having to hand-dig roots. Older models are relatively common and affordable at agricultural auctions or lying around old potato farms. The challenges with chain diggers are that they are prone to breakdowns from the amount of stress placed on them as they are pulled through the soil and they tend to clog in stony soils or in fields with excessive crop residue, which can tangle in the chain. We use a 1940s International Harvester potato digger for harvesting all our deep-rooted crops, such as angelica, burdock, and yellow dock. When we bought it at auction it was ground driven and was inefficient because of its poor traction, so we welded the rear differential from a 1974 Jeep onto it. Now it runs off the tractor PTO and works incredibly well. Although it breaks down from time to time, the savings in labor we have accrued from its use eases the frustrations of having to repair it.

## Bed Lifter

Bed lifters are some of the simplest yet most effective implements used for harvesting vegetable and herb root crops. They consist of a heavy-duty square steel frame, weighted and attached to a tractor's three-point hitch. Attached to the bottom of the frame is a four-foot-wide beveled and angled flat steel bar that undermines and lifts the entire bed of roots above it as it passes under them. The angle of the flat bar as it is pulled through the soil provides the leverage to simply break the bond between root and ground and gently pry the roots high enough so that farm personnel can lift the roots up and out of the soil by hand without having to dig them. Roots are then placed in bins to be hauled back to the farm. We built our bed lifter out of used steel grader blades that our town road crew gave us. We use the bed lifter for all of our shallowly rooted crops, such as echinacea and dandelion root.

## Sickle Bar Mower

The sickle bar mower is the grandfather of the modern mowing machine. Once commonplace on farms across the landscape, these implements have become harder to find and have fallen into more of a "specialty item" niche for the manufacturers that still produce them. As a result their value has increased, and even the older models hanging around barnyards from coast to coast are being reconditioned or sold "as is," often for far more than their original purchase price.

Modern agricultural technology has advanced mower designs that are far more efficient and practical than the sickle bar mower, yet there are field applications where the sickle bar still reigns as the tool of choice. Unlike most types of mowing and harvesting machines that are pulled along behind or beside a tractor on wheels or mounted underneath or behind the tractor, the sickle bar mower consists

**Figure 6-20.** Chain digger.

**Figure 6-21.** Bed lifter. Photograph courtesy of Bethany Bond

**Figure 6-22.** Sickle bar mower.

of a six- to eight-foot-long bar with a reciprocating set of notched knives that is either mid- or rear mounted to the tractor and is extended to one side of the tractor by way of hydraulic cylinders or manual levers. The knives are powered by the tractor's PTO and slice through grasses, plants, small shrubs, and thin saplings.

One of the greatest benefits of the sickle bar mower that other types of mowers don't offer is their versatility with mowing hillsides and underneath fences and other objects that most mowers would not be able to reach. As long as the tractor is on stable ground, the sickle bar can be angled either up or down to mow areas such as roadside banks and ditches and steeply angled hillsides. One challenge with these implements is that they tend to clog if they encounter material too thick to cut or if they are operated at ground speeds too high for the cutter bar to keep up with. They also require a bit more maintenance and sharpening than many other types of mowers.

Herb farmers have discovered the utility that sickle bar mowers have to offer with harvesting their leaf crops. As long as the crops are standing relatively upright and are not too thick or woody, sickle bar cutters can make short work of mowing down

such crops as stinging nettle, lemon balm, and other herbaceous plants without compromising the quality of the plants. The cutter bar is simply pulled through the crop, and the notched teeth cut through the stems at a height controlled by the operator; the machine lays the plants down in a swath, to be picked up by hand and brought back to the farm for processing. There are some large-scale commercial models that are equipped with conveyors that automatically load the harvested material onto trucks or wagons, but they are expensive to purchase and maintain. We use a six-foot sickle bar mower at ZWHF to harvest many of our leaf crops.

## Miscellaneous Implements

Miscellaneous implements compose the remainder of our "mechanized toolbox" and perform a variety of useful tasks.

### Manure Spreader

Manure spreaders come in a variety of sizes and shapes. They are basically a wagon attached to a tractor's draw bar or trailer hitch to hold manure or compost with an open rear end and rotating

**Figure 6-23.** Manure spreader.

"beaters" to distribute it. Some are ground driven, meaning that a gear attached to one of the wheels is connected by a chain to a smaller gear on the beaters. The beaters are serrated, angled thin steel blades attached to an axle on the back of the spreader that spin rapidly, spreading the manure out behind the spreader as it is towed. PTO spreaders work by having the beaters attached to gearing driven by the tractor's PTO. Both types have chain-driven rows of steel bars on the floor of the spreader that continually move backward, pushing the material into the beaters so it can be continually spread.

We use a ground-driven manure spreader that we adapted for spreading finer composts by welding additional steel fins onto the beater axles. There are a lot of old spreaders lying around barnyards and they are relatively easy to find at a reasonable price. Maintenance can be more challenging and expensive with the PTO-driven spreaders than with ground-driven spreaders. The advantage of PTO spreaders is that the beaters can be driven fast or slowly, independent of the ground speed of the tractor. You can even stop at the edge of a field and sling manure onto it without driving into the field. Ground-driven spreaders, on the other hand, are only spreading manure as the spreader is being pulled along. A lever on the spreader engages or disengages the beaters as needed.

## Cone Spreader

Cone spreaders are great for distributing granular fertilizers and mineral powders. They consist of a large hopper, made of steel or poly, mounted to a steel frame

**Figure 6-24.** Cone spreader.

**Figure 6-25.** Rotary cutter. Photograph courtesy of Bethany Bond

attached to the tractor's three-point hitch or towed on wheels behind the tractor. A PTO shaft connects to a spinning plate with fins attached that distributes material as it flows down through the hopper and onto the plate. Cone spreaders are inexpensive and easy to operate and maintain. The biggest drawback is that they don't have a large capacity and require frequent refilling as the operator spreads material over larger tracts of land. We use a Vicon spreader to distribute mineral powders on our farm.

## Rotary Cutter/Brush Hog

Rotary cutters, often referred to as brush hogs, are rugged mowers designed for rough mowing work, such as clearing small trees and saplings and mowing overgrown pastures. They are affordable, easy to maintain, and extremely useful. Rotary cutters are made of a heavy-duty sheet metal shield covering a set of heavy-duty steel blades that rotate by way of being attached to the tractor's PTO. Gauge wheels and three-point lift arms adjust the height that the mower operates as it is pulled along behind the tractor, attached to its three-point hitch.

The drawbacks of rotary cutters are that they are not well suited for fine mowing or harvesting because of their aggressive nature. The benefits are their durability, affordability, and utility. We use a six-foot-wide rotary cutter to maintain fields and pastures and to cut the tops of root crop plants before we dig the roots.

# CHAPTER 7

# Field and Bed Preparation

*Round and round, the cut of the plow
    in the furrowed field,
Seasons round, the bushels of corn
    and the barley meal,
Broken ground, open and beckoning to
    the spring,
Black dirt live again!
The plowman is broad as the back of the land
    he is sowing,
As he dances the circular track of the plow
    ever knowing
That the work of his day measures more
    than the planting and growing
Let it grow, Let it grow, Greatly yield.*
— LYRICS FROM THE GRATEFUL
DEAD SONG, "LET IT GROW"

**Figure 7-1.** Beginning spring tillage.

Preparing fields and beds for planting requires a thoughtful, well-planned, and well-implemented approach to provide plants with a good foundation in which to grow while at the same time mitigating the damage caused by poorly timed or poorly executed tillage. We have already discussed the choices between till and no-till farming methods. In this chapter we will discuss types of tillage performed during various times in the farm season, as well as the addition of soil amendments during tillage.

There is something magical about opening the earth in spring that never loses its luster no matter how many times we do it. At first the rich soil smells a bit foreign to us after a long winter deprived of its heady-earthy aroma, yet in seconds the familiarity returns—like seeing an old friend. As we inhale deep breaths full of molecules of decomposing organic matter and our senses are awakened from winter's slumber, our minds recall this experience on the conscious level as far back as childhood springtimes spent digging in the dirt. Perhaps we also recall this experience subconsciously, maybe even genetically, from our ancestors' footprints in the fresh spring earth.

The robins show up and begin plucking earthworms from the newly turned sod, reminding us of the cycle of life that stems from this medium, and we bask in the comfort and knowledge that hope springs eternal. Lengthening daylight illuminates our way as we scatter dormant seeds into this rich, dark earth, confident that life will emerge from their shiny seed coats and grow to nourish our bodies and minds as it has again and again.

For the farmer the anticipation of spring tillage is a feeling like no other. Our patience has worn thin

**Figure 7-2.** Waiting for spring in Vermont.

**Figure 7-3.** Black cohosh rootstock unfurling with spring energy.

after waiting and planning all winter for this moment. We are filled with anticipation, yet we have to remain mindful to not let this feeling cloud our judgment. Is the ground ready to be tilled yet? Is it dry enough? Are we jumping the gun? The tractor, plows, and harrows are all tuned up and greased; the seedlings are bursting forth from their plug trays in the greenhouse; employees are eager to work; the birds are cheering us on to expose worms for them to gorge on. There is so much to do, so much happening. When can we begin?

## Spring Tillage

In the frenetic energy that rules spring we need to remember to think and act rationally when performing early tillage. In some regards, the earlier we commence tillage, the better off we are. Getting an early start in spring can help set the tone for the entire growing season. Early tillage means aerated soils dry more quickly, we get a jump on weed growth, we can get plants in the ground sooner—which in turn can potentially increase yields. We have more flexibility

**Figure 7-4. Spring rains.** Photograph courtesy of Taylor Lee Chapman

and don't feel as rushed, the plants don't get root-bound in their pots, we can sleep better at night, and on and on.

On the other hand, if we let enthusiasm or stress cloud our judgment and we till too early, this can have dire and long-lasting consequences on the health of our soils. Good soils are very complex ecosystems made up of a fine balance between particles and the space between those particles. The particles consist primarily of minerals condensed from erosion of the parent rock found nearby. This mineral base makes up 90 to 95 percent of the soil particles. The remaining 5 to 10 percent of these particles consists of decomposing organic matter, along with a diverse community of flora and fauna, such as bacteria, fungi, algae, nematodes, insects, and earthworms. This space between the particles, called the pore space, consists of water and air. More air than water is found in these spaces when the soil is dry, and more water than air is found when it is wet. The ratio of pore space to particles in good soil is generally about 50:50.

Maintaining the equilibrium between soil particles and pores is absolutely crucial in keeping our soils healthy and productive. Unfortunately, this balance is extremely delicate and can be upset by literally one poorly timed tillage pass. When excessively wet soils are tilled, the pressure and vibrations exerted by tractors and tillage equipment compress soil particles together and reduce the pore space between these particles. Reduced pore space means that the things that normally occupy these pores, such as oxygen, living organisms, and plant roots, have trouble finding space to occupy. When plant roots can't find oxygen, plants suffer; when the soil lacks oxygen the biological life normally teeming within its structure can't breathe, and the whole system falters.

The damage isn't usually irreversible, but it can take years of leaving the ground fallow and planting cover crops to help it heal itself of the damage done to it. Soil can usually do a better job repairing itself gradually than we can do rapidly. We can certainly assist the process of remediating damaged soils by adding more organic material and remineralization, but the delicate balance of minerals, organic matter, organisms, water, and air found in healthy soils takes time and effort to repair.

## How Wet Is Too Wet?

There is no hard-and-fast rule that says soil is either too wet or too dry. There are so many different soil types on this earth, and each of them is unique. In general, clay soils take the longest to dry out and are the most susceptible to compaction. Sandier soils dry out more rapidly and are more resilient to compaction. The best approach is a cautious one. When you enter the field to evaluate soil conditions, does the soil stick to your boots or tractor tires? If so, that is an early warning that it is too wet to till. If the surface of the field seems dry but you are unsure about the bigger picture below, another effective method that farmers employ to determine soil moisture content is to probe into the soil approximately six inches deep in several locations in the field you are evaluating. Grab a handful of soil from this depth, and gently squeeze it into a ball. Bounce the ball in your hand a couple of times. Does the ball crumble easily and break apart into several multisize pieces? If so, your soil is probably good to go for tillage.

If your ball just breaks in half or into a few big chunks and the big pieces seem sort of "glued together with moisture," that is cause for further evaluation. If your ball stays put firmly together and doesn't break up at all, wait a day or two and try again, provided it doesn't rain. All forms of tillage, even in the best conditions with the best choice of tools, will compact the soil and reduce organic matter to some degree. We can mitigate this impact by making wise choices when our steel meets our soil.

# Primary Tillage

Now that we have determined whether or not our soils are dry enough to till (remember, "till" does not necessarily mean "rototill"), it's time to fire up the tractors, harness the oxen, or break out the broadfork and get to work. Let's hope you have already mapped out your fields, taken soil tests to determine what if any amendments may be needed (we discuss soil fertility with medicinal herb crops in detail in chapter 9), and are ready to get to work breaking ground. It is wise to look at the weather forecast and, if possible, perform tillage during a dry, sunny stretch of weather so that the weeds and sod you are tilling are desiccated by the sun and become easier to kill and incorporate.

## Choice of Tools

The next thing you need to decide on is what tool to use. If this is virgin ground that you are working with, are there lots of whip trees, roots, and stones on or just below the surface? If so, you have some prep work to do to remove these obstacles or plan to work around them. How much crop or weed residue is on the soil surface? Lots of long fibrous plant matter can complicate things, so you may want to consider disc harrowing first to cut that stuff up a bit and avoid having it clog equipment.

If the residue is light and the soil is dry enough, go right for the primary tillage with the deepest tillage tool you have available. This will open up the soils for further drying and loosen everything up a bit to ease secondary tillage. Crop residue such as that from cover crops left on the soil's surface can be beneficial for preventing erosion and preventing

**Figure 7-5.** Disc harrowing helps to break up sod.

weed seeds from germinating. As long as it isn't interfering with tillage and planting, you may want to consider leaving residues on the surface when possible instead of incorporating them, as they can help smother weeds and conserve soil moisture.

For tilling new ground with a dense mat of sod or a field that has been cover cropped, our preferred approach is to start out with the three-bottom mold-board plow. This initial plowing turns the sod over and helps dislodge any large stones, which we load into the tractor's bucket loader and haul away. After plowing we like to let the field "rest" for at least another day or two before we come in again with the tractor. This pause allows the soil to dry out a bit, and the sod begins wilting, which eases incorporating it. We try to till all of our fields at the same time as early as possible in the spring. By prepping all of the fields at once systematically, this lessens the amount of time involved with switching tractor implements back and forth and makes the process more efficient. After plowing we usually come in with the disc harrows and make a couple of passes over each field, chopping and incorporating the sod and residue.

If we have a good window of sunny, dry weather at this point, we like to let the fields rest for another day or two to encourage further desiccation of sod and weeds. During this pause in tillage, we take the time to spread composted manures and any other soil amendments, such as mineral powders, that are needed based on soil tests. We use our ground-driven manure spreader for this and usually spread between one and three tons per acre of composted manure, depending on soil and crop needs, along with the type of compost we use. For mineral powders we use our cone spreader. After fertilizing we go back and disc harrow again as needed to incorporate the soil amendments.

After two to four passes with the disc harrows, the soil and residues will be starting to homogenize. Remember, the less you till, the better off the soil is. It is always challenging to find the balance between thoroughly incorporating weeds, crop residue, and soil amendments without pulverizing the life out of the soil. We try to maintain the "less is more" approach, and if we have to do a little more cultivation of weeds in the short term, it is usually worth the long-term savings in soil biota and tilth.

**Figure 7-6.** Removing a large rock from our upper fields.

# Secondary Tillage

After primary tillage is completed it is time to move on to secondary tillage and bed preparation. At this point you should already have a plan for where crops are going to be planted. There have been times on our farm when the high water table in spring has delayed tillage in some of our fields that are not as well drained as others. When this happens, as it often does, we have to make adjustments on the fly to our planting schedule and map. For example, some of the transplants that we seeded early in March may be starting to become root-bound in their plastic cells. We try to plant these first, even if it requires completely rewriting the planting map.

We also try to get a cover crop of oats and red clover direct-seeded as early as possible. Oats grow best in the cooler months, and since the dried milky oat heads are one of our top-selling items, we often need to get a crop planted as soon as possible to replenish our inventory. What this means is that often the first field that is dry enough to till and plant is divided into one section for the row crops that are outgrowing their cells and another section for oats and clover. We will discuss cover crops in further detail in chapter 9.

This example of row crops and cover crops growing together in the same field illustrates our two separate approaches to secondary tillage. For the row crops we are often going to be preparing raised beds using our bed former, and we may be mulching some of these beds with polyethylene mulch. In these fields we focus our tillage efforts primarily on the bed itself without worrying about the pathways between the beds. At this point we have gone from using "broad strokes" across the whole field with the disc harrows to narrowing our focus and only tilling the soil where the plants are actually going to be growing, not necessarily in the paths between them.

The first step we take in this final secondary tillage for beds is chisel plowing. Although the mini-chisel plow is considered a "primary tillage" tool, we can also use it for secondary tillage by deeply loosening the soil under each bed. This breaks

**Figure 7-7.** A cover crop of red clover. Photograph courtesy of Bethany Bond

up any hardpan that may have formed from our primary tillage with the moldboard plow and disc harrows. It also aerates the soil, which will help promote good root penetration and water infiltration.

The final step of our secondary tillage method is to rototill the beds prior to forming them. Rototilling breaks up any remaining large clods left after primary tillage and chiseling and loosens the soil enough that the bed former has plenty of material to form nice raised beds with good planting and seeding surfaces. Generally one pass with the rototiller at low rpms and relatively high ground speed does the job really well, but sometimes we have to do a second pass or slow the ground speed down with this tool if the soil is still excessively cloddy. At this point we are ready to form beds with our raised bed former.

## Tillage in Preparation for Raised Beds

Proper tillage is an absolute prerequisite to forming raised beds with a bed former, and there is usually a direct correlation between the depth and quality of tillage and the quality of the raised beds that can be formed after that tillage is performed. Bed formers are designed to pull loose soil into their pans with discs and wings and to extrude that loose soil into a firm raised bed resistant to erosion from the elements and to crumbling when being cultivated or planted into. They are not designed to loosen the soil or perform any type of tillage as they are pulled along through it. Improper tillage can result in difficulties forming beds as the bed former encounters compacted soil, ridges, clods, or large stones. For applications where plastic or bioplastic mulch is applied to raised beds, improper tillage can result in uneven bed tops, which can stretch or tear the mulch or cause difficulties with using automated transplanting equipment.

By utilizing thorough secondary tillage we can ease the process and improve the overall quality of our raised bed culture. Raised beds can be formed any time that growers wish to plant during the growing season. Some farms even form raised beds with or without plastic mulch on them during the fall and

**Figure 7-8.** Jeff tilling cover crops in preparation for spring planting. Photograph courtesy of Bethany Bond

**Figure 7-9.** Mulched, raised beds ready for planting.

winter to get a head start the following spring. Since we are primarily growing perennials on the medicinal herb farm, we have the benefit of being able to plant perennials at any time during the growing season to be harvested in subsequent years.

## Secondary Tillage for Cover Cropping

For secondary tillage in preparation for planting cover crops, we use a different approach. Since the field will be direct-seeded with a broadcast seeder, we till the entire surface. The first thing we try to do is determine how compacted the soil is in the field we plan to cover crop. We conduct this test by shoving a spading fork with long tines into the soil in several places within the field. If the spading fork tines penetrate the soil without excessive effort, that is a fairly good indication the soil is probably not excessively compacted. If we have to step on the spading fork with our feet and apply a lot of pressure to get the tines to penetrate, we know we likely have some compaction. This is a pretty subjective and basic test that will definitely vary, depending on a lot of factors, but when used in tandem with our instincts and experience, it gives us a good general indication of soil density.

Some people use a "penetrometer" to test for compaction, which is definitely more accurate than using spading fork tines. This instrument is basically a long steel rod with a spring gauge. The rod is inserted into the soil, and the gauge gives the user a reading on the density of the soil below.

If the soil is compacted in our cover crop fields, we do a single pass of the entire field with the mini-chisel to break up any hardpan (provided there isn't excessive crop residue on the surface), and we follow that by disc harrowing. If the soil is not compacted we go straight for the disc harrows or rototiller, depending on the amount of clods and residue on the surface. Since the entire surface of this field will be direct-seeded, we try to get it as weed-free as possible.

One of the most effective approaches we have adopted for creating a weed-free seedbed is the "stale seedbed" method. A stale seedbed is a seedbed that is not only free of weeds but also relatively free of weed seeds that can germinate later on. To create the stale seedbed, we use our tine weeder to make two shallow passes on the field approximately one to two weeks after our initial tillage. This one- to two-week window gives time for the weed seeds to start germinating and develop into the "white root stage," which is the time when they are most vulnerable to being killed by the tine weeder. Some farms use flame weeding with tractor- or backpack-mounted propane burners for this purpose, which is also effective. The stale seedbed technique is applicable and highly recommended for all direct-seeding and transplanting applications, not just for cover crops.

## Midseason Tillage

Throughout the growing season, with so many species of herbs in cultivation, there are times when we have to perform midseason tillage. Chamomile, for example, grows best during the cool spring and early summer months; then blossom production usually starts to wane in late summer when the heat cranks up. We try not to let the chamomile go to seed at this point—it can become a weed the following year by putting so much seed out into the field. Instead of letting it go to seed, we mow the chamomile down with the rotary or sickle bar cutter, till the bed, and plant a cover crop in its place to hold the soil and restore biomass. Some people prefer to let the chamomile go to seed so they don't have to replant it the next year, but we have found that weed pressure within beds of chamomile that have reseeded themselves can negatively affect harvesting efficiencies. When we till during midseason we are mainly concerned with incorporating crop residues and preparing a seedbed for cover crops.

Our primary tillage tool for midseason cultivation is the disc harrow. We don't need deep tillage at this point so we don't chisel plow, and we try not to use the rototiller because crop residues tend to get wound around the tines and because we don't like to pulverize our soil if we can avoid it.

## Fall Tillage

Fall tillage is primarily focused on preparing seedbeds for planting winter cover crops such as winter rye or oats and clover. Incorporating weeds and crop residues isn't as important to us this late in the season; in fact, the more residues we leave on the soil surface, the better protected our soils are from erosion by wind and rain. For the last few months during the growing season often we have driven through the fields with trucks

**Figure 7-10.** Mowing oats for a winter cover crop.

and tractors numerous times to cultivate weeds and to harvest crops. This traffic causes soil compaction; this may be a good time to consider chisel plowing to break up the hardpan formed by all that traffic. Chiseling at this point can help tremendously by improving soil drainage during the wetter winter and spring months.

Early fall is also the time of year that, if you have open soil on sloping ground, it would be wise to consider where the water is likely to go during torrential rains and attempt to mitigate the potential for erosion. Erosion can be avoided by tilling small ridges perpendicular to the slope of land to encourage water to disperse sideways as it flows downhill, lessening its kinetic energy and erosive force. We often like to set ridges with our disc harrows by extending the toplink on the tractor's three-point hitch. This angles the harrows rearward and causes the outer rear discs to leave furrows in the soil that when set perpendicular to the slope of a hillside help to divert any water flowing downhill from gaining momentum and carrying soil with it. After this we sow oats or winter rye as soon as possible to get the cover crop established well enough before winter to help hold the soil in place during torrential rains. The furrows set by the disc harrows, which are then held in place by the roots of a cover crop, do an excellent job diverting any potentially erosive runoff.

**Figure 8-1.** Glorious fields of dandelion in bloom. Photograph courtesy of Larken Bunce

# CHAPTER 8

# Plant Propagation

Many of the medicinal herbs we grow and use are considered weeds by most people; in fact, one of the most entertaining aspects of our work is seeing people's reactions when they ask what we grow. Often they think we're joking when we tell them we are growing dandelion, burdock, and stinging nettles. "You mean you're *purposely* growing these plants?" they ask incredulously and usually follow that with, "Why don't you just come over and dig them out of my garden?" They laugh at the thought of our growing acres of these "weeds," then inevitably ask, "So how much do you charge for those weeds?" The answer to that question raises more laughter and is often followed by, "And people actually pay you to buy those weeds?"

## Growing Weeds from Seeds

Yes, as humorous as it may sound, it's true that we are actually growing and selling many species of plants that some people are desperately trying to eradicate from their lawns and gardens, and people certainly do buy these weeds from us. One would think that these weeds must be some of the easiest plants in the world to grow since they seem to grow everywhere we don't want them to grow, but the truth of the matter is that things are often not as easy as they may seem.

Each spring we start planting seeds in flats in the greenhouse in early to mid-March. Along with some of the medicinal herb species that we seed early, we will also seed a few flats of edible greens, such as kale, spinach, and mesclun mix, for our home vegetable garden. It is astonishing to watch how quickly

and uniformly those vegetable seeds germinate and grow, especially when observed in comparison to the flats of herbs sitting beside them.

As we mentioned in a previous chapter, most vegetables have been selected, bred, and hybridized by humans for hundreds or even thousands of years to improve both the plant's efficiency of growth and the quality of the desired plant parts. Wild plants, on the other hand, which almost all of our medicinal herbs still are, haven't been bred as much and have retained their wild nature in spite of the fact that we have domesticated them.

**Figure 8-2.** Jeff and Lily planting seeds in early spring.

**Figure 8-3.** Medicinal herb seeds: (*from bottom left going clockwise, ending in the center*) angelica, calendula, castor, Saint John's wort, moringa, and dandelion. Photograph courtesy of Bethany Bond

Out in the "real world," where plants grow unassisted by humans, there are countless challenges and opportunities for these plants to do what the universe has designed them to do—to grow and to procreate. All it takes is the ideal combination of a suitable medium in which to grow; nutrients, oxygen, water, and sunlight to fuel that growth; tolerable temperatures; and a niche to occupy. The challenges these wild plants face in growing are numerous, and many of them have developed their own form of "life insurance" to protect themselves and their "families" from catastrophic loss.

Imagine what would happen if a plant such as goldenseal (*Hydrastis canadensis*) were to go to seed in the fall and drop those seeds into the earth, where they lay dormant all winter; then, during the first warm, sunny days of spring, those seeds all germinated at approximately the same time. Pretty efficient procreation, right? Now imagine what would happen if within days of all of those seeds germinating, some ignorant driver threw a lit cigarette butt out the window of his or her car and all of the dry, dead brush and leaf litter became engulfed in flames and spread through the forest understory. Bye-bye, goldenseal babies!

Many species of native plants have adapted themselves to threats of this nature by ensuring that their germination is sporadic by way of germination-inhibiting compounds found within the seeds. By germinating sporadically, the plants ensure that if the first seeds to sprout get wiped out by such calamities as wildfire, drought, flooding, browsing by critters, or perhaps even humans with lawn mowers, there are still some dormant seeds waiting to germinate in order to ensure their survival. Some seeds stay dormant for years before they germinate, which could be thought of as sort of a "whole-life insurance policy" for the plant's progeny.

## Stratification

Stratification is the term we use to describe attempting to break seed dormancy by imitating natural forces that stimulate germination in the wild. Basically this means exposing the seeds to fluctuating cold/warm temperatures, light, and moisture, we do this by sowing them in flats, gently watering them in, and leaving them in the unheated greenhouse for a few weeks before we begin turning the heat up. This is our attempt to mimic what goes on outside naturally in the spring when the nights remain cold, the days grow warmer, and the rain falls. Some growers prefer to simplify this process by mixing the seeds in a damp medium such as sand or peat moss and putting them in the refrigerator for a few weeks. We feel that the more natural temperature fluctuations experienced in the greenhouse, along with increasing sunlight, helps to break dormancy faster, more naturally, and more

**Figure 8-4.** Ashwagandha seeds in a flat, showing sporadic germination. Note that all seeds in this flat were planted on the same day. Photograph courtesy of Bethany Bond

**Figure 8-5.** Melanie scarifying seeds with sandpaper. Photograph courtesy of Bethany Bond

effectively than the refrigerator method. Many of the perennial herb species that we grow benefit greatly through germination enhancement by this treatment. When the seeds start germinating it is often very sporadic, but in time they wake up one by one and usually fill the flats eventually with healthy transplants.

# Scarification

There are so many different ways that plants have adapted methods of ensuring that their progeny survive to keep spreading their genes far and wide. Take astragalus, for example. This effective and popular tonic herb is a member of the pea and bean (Fabaceae) family, and its seeds are tough little beans that require scarification to germinate. To scarify a seed means to penetrate its seed coat and expose its endosperm by way of piercing, nicking, or abrading. This scarification allows water and air to enter the seed coat and begin to stimulate the process of germination. Here in our greenhouse we scarify seeds that require this treatment by rubbing them over medium-grit sandpaper until we see a bit of the endosperm peeking through the abrasion; then we often soak them in a diluted seaweed solution overnight until they swell, at which point we sow them into flats.

But how does this scarification process occur in nature? As far as we know, garden fairies aren't standing by with sandpaper waiting for the seeds to drop so they can begin sanding them. Instead, it seems that the fairies have recruited birds and mammals to help them scarify the seeds, and they don't even need to use sandpaper. All these birds or critters have to do is ingest the seed, either purposely or incidentally, and the hydrochloric acid in their stomachs dissolves just enough of the seed coat so that by the time they excrete this bolus of biota the germination process has already begun. With this method, the seed even comes complete with its own starter fertilizer. Now that's what I call ingenuity!

At ZWHF we scarify the seeds of the leguminous plants *Astragalus membranaceus,* licorice (*Glycyrrhiza glabra*), and *Baptisia australis* as well as some others by hand with sandpaper. Some people apparently use sulfuric acid to do this, but sandpaper works fine in our experience.

# Light-Dependent Germinators

Some seeds will germinate only when exposed to light. This seems to be the plant's method of ensuring that its seeds don't germinate too deeply in the soil, where they may not have enough energy to reach up to the surface and grow. Light dependency is most common with the tiniest seeds that we sow, such as *Lobelia inflata*, tobacco, stinging nettle, Saint John's wort, wormwood, mullein, and many others. We plant these seeds by sowing them on the surface of the soil or potting mix, very lightly covering them, and gently watering them in with a misting nozzle on the watering wand to avoid washing the seeds away.

# Heat-Dependent Germinators

Heat-dependent germinators won't germinate until the soil warms up. Most of the plants whose seeds bear this characteristic are species that grow in tropical climates. Some excellent examples of this are ashwagandha, gotu kola, and tulsi. We start these seeds by sowing them in early March in flats of

**Figure 8-6.** Astragalus seedlings.

potting soil under T-5 flourescent grow lights in a warm room. Then we move them to the greenhouse and keep them warm by putting them close to the heater while they grow on.

# Direct Seeding

In an ideal world on an ideal herb farm, we would be able to direct-seed all of the plants we grow by preparing the fields, sowing some seeds, then watching them germinate and grow. The problem is that, as far as I can tell, there is no such thing as an ideal world, and although I know of some herb farms that could possibly qualify as ideal, I have yet to see one that has mastered the art and science of direct seeding. Again, the challenge here is that most medicinal herb seeds are from wild plants, and they want to grow how, when, and where they darn well please, while paying little regard to our grand master plan. Don't get me wrong—success can be and has been achieved with direct-seeding herbs, but it requires a very concentrated effort, along with a few ounces of patience and a splash of luck.

The biggest benefit to direct seeding is that it saves the grower a giant step in the growing process. When transplants are grown in a greenhouse, for example, the farmer has to plant twice, once by putting a seed in a flat and again when putting that transplant into the ground. Direct seeding halves that process; saves a whole lot of time, space, and materials such as plastic flat trays and potting soil; and generally makes for a more well-rooted and healthy plant. Some plants that we direct-seed germinate and grow easily and dependably. These are our medicinal cover crops such as oats and red clover, which we broadcast-seed to create dense, full field plantings. The culinary herbs that we grow for market, such as cilantro and parsley, are also very dependable for direct seeding.

We have made some interesting observations regarding direct seeding in the past fifteen years of

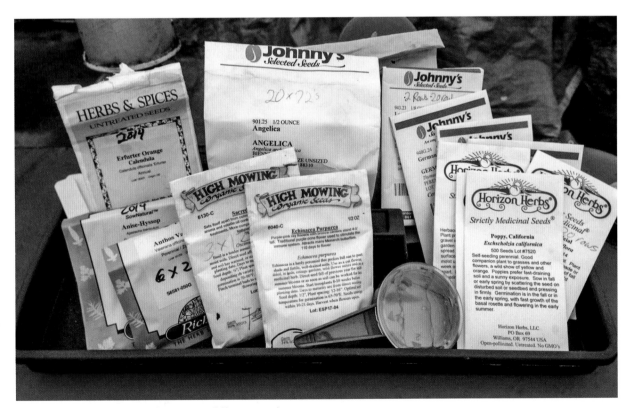

**Figure 8-7.** We source seeds from many different seed companies.

**Figure 8-8.** Ashwagandha seed beginning to germinate.

**Figure 8-9.** Jeff direct-seeding cilantro with an Earthway seeder.

farming herbs. In fact, the most consistently reliable method of direct seeding that we have observed is that of the "incidental method." An example of this

is that reliably every year one or more species of herbs that we seed into flats in the greenhouse will fail to germinate well, if at all. No matter how long we wait or how much we water, no matter the intensity of mantra chanting, calling on spirit guides for assistance, or leaving offerings to the plant gods and goddesses, the seeds just won't grow.

After a couple of months algae sometimes starts to scum over the top of the potting soil surface, and weeds start to pop up. At this point we will often throw in the towel and dump those flats into the compost pile in order to make room for more seedlings. Then, almost without fail, the dormant seeds begin reliably germinating at an astonishing rate within weeks. This is the nature of farming with native plants. Just when you think you have provided the perfect environment for germination and growth, the plants show you what their idea of the perfect environment really is.

Another observation we have made is that the perceived ease with which a particular plant should grow by direct seeding is often directly proportional to the difficulty achieved at getting said plant to germinate and grow. Burdock is a good example of this. I mean, how could you *not* get burdock to grow, right? They grow everywhere! Yeah, that's what we thought, too, until we tried to direct-seed it. Burdock did well the first year we direct-seeded it, but it didn't germinate well at all the second year. The third year we got fancy and used an Earthway seeder instead of putting the seeds in the ground by hand, but germination was still spotty.

What we eventually learned is that burdock needs a consistently moist but not sodden soil during the two weeks or so it takes to germinate, and if the seed dries out at all we see poor germination. Sowing extra seed and installing T-tape irrigation seems to have solved that dilemma for us, but we often wonder why can't it be as easy as it is in the wild, where the plants seem to grow everywhere. The answer is, because as much as we try, we can't reliably control nature according to our needs and desires, nor should we. As native plant growers we just have to accept that fact and do what we can to face the challenges.

# Taking Advantage of Cosmic Influences

Another way that we can help to meet the challenges of growing wild plants in a managed environment is to take advantage of the influences that the moon, sun, planets, and stars exert on natural processes here on Earth. We all know that the moon in particular has some very obvious physical effects on the earth as her gravitational pull causes the ocean's tides to rise and fall. There are also numerous biological phenomena, such as human, animal, and plant reproductive cycles, that are thought to correlate with the moon's influence. The sun certainly has some very obvious effects on the earth and its inhabitants that are too numerous to list here.

What about the stars and planets? Perhaps their effects aren't as obvious to us as the moon and sun, since they aren't as prominent from our limited worldview, but just because we may not be able to see them or to objectify them in scientific terms, that doesn't mean they don't exert influences. Many people understand and agree with our sense that there is a lot more going on than meets the eye when it comes to ways that various components of the universe outside our scope of reason affect us.

One way we try to take advantage of these cosmic influences is to plant according to recommendations in the biodynamic calendar. This calendar was established based on biodynamic methods developed in the 1920s by Rudolf Steiner and later refined by Maria Thun in the 1950s. Their work promoted the effect of cosmic influences on agricultural practices and gave recommendations on how to integrate agricultural work with these influences. In our experience Melanie and I feel that planting and harvesting by the phases of the moon has had a tremendous influence on the quality of our work with the plants. Sowing seeds around the time of the full moon, for example, seems to enhance germination. Harvesting, however, is best done closer to the new moon's phase.

Scientific studies have demonstrated that around the time of the full moon, plant metabolism speeds up, which can enhance growth and water absorption.

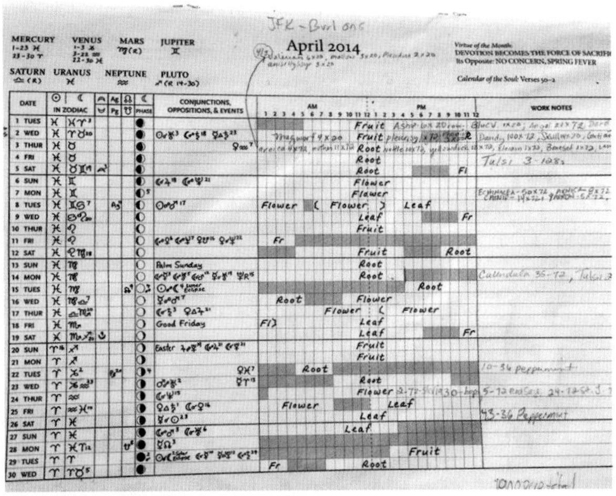

**Figure 8-10.** Keeping track of seeding on a biodynamic calendar.

This increased moisture content brought on by the full moon can make it more challenging to dehydrate herbs well and can even lessen the overall product quality. It is interesting to go back and look at the data we have recorded on our biodynamic calendars from year to year and to correlate it with our own observations.

# Growing Herbs from Transplant

Of the more than fifty species of medicinal herbs that we grow at ZWHF, fewer than ten of those species are direct-seeded into the ground. That leaves more than forty species of plants that have to be grown from either seed or cutting in potting soil in plastic seed trays in the greenhouses and cold frame. The primary reason we choose to transplant rather than direct-seed the majority of our crops is the challenges, not only with sporadic germination but also the fact that the weed pressure often builds while we are waiting for those long-germinating direct-seeded plants to grow. We can counter both of these challenges by producing healthy transplants early on in our greenhouse and densely planting them into weed-free flat or raised beds. We use mechanical transplanters to speed up this process

**Figure 8-11.** Eric and Brian transplanting blue flag iris rhizomes.

and to ease the physical effort of putting thousands of plants into the ground.

# Choosing Your Potting Soil

The first and most important aspect of producing high-quality transplants (aside from the seeds themselves) is the potting medium. We use a soil-based potting mix called Fort Vee from Vermont Compost Company—made with compost, peat moss, mineral powders, bone char, kelp, and blood meal—on our organic farm, whereas most conventional (nonorganic) farms and greenhouses use what is referred to as a "soilless" mix. Soilless mixes are sterile and don't contain the living organisms found in soil-based mixes. The biological life thriving in our potting mix helps us produce healthy, disease-resistant transplants.

In the early years when our farm was smaller, we made our own potting soil in small batches by mixing ingredients in a wheelbarrow or in the tractor's bucket loader. We soon realized that the amount of time we were spending mixing the ingredients well with a shovel was excessive, so we began buying potting soil by the yard, which has saved money as well as our backs. A well-made potting soil should ideally

be able to supply the necessary nutrients to plants until their roots start to fill the cell they are planted in. If rain or other factors prevent us from getting the plants into the ground in time, we can supplement fertility by feeding the plants with a combination liquid fish/seaweed solution.

# Vegetative Propagation

In addition to starting plants from seeds, we use vegetative cuttings to make new plants. Vegetative propagation is a form of asexual reproduction in which a piece or pieces of a plant are removed and placed in a suitable medium in which to form adventitious roots (roots that can grow from plant parts other than the primary root system). Once these adventitious roots form and the plant starts to grow on its own, it becomes a genetically identical clone of the plant the cutting was taken from. In order to encourage genetic diversity when we take cuttings, we try to take them from as many different plants as possible that were originally started from seed.

This type of propagation has become our preferred method of starting several plants in the mint family that we grow, such as peppermint, lemon balm, nettle, and skullcap. We take thousands of cuttings from these plants in early spring (sometimes while there is still snow on the ground) by snipping pieces of soft, fleshy, laterally growing shoots approximately two inches long from plants growing out in the fields that are just starting to break winter dormancy. We then insert these root cuttings into cell trays filled with a special well-drained potting mix that we custom make using peat moss, perlite, sand, compost, and a bit of wood ashes. Then we thoroughly water the cuttings in and place them under the benches in our greenhouse to give them protection from the hot sun while they start to grow. We mist them with water once or twice a day to keep them moist in the early stages of rooting.

Once they become established in their cells and start to grow vegetatively (generally two to three weeks), we move them out from under the benches into the full sun in the late afternoon so they have a chance to adapt gradually to the intensity of the sun-light. We then apply a foliar application of seaweed spray to give them a little fertility boost. Within a week or so of growing in the full sun and once their roots have reached the bottom of the cells, they are ready to be transplanted into the field. We have a very high success rate with all the plants we start using this method.

Another method of vegetative propagation we use is taking hardwood cuttings from dormant woody shrubs and plants such as elderberry (*Sambucus canadensis*). In the early spring, before the plants have started to grow, we use sharp pruning shears to cut off twelve- to sixteen-inch growing tips from the dormant plants. We then insert these hardwood cuttings into pots filled with our custom-made prop-agation potting mix, water them in well, and give

**Figure 8-12.** Making hardwood cuttings of elderberry.

**Figure 8-13.** Propagating lemon balm from vegetative root cuttings.

**Figure 8-14.** Elder cuttings maturing.

them the same treatment as the root cuttings in the previous paragraph. Hardwood cuttings take longer to root, and we don't have quite as high a success rate as we do with herbaceous cuttings, but this is still an amazingly effective way for us to increase our planting stock affordably and efficiently.

Some people use rooting hormones, both natural and synthetic, to help speed up the rooting process and prevent fungal diseases, but we have found that we have good success without this treatment. Foliar fish and seaweed sprays seem to do wonders for helping to keep the cuttings healthy and free of fungal diseases while they grow into healthy plants.

## Choosing the Right Containers

In addition to choosing the proper medium in which to grow plants, choosing the right container in which to grow them is also a very important consideration.

Some farms choose to use soil blocks, which are basically individual blocks of soil formed in groups separated by a thin air space. These blocks are formed with mechanical soil-block makers that can create several blocks at a time by compressing the soil into a metal mold and releasing the blocks into trays or onto a solid surface, where they are then seeded or planted into. The empty spaces between each block "air-prune" plant roots so they don't grow together.

We choose to use plastic cell trays on our farm because after trying both methods we feel we can produce more plants in a shorter time while taking up less space with cell trays than with soil blocks. Our cell trays measure ten inches by twenty inches and hold 36, 72, or 128 individual cells, which we fill with soil and plant seeds or start cuttings in. We also use "20 row" seed trays for a lot of the tiny seeds that we start, then "prick out" any seeds that germinate into 72- or 128-cell trays to grow on. This saves us the potential of wasted space in cells that don't germinate. All four types of plastic cell trays we use are very affordable and last a long time if they are stored out of direct sunlight when not in use. We can use many of them for up to four years or so before they begin to break down and we have to replace them.

## Planting Seeds in the Greenhouse

Each spring we eagerly anticipate the time when we can begin planting the season's first seeds. We start preparing the greenhouse in late February or early March by setting up growing and potting benches and watering systems, preparing cell trays, and stocking up with potting mix. We have tested the germination rates on any leftover seeds and have calculated how many of which species we will seed into which size cell trays at what date. There is usually still snow on the ground outside, yet on clear, sunny days, temperatures inside the greenhouse can easily reach 80°F, and we often work in shorts and T-shirts. The greenhouse is a pleasant working environment and respite from the cold in what is more or less still winter here in Vermont.

Seeding begins shortly before the full moon in March every year with the heat-loving crops such as ashwagandha, gotu kola, and tulsi. These are started inside using bottom heat at a constant 80°F under efficient T-5 fluorescent lights. We then start seeding all the perennial herbs that can benefit from stratification. These stratifying seeds are lightly watered in and placed on benches in the unheated greenhouse, where they experience large fluctuations in temperature for about three weeks. We have to be careful not to let them completely dry out during this time, so we give them an occasional misting with water to keep the soil surface slightly moist.

After three to four weeks of stratification we begin heating the greenhouse, and we plant the rest of the seeds that don't need stratification, such as calendula, spilanthes, astragalus, and many others. The greenhouse is heated via Modine heaters with forced hot water from an efficient wood gasification

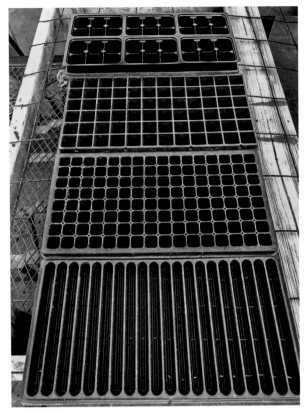

**Figure 8-15.** The different seed trays we use to make transplants. Photograph courtesy of Bethany Bond

**Figure 8-16.** Filling the greenhouse with seeds. Photograph courtesy of Bethany Bond

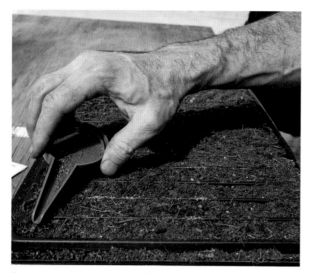

**Figure 8-17.** A hand seeder. Photograph courtesy of Bethany Bond

**Figure 8-18.** Healthy seedlings come to life. Photograph courtesy of Bethany Bond

boiler that also heats our house and domestic hot water. We set the nighttime temperature at 60°F and the daytime temperature at 75°F. If the sun heats the greenhouse above 85°F during the day, an electric thermostat opens shutters and activates a fan to cool the house. As seedlings grow on, many are moved into our drying shed, which serves as a cold frame in the spring, in order to begin preparing them for transplanting and to make room for more seedlings in the greenhouse.

## Seeders

There are many ways to sow seeds into flats, the simplest method being using your fingertips to sprinkle or drop seeds onto the soil surface. For years we used small plastic seed-sowing devices that held seeds in a clear compartment and dropped them down a small chute while we gently tapped it with our fingers. There were many years that we said to ourselves, "There has to be a better way to do this," and lo and behold, there was. We finally built a vacuum seeder,

after realizing how much time we were spending seeding the slow way. This device has made a dramatic improvement in the speed at which we can seed flats.

It works by utilizing the suction from a small vacuum motor to pull seeds onto holes drilled into a thin plate of aluminum. The holes are slightly smaller than the seed itself so the seeds stay sucked to the hole but don't get pulled through it. These holes line up with the cells of our cell trays. When we place the aluminum plate with all the seeds onto the cell trays and release the vacuum, voilà! The seeds drop into their cells. We custom-drill our own aluminum plates to match the different-size seeds we use. There are many different types of "automatic" seeders, ranging from simple and inexpensive to elaborate and expensive.

Now that we have discussed the fascinating world of seeds and how best to start and grow them into healthy transplants, let's step out of the greenhouse and into the fields to get these transplants growing outside, where they belong.

# CHAPTER 9
# Considerations for Growing Medicinals

As I mentioned at the beginning of this book, Melanie and I were pretty young and inexperienced when we bought our farm and began growing medicinals. We started our seedlings in a little makeshift lean-to–style greenhouse stapled to the side of our garage that first year. I can remember lying awake many nights that spring, listening to make sure that I could still hear the heater blowing warm air across the plants on chilly nights. We would check the greenhouse obsessively all day long to make sure the seedlings were well watered and still alive. We were so cautious in the way we and others handled the plants, and when we finally got ready to transplant them into the field, our anxiety levels soared.

Now that we have raised our daughter Lily for the last eight years, it is humorous to consider the parallels in our approach to raising her the first year of her life with raising our plants the first year on our farm. Shortly after Lily was born we spent many restless nights, listening to be sure she was still breathing. We checked on her constantly and fed her incessantly, and we were so cautious in the way we and others handled her. Lily is eight years old now, and we realize that it won't be long before she goes out into the world on her own. The thought of that alone makes me anxious.

So there we were back in 1999, two young, somewhat idealistic new farmers setting our transplants out into the field for the first time, not knowing whether they would thrive, barely survive, or succumb to the elements and crush our hopes and dreams. Of course the plants survived. They actually thrived, and we finally breathed a tremendous sigh of relief. Friends of ours who have had multiple children describe how they too were paranoid with their first child, but by the second or third child they had adopted a more laissez-faire approach, and the kids pretty much took care of themselves.

**Figure 9-1. Medicinal harvest.** Photograph courtesy of Kate Clearlight

Okay, that's a bit of an exaggeration, but again, the parallel is worth noting, because now that we have produced fifteen generations of offspring from our greenhouses, the way we manage them is much more hands-off. I'm not saying we are careless—our livelihood depends on having strong transplants develop into strong field crops. The difference is primarily that we trust that the plants have a will to grow, and if we give them a little TLC and provide them with a good home and some nourishment, they will grow up to be fine, upstanding members of their community.

## Planting Time

It is late spring in Vermont, and our greenhouse is bursting at the seams with transplants. The fields are all tilled, beds are prepped, irrigation headers are set, phone is ringing off the hook with herb orders, and the farm crew is chomping at the bit for more hours. What more could we ask for at this point in the game? This is definitely an exciting time on the farm, as it truly signifies the real start of the growing season.

But before you start putting plants in the ground, it is important to ask yourself: Are you really ready? Is the threat of frost past? If not, have you hardened your transplants off well enough or stocked enough row cover to keep all those plants alive if temps dip below the freezing mark? Did you spread and incorporate enough composted manure and rock powders to grow healthy plants? Will you mulch beds or cultivate them? Have you strategically planned where all these plants will grow? If the answer is "yes," good luck and happy planting. If the answer is "no" or "maybe," here are a few things to consider before you set out to plant.

## Hardening Plants Off

Springtime in most northern climates is a time of flux. The angle of the sun increases, providing intense solar radiation by day, but clear nights can promote rapid radiational cooling and allow temperatures to plummet below the freezing mark. Here in Vermont we aren't really considered "in the clear" as far as

our last average frost date until around Memorial Day. However, like many farmers, we calculate the risk/reward factor while putting plants in the ground as early as we dare and hope that our gamble pays off. When it doesn't pay off, we have to scramble to break out row covers along with all manner of tarps, bed linens, shower curtains, tablecloths, or what have you in a race to minimize frost damage.

One of the best ways that we can try to tilt the odds in our favor when it comes to late spring frosts is to "harden the plants off" before we put them out into the field. Hardening off basically means getting them used to being out in the real world after they have been pampered in the greenhouse for so long. Those of us who live in cold climates know that in the fall when the mercury starts dropping we feel very sensitive to the change in temperature, so we add layers of clothing or seek warm shelter. In September a 50°F day can feel pretty darn chilly. We aren't used to its being that cool yet, but by the end of January we have experienced so much cold weather that a 50°F day can feel downright balmy. Plants are similar; it takes them time to get used to changes in their environment. Like humans, though, plants are adaptable if we approach changes in their environment gradually and thoughtfully.

The first thing we do to harden our plants off is to roll up the sides on the greenhouse for a few days and nights as long as the mercury doesn't dip below freezing. This is also the time when we decrease or completely withhold the amount of water we are providing them so they experience a little bit of drought stress. After they have started to get used to more natural conditions, we bring them outside during the morning or evening when the sun is low so as not to shock them, and we put them on benches out in the open. Now they really start to get used to the full sun and the elements. If it gets too cold at night we can cover them with Reemay, but ideally they get to experience temps close to but not quite freezing, which helps strengthen them to withstand a light frost if it happens.

We are still reducing water to the plants all through this time, and we actually like to see them

**Figure 9-2.** Plants outside to harden off.

wilting slightly because that means we are really test-ing them. After several days and nights of this, they are much stronger, prepared to face the elements and ready to go out into the field. At this point we water them in really well to prepare them for their journey.

# Fertility: Less Is Often More

One of the most challenging aspects of our work with medicinal herbs has been figuring out how to address soil fertility. With so many different species of plants growing simultaneously, how do we please them all? Everything we have learned about growing plants in general tells us they grow best when we provide them with optimum fertility levels and ideal growing conditions. When we follow these recommendations our plants are healthy, and they produce high yields. However, our own instincts combined with what we have learned from our background as herbalists tells us that with certain species of plants, perhaps less is more when it comes to fertility. The "less is more approach" closely follows the way we see many of these plants growing in the wild, where fertility levels are usually far from optimum and growing conditions are less than what one would consider ideal.

We have learned that many of the medicinally active compounds that we seek from plants come from secondary metabolites. Some of these second-ary metabolites are purportedly higher in plants that have experienced stressful conditions and have grown in what we would consider nutrient-deficient soils. Many herbalists that we know and trust tell us this same thing, that medicinal herbs found growing in seemingly harsh conditions in the wild are often found to produce more potent medicinal compounds than those cultivated in rich soils and protected from stressful conditions by way of irrigation, weeding, and pampering.

**Figure 9-3.** Field-grown rhodiola.

So how do we approach this dilemma? Do we purposely refrain from increasing fertility in our soils instead of enriching them? Do we drought-stress the plants by withholding irrigation and growing them in harsh conditions? The answer usually falls somewhere in between the two extremes. First of all, not all medicinal plants found growing in the wild like to grow in harsh conditions and poor soils. Take stinging nettle, for example, one of the most nutritive vitamin- and mineral-rich tonic herbs we know of and also one of our best-selling herbs year after year. When we see nettles naturalized in the wild, they are usually growing out of old manure piles in barnyards or in pockets of rich, moist soil in woods and

meadows. That is where they like to grow, and they seem to produce an abundance of readily available vitamins and minerals when grown in rich soils.

Skullcap, one of our premier nervous system herbs, also doesn't seem to like to grow anywhere but in rich, moist soils. Arnica and Saint John's wort, on the other hand, struggle to grow in soil that is too rich, and they seem to make stronger medicine when grown in poor to average soil. Lobelia grows abundantly here in the wild on compacted logging roads in the woods and in gravel deposits on the edge of meadows. This wild lobelia produces some of the strongest alkaloid-rich medicine we know of. However, when we planted lobelia in our rich farm soils, the plants got twice as big and green as the pale-green stunted ones we found in the wild, yet the alkaloid concentrations weren't anywhere near as strong as in the wild plants.

We have found that the key to this dilemma is to plant our crops in soils that are similar to the type of soils we find them naturalized in or that are found in their native habitats. Before we started farming this land our soils were naturally low in fertility and very acidic. This is the perfect condition for growing plants such as arnica, lobelia, rhodiola, Saint John's wort, and yarrow. All we have to do to provide good growing conditions for these plants is to till the soil and prevent excessive competition from weeds. We don't usually add much if any soil amendments to the beds these plants are planted in, and we usually don't irrigate them, either. They grow well in these conditions and make strong medicine.

The majority of crops we produce are species of plants that like to grow in moderately fertile soil that doesn't contain high nutrient levels such as nitrogen and phosphorous but does contain ample quantities of soil organic matter. This is fairly simple for us to replicate. When we take our soil samples and send them off to the lab for analysis, the results come back with specific recommendations as to fertility and pH levels for growing commercial field crops. If our soils are a bit low in primary nutrients such as nitrogen and phosphorous, for example (which they commonly are), there may be a recommendation to apply up to two hundred pounds per acre each of nitrogen and phosphorous, for example. In that case, we may apply between fifty and one hundred pounds instead of the two hundred pounds recommended. The composted chicken manure we use is relatively high in nitrogen and phosphorous, so we have to be thoughtful as to how much is enough without going overboard. Our typical application rate for "maintenance fertility" for crops that prefer soils with average fertility is generally one-half ton per acre of composted chicken manure. This composted manure that we get from a local egg farm has sawdust bedding mixed in with it, which helps the compost release its nitrogen at a slower rate, increases soil organic matter, and is an effective and affordable source of fertility.

For crops such as angelica, burdock, elecampane, comfrey, and some of the "mint family" plants such as nettles and skullcap that prefer a richer soil with more readily available nitrogen, we follow the soil test recommendation data forms that the soil labs give for "field crops." (Incidentally, I have never seen "medicinal herb crops" listed on these soil test recommendation data forms. We hope this fact changes in time as medicinal herb production becomes more common, but again, fertility needs are so variable with medicinals that it would probably be difficult to establish baseline recommendations.) The fertility recommendations from the soil testing labs for field crops generally recommend one to one-and-a-half tons per acre of composted chicken manure.

The mint family crops such as peppermint, lemon balm, and spearmint are prized for their production of highly aromatic essential oils, which tend to increase when the plants experience stresses such as droughty conditions. We like to encourage early leaf growth on these crops by providing them with good (not excessively rich) soils and irrigation. Then a couple of weeks before they begin flowering we shut the irrigation off and hope that we have hot, dry, sunny days that help to stimulate increased production of these essential oils. The weather doesn't always cooperate this way, but when it does, some of our most potent aromatic herbs are produced.

**Figure 9-4.** Pleurisy thriving in calcium-rich soil.

# Soil Remineralization for Medicinals

Most of the soil types found on our farms and in our gardens are lacking the complete range of soil minerals and trace elements that are necessary for optimal plant growth. These deficiencies vary from region to region and are primarily determined by type and solubility of the parent rock (bedrock) that our soils are composed of. Secondary nutrient recommendations from soil tests, along with soil maps from the Natural Resources Conservation Service (NRCS), are useful in that they can help to guide us in adjusting mineral levels such as calcium and magnesium, which help determine our soil pH.

Some of the medicinals we grow, such as echinacea, garlic, ginseng, pleurisy, and red clover prefer soils containing more readily available calcium. For those crops we may add a little extra lime or wood ash to provide that calcium. It is important to note that timing and the solubility rate (fineness of particle size) of the product being applied are important considerations when applying lime because of its slow-release tendencies. Lime applied in the spring will generally not have enough time to become available to plants during critical growth periods, so we apply lime in the fall.

When applying lime to raise soil pH and provide calcium, the percentage of magnesium in the soil, as well as in the type of lime, plays a critical role in determining which type of lime you should apply. Whereas plants absolutely need magnesium to undergo photosynthesis, excessive magnesium levels can negatively affect soil biological activity and interfere with the plant's ability to utilize other important nutrients. Striking a balance with magnesium levels is extremely important. Dolomitic lime, which contains up to 11 percent magnesium carbonate, is generally recommended for soils low in magnesium. Calcitic lime, on the other hand, contains on average 40 percent calcium carbonate and only around 2 percent magnesium carbonate, which is a better choice for soils with sufficient magnesium levels.

Other plants, such as angelica, blue vervain, and valerian, for example, seem to prefer a slightly more acidic soil with less available calcium. For these crops we will hold off on the calcium application and focus on adding plenty of organic matter via composts and cover crop residues. Our composted chicken manure is relatively high in calcium, so we always factor that into the equation and try to use a different type of compost, such as that made from cow manure and decomposed plant material, on crops that prefer less calcium.

Our biggest goal in keeping our soils and thus our plants healthy, in addition to remineralization with mineral powders, is to provide the soil with as much organic matter as possible. Organic matter is something we never really have to worry about applying too much of because our soils and plants need it, and we burn some of it up every time we till or cultivate. Our approach to supplying ample amounts of organic matter is to spread compost, cover crop our soils, and incorporate plant residues into the topsoil. In addition to increasing soil organic matter, we also have to do what we can to ensure that it stays in the soil where it belongs, rather than being released as gaseous carbon or eroded into our watersheds.

## To Mulch or Not to Mulch?

Most people who grow plants outdoors seem to have a strong opinion one way or another about whether or not to mulch plants. Also noteworthy is that many of those who have an opinion on mulching have changed their opinions through the years and are just as passionate about their new opinion as they were about their old one, even if those opinions run counter to each other. I have to admit that Melanie and I fall into the category of the newly converted and still opinionated.

We spent fourteen years farming herbs without mulching anything, with the exception of a light layer of bark mulch on our perennial beds in the front yard. My arguments against mulching were that straw (which is the mulch of choice of most commercial growers around us) would be too expensive because of the amount of material it would take to mulch several acres of herbs. Good straw can also be difficult to source from a certified organic grower because of the high demand. There isn't much grain grown where we live, so straw often has to be sourced from afar. I also felt that the labor costs involved with applying enough mulch by hand for it to really be effective at weed control would be exorbitant.

I wasn't even convinced that having a layer of straw around the plants and in the paths would really keep the weeds down all that well, and if it didn't would the straw get in the way when we had to come in and weed it with tools? Could the straw create its own weeds by bringing new seeds into the fields? Hay was another more affordable option that

**Figure 9-5.** Tulsi growing on mulch to help increase yields, reduce weed pressure, and prevent soil contamination on leaves. Photograph by Alden Pellett

we considered, but we figured the amount of weed seeds in that hay was almost certain to add to rather than subtract from the weed problem. Why would we do such a silly thing as mulching when we could cultivate weeds with the tractor, then hand-hoe any weeds that the tractor couldn't get?

Lots of farmers around us who grow garlic swear by their mulching and say that it saves their crops during the winter if there isn't snow on the ground for insulation and they get a deep, hard freeze. Our farm is at eleven hundred feet above sea level in northern Vermont, and we consistently have had between three inches and three feet of snow on the ground all winter long without fail. Our thinking was that snow makes an incredibly effective insulating barrier against deep-freezing temperatures penetrating into the soil and damaging plants, so

**Figure 9-6.** Beautiful and aromatic dried tulsi. Photograph courtesy of Bethany Bond

why would we ever mulch our garlic (or anything else for that matter) when there is no need to? Basically what it came down to was that for years we saw absolutely no compelling reason to start mulching crops, so we didn't mulch.

Then things started to change. The first thing to change was that more wholesale herb buyers were requiring our herbs be analyzed for microbiological contamination, and that they fall within acceptable levels, before they would buy them. This microbiological testing is part of the Good Manufacturing Practices (GMPs) we have mentioned in previous chapters. What on earth does this have to do with mulching, you might ask? Good question. Basically what it means is that when it rains hard or the wind blows and soil gets splashed or blown onto the leaves of plants they become increasingly susceptible to fungal contamination.

Our soil and the air around all of us contain microscopic yeast and mold spores, regardless of how good the soil or how clean the air is. When those spores land they often remain dormant and benign for the most part while the plants are still growing. Once we harvest the plants, however, and they go into the drying shed, that humid environment created from removing tons of water from the plants drying inside it can cause those yeast and mold spores to multiply rapidly. In all the thousands of pounds of herbs we have dried and had tested for microbiological contamination, we have had a few microbiological tests come back positive for elevated yeast and mold levels. But that was a few too many, and we have realized that we have to do everything we can to keep the plants' leaves cleaner while they are in the field by preventing soil splash, especially with increasing regulations ensuring that raw materials used in herbal products are free from microbiological contamination. Part of the solution to maintaining sanitary conditions with our herbs is to use mulch to prevent soil deposition on the plants' leaves.

The second thing to change regarding our mulching evolution was our weed pressure. We had a couple of back-to-back years with more rain than we ever imagined possible in this part of the world.

When it rains for days and days, the weeds grow and grow, and we can't get into the fields to cultivate or hoe because of the muddy conditions. When we can't cultivate weeds they grow large and rapidly go to seed. When weeds are allowed to go to seed, even for one season, that can set back a majority of the work we have done in previous seasons to stay ahead of weed seed populations.

Not only did our existing weeds go to seed during those excessively rainy seasons, but we also started noticing a new species of weed creeping onto the scene. The dreaded galinsoga! Hairy galinsoga (*Galinsoga ciliata*), or "quick weed" as it's often called, is the bane of existence for many growers all across its viable range. It is an annual weed that goes to flower and produces an enormous amount of viable seed before the blink of a farmer's eye. With most weeds we have a visual warning that the plants will be going to seed soon after they begin flowering. With galinsoga, as soon as it starts flowering it is able to start producing viable seed almost immediately.

It's almost hard to believe we went as long as we did here before this new weed entered the picture. Anyone who farms in this region will tell you that if you don't have galinsoga yet, do everything you can to prevent it because it's only a matter of time before it comes in. It travels via bird poop, deer poop, someone's farm boots, "imported" manures, and so on. There are endless possibilities for this opportunistic plant to find its way into your land. The solution to this weed problem, as I learned from so many who have experienced its challenges, is to cultivate it or flame-weed it before it goes to seed or smother it with mulch. Once galinsoga is on your farm you basically have to accept that it is there to stay and learn to live with it because it will likely never completely go away. We are going to try all of these methods, including mulching, and yes, we will have to learn to live with it.

The third and final thing to change in our mulching paradigm shift was that we finally did a business plan, and in that business plan we uncovered the fact that we were spending way too much money paying people to hoe and hand-weed. Tractor cultivating can be very effective at controlling most of the weeds, especially early in the season, but since we are planting acres of long-term perennial beds, once the plants get large it can become difficult or impossible to tractor-cultivate, and the weeds can rapidly grow, especially when we harvest the plants' leaves two or more times per season, which gives weeds an opportunity to take hold. Also, running tractors is costly on many levels, and the less we have to use them the better off we all are.

Many people have asked why we don't use cover crops as a "living mulch." We tried this with white clover in the paths one season, and it worked great until the living mulch became a pernicious weed in and of itself. It was also no match for some of the perennial grasses that got established while the clover was just starting to take hold. White clover has a vigorous lateral-growing root system that seems to go on and on, and in some crops it started outcompeting our plants for water and nutrients. Living mulches may be practical for small-scale home gardening, but we have found them impractical for medium- to large-acreage production.

## Types of Mulch and How to Apply Them

As it became obvious to us that we needed to consider mulching, the next step was deciding how to do it. Melanie and I spoke with several growers for advice, and they all told us basically the same thing, that straw mulch is definitely an option, but it is very expensive to buy if you can't produce your own or don't already have a good local source. It can be very labor intensive to apply, and it's also difficult to weed around it if it fails in its role of prevention. Mulch hay was recommended by some, but with that recommendation came warnings about using hay cut from flowering grasses because of the amount of weed seeds it contains. These things we knew already, so we considered the next and more highly recommended option, which was to buy a combination bed former/mulch layer that attaches to our tractor's three-point hitch and apply either polyethylene mulch or biodegradable plastic mulch

made from non-GMO cornstarch, which as of the writing of this book is still in the process of becoming allowable for use on certified-organic farms.

We had seen lots of vegetable growers over the years using plastic mulch on raised beds, and we told ourselves over and over again that we would never resort to that method on our farm. Purposely laying a petroleum-based product on organic soils seemed to run counter to the very ethos of organic farming. I spoke to more growers that were using poly mulch, and they all expressed the same moral dilemma but felt that there was no option at the scale they were farming. The most challenging aspect of the poly mulch, they said, was in having to remove and landfill it after it is removed. Other than that they seemed unanimously to say that the benefits far outweighed the challenges, and some even said they didn't think they would still be farming without it.

Biodegradable mulch seemed like a viable alternative that we had to consider. At least we wouldn't have to haul it off to the landfill in the fall. When we factored in all the diesel fuel and labor that we use cultivating, plus the loss in soil organic matter, plus the expense of paying people to hand-weed, plus the risk of our herbs failing microbiological tests from soil splashing on plants, we had to make the right decision to keep our farm going, which was to start mulching. The only problem with using bioplastic mulch is that it is currently stalled in the process of becoming approved for use on certified-organic farms. While we await the approval of bioplastic mulches, we are using black plastic mulch with great success.

Mulching raised beds has been one of the most cost-effective improvements we have made, and although we wish we had never come to the point where we felt like we had no other choice but to start mulching, we have never doubted our decision to make that choice. We only currently mulch ten of the fifty-plus species of crops that we grow, and most of these are either species that grow close to the ground and are susceptible to soil splash, such as lemon balm and anise hyssop, or tropical plants such as ashwagandha and tulsi that benefit from the fact that the black-colored mulch warms the soil earlier in the spring and provides additional heat throughout the season, which these plants thrive on.

After we made the leap to mulching our raised beds we were faced with the dilemma of how to control the weeds in the "middles," or pathways between the raised beds. We consulted with our local community of vegetable growers and found that most of them were having great success with using spoiled hay from local dairy farms that had been cut from fields of nonflowering grasses. Thankfully, our neighbor's dairy farm produces a surplus of round bales that are very affordable and surprisingly free of weed seeds. We now use this hay to mulch almost all of the middles. There is an initial labor expense for spreading the hay by hand at a depth of three to four inches, but this expense is minimal compared with the labor

we were paying for hand weeding. Plastic-mulching raised beds with hay mulch in the middles has proven to be a winning combination on our farm, as well as on other farms, and we highly recommend it for producing sanitary herb crops in weed-free fields.

## Transplanting: By Hand and Machine

For the small-scale herb farmer, transplanting by hand is a great way to get plants into the ground. We transplanted by hand for the first few years we farmed, and we got pretty good at it through plenty of practice. Since we were cultivating with a tractor we had to have straight rows, so we would set a long string line between two stakes, one at each end of

**Figure 9-7.** Lily, Bethany, and Melanie transplanting tulsi with the water-wheel transplanter.

the row, and hoe a four-inch-deep furrow under the length of the string line. We would then walk along with five-gallon buckets and fill the furrow with compost and mineral powders when needed. Then we would cover the furrow with a steel rake and level it out. One person would then walk along with a trowel or narrow pointed shovel making holes for the plants to go into while a second person pulled plugs from flat trays and dropped them into the holes. A third person walked down the row with a rake or hoe, covering the plants with soil. With three people it took us approximately two days to plant one acre at this rate. Let's say two ten-hour days = 20 hours × 3 people = 60 hours × $12 per hour average = $720 just in labor to plant an acre.

As our acreage increased and we started doing the math on our cost of production we realized that we desperately needed to automate that process. We bought a used Powell mechanical disc-type transplanter from a local vegetable grower for $500. It took us a day or two to get it adjusted to our plug size and row spacing, but when we did it was obvious that this was the way to go. We figure that this piece of equipment probably paid for itself within a few days of use by the labor savings alone. Now, with two people riding on the transplanter feeding plants into it and one person driving the tractor, we can easily plant two acres or more per day.

When we started working with raised beds and plastic mulch, we bought a water wheel–type transplanter. This machine allows us to transplant into our mulched raised beds as it punches holes through the mulch and into the bed to plant into at a predetermined spacing. A large water tank attached to the planter fills each hole with water while two people riding on the planter push a plant into each hole. When the water slowly drains out of the hole, it pulls soil around the plant, giving us excellent soil-to-root contact, which helps ease transplant shock. This transplanter allows us to simultaneously plant two rows at a time in our double-row mulched raised beds. We are also experimenting with using our water wheel transplanter to perform no-till planting directly into cover crop stubble.

# Establishing Long-Term Perennial Beds

Since we are primarily growing perennials on the organic medicinal herb farm, we have the benefit of establishing long-term plantings instead of having to replant each crop every single year, as we would have to do with annuals. Many species of perennials will last ten to twenty years or more, even decades, with little maintenance if left to grow fully each year. However, since we are harvesting most of the perennials that we grow for leaf harvest once or twice each season, their vigor starts to wane after three to five years of repeatedly being harvested, and then it's time to replant them. This isn't true with all perennials, of course. Some, especially the shrubby and woody types, such as elderberry bushes, will last for decades even with being repeatedly harvested.

One of the biggest challenges we face when trying to keep perennial beds going long term is weed control. When we harvest the perennial leaf crops we are generally harvesting approximately 75 percent of each plant. We leave about a quarter of the plant intact so its leaves can keep photosynthesizing and feeding the plant's root system. However, even when we leave this 25 percent intact, we are decreasing the leaf canopy so much that increased sunlight penetrates the canopy to the soil below the plant. When this happens, no matter how weed-free we thought the bed was, weeds inevitably start to pop up, especially perennial grasses. When the grasses come in, our labor costs increase from having to pull grass out of the herb plants we harvest. When we harvest, not only are we letting more sunlight penetrate the leaf canopy, we are also diminishing the strength of each plant by cutting most of it off. When we continue to diminish the strength of the plant, eventually we see diminishing yield returns. We have experimented with harvesting 50 percent of each plant instead of 75 percent, but when we do the math on that, the losses in yields do not make up for gains in the amount of years we can harvest those plants. It is more cost effective to replant these perennials when their vitality starts to fade, and this is a fact of life we have come to accept.

**Figure 9-8.** Fields of peppermint.

Perennials such as echinacea, elecampane, and marshmallow that we grow for root harvest, on the other hand, are only grown for two to four seasons before we harvest them. Beyond four years, the increase in quality and yield of those roots usually either levels off or starts to wane, which can mean decreasing return on the costs of maintaining them and occupying valuable field space. Woodland perennials such as goldenseal, ginseng, and black cohosh, on the other hand, seem to increase in size as well as potency each year and can live for a very long time.

Some of the mint family crops that we grow, such as peppermint, spearmint, and skullcap, spread via laterally growing shoots. These plants are considered invasive weeds by some because they can spread aggressively in a relatively short period of time and outcompete other plants. We can take advantage of their vigorous growth habits by utilizing these laterally growing shoots to establish new plantings. In addition to taking cuttings from dormant mint family plants in the spring, as we discussed earlier, another method we use to propagate these types of plants is to divide and move them in the fall. We do this after three or four years of growing in one place when the weeds start to infiltrate the crop and we see yields declining. After we have made the last leaf harvest in late summer or early fall we go into the field with the rotary cutter or sickle bar mower and mow the plants down to ground level. We then run the disc harrow over these plants several times, which chops them into pieces. Then we go into the field with bins and fill them with plant divisions. We take these plant divisions to another freshly prepared field and plant them using a transplanter.

This has become a very efficient way of "recycling" older plants for us and restoring their vitality by moving them and giving them a fresh start in new soil. They really take on a new, youthful vigor after being divided. After we have removed the plants from the old field, we disc harrow that field a couple of more times during a dry spell to ensure that the plants we are trying to remove don't turn into weeds. Then we plant a dense cover crop, usually consisting of oats and red clover.

# Crop Rotations

Crop rotations are an integral component of the medicinal herb farm plan. Although there is no single method or system that is universal for every farm, there are some key principles that can be broadly applied when planning how, when, and where to rotate crops from year to year on your own farm. Following are the benefits that can be had by thoughtfully planning and executing crop rotations on your own farm.

## Disease Prevention

By moving plants that are susceptible to disease as often as possible we can help limit the introduction and spread of plant pathogens. Soilborne diseases can overwinter and remain in infected fields for years. By moving infected host plants out of contaminated fields for at least three years we can mitigate these diseases by depriving them of a host. We can also avoid disease transmission by choosing crops that are not susceptible to the same pathogens that may be present in certain fields. Saint John's wort is a good example of a plant that is susceptible to fungal disease on our farm, and we take great precautions to rotate this species in fields located far away from fields it has previously occupied. Well-balanced soils and good growing conditions also help to limit disease transmission. We discuss the specifics of plant diseases in greater detail in chapter 11.

## Insect Control

Insects can be detrimental to crops not only by the damage caused by feeding on them but also by spreading disease as they do so. Most common plant diseases are spread by insects that serve as "vectors," or agents of transmission. These vectors transmit pathogens when they become contaminated from an infected organism (usually a plant) and transmit it to another. When crops that are prone to disease are moved as often as possible, we can help to limit the spread of pathogens by limiting access to the hosts that vectors feed on. Echinacea is a good example of a species that is prone to being infected with aster yellows disease by leafhopper insects, so we do everything we can to rotate this crop in a thoughtful way to prevent leafhopper damage. We discuss pest control in greater detail in chapter 11.

## Weed Control

By moving crops around from year to year we can help to control weed problems by using good tillage and cover cropping practices. Perennial weeds, such as various species of grass, often pose the biggest challenge, as they can be difficult to eradicate without excessively cultivating the soil.

## Soil Health

Crop rotations give us a chance to restore soil tilth and fertility by planting cover crops in between years of having herb crops grown on them. We can also help remediate soil compaction issues caused by cultivation and harvesting traffic by using deep-rooted cover crop species such as alfalfa or red clover.

With so many species in rotation consisting of both annuals and perennials on a typical medicinal herb farm, it requires thoughtful planning and good record keeping to keep track of yearly rotations. At ZWHF we have come up with a pretty simple formula that guides us in our process. Everything we do is recorded from year to year so we can reference what was growing where, how, and when.

**Figure 9-9.** *Echinacea purpurea* infected with aster yellows.
Photograph courtesy of Kate Clearlight

## Crop Rotations for Annuals

Annuals that we grow for aerial (leaf or blossom) harvest, such as calendula, chamomile, and tulsi are planted in spring, harvested multiple times during the growing season, then tilled under by early October. We immediately follow these with a fall and winter cover of either winter rye or oats. The following spring this field is usually planted with root crops that we grow for one season, such as burdock, dandelion, and yellow dock. After these roots are harvested that fall, we like to plant that field with a mixture of oats and clover. The annual oats are winter-killed, and the perennial clover emerges in spring as an excellent nitrogen-fixing green manure and medicinal herb crop (red clover blossoms) for the entire following season.

## Crop Rotations for Perennials

Perennials are generally grown for anywhere from two to five years in the same bed and are usually followed with a year or two of oats and clover. If perennial weeds such as grasses have become a problem in these perennial beds, either we will do two "quick crops" of buckwheat the following spring and summer followed by oats and clover or we may opt for planting sudangrass × sorghum (Sudax) for its weed-smothering, biomass-producing effects.

There are three major disease issues that affect perennials that we have dealt with periodically on our farm: aster yellows disease, anthracnose, and powdery mildew. Aster yellows is spread by leafhoppers and occasionally affects our calendula and echinacea crops. We have had success mitigating the effect of this disease by moving these crops as far as we possibly can from season to season and by attempting to control the leafhopper population.

Anthracnose is a fungal disease spread by windborne spores that causes rustlike lesions on our Saint John's wort. This disease can overwinter in the soil, so we move this crop as far as we can from season to season and try to plant it in areas with good air circulation to keep the spores at bay. Powdery mildew is another fungal disease that can be spread by

**Figure 9-10.** Burdock flowering. Photograph courtesy of Larken Bunce

insect vectors and wind and can remain in and on the soil. Our skullcap has been susceptible to this disease, so we try to prevent it by rotating the crop into a new bed with good air circulation each season that we plant it. We will discuss plant diseases and prevention in greater detail in chapter 11.

## Using Cover Crops to Increase Soil Health

*The farmer can ameliorate 100 acres with clover more certainly than he can 20 from his scanty dung heap. While his clover is sheltering the ground, perspiring its excrementitious effluvium on it, dropping its putrid leaves, and mellowing the soil with its tap roots, it gives full food to the stock of cattle, keeps them in heart, and increases the dunghill.*

— J. B. BORDLEY, *ESSAYS AND NOTES
ON HUSBANDRY AND RURAL AFFAIRS,*
PHILADELPHIA, PENNSYLVANIA, 1801

**Figure 9-11. Buckwheat.** Photograph courtesy of Dalgial, Wikipedia

Cover crops are, dollar for dollar, one of the most cost-effective ways of building and maintaining soil health and fertility. The simple definition of cover crop means a crop that is grown or left on the soil surface to prevent or smother weeds and prevent erosion from wind and water. That definition merely scratches the surface of what these plants truly offer, however. In addition to preventing erosion, and suppressing weeds, cover crops can add significant amounts of biomass to the soil when their residue is incorporated; they can help decrease soil compaction and mine minerals with their elaborate root systems; and they can help to provide habitat for beneficial insects.

Some species of cover crops can serve as "green manures" by taking gaseous nitrogen from the air we breathe and converting it to a solid nitrogen fertilizer that can help to enrich the soil that cover crop was grown in. Many cover crops are also efficient at "capturing" nutrients while growing and temporarily storing them until they are incorporated into the soil, at which point they can slowly release those nutrients

to feed the soil and any crops growing in it. One of the biggest benefits of using cover crops for the medicinal herb farmer is that we have the added benefit of being able to market some of these cover crops, such as oats and red clover, for their healing attributes.

On the other hand, there can actually be some drawbacks to using cover crops in certain situations. For example, cover crops provide nectar and cover for beneficial insects, but they have also been known to attract and harbor harmful insects and diseases. Deer and other browsing animals may also be attracted to cover crops, which can lead to increased pressure on commercial crops nearby. Another challenge, which we've mentioned previously, is that cover crops can become weeds themselves if they are not managed properly. A crop of buckwheat or winter rye left to go to seed, for example, can germinate and come back when and where it isn't necessarily wanted.

Incorporating cover crops can actually be more detrimental to the soil than beneficial if it isn't done correctly. Take winter rye, for example: We seed this crop in late summer or early fall, and it tends to become well established in a relatively short period of time. It is best to wait until the rye starts to flower the following season, then mow it down with a sickle bar or flail mower. After it is mowed it is much easier to kill by tilling it in. If the grower is in a hurry to get the field prepared before rye flowers, killing and incorporating it can take a toll on the soil when it is excessively tilled, thus providing a negative return.

Some cover crops, especially winter rye, have what are known as allelopathic properties. Allelopathy describes the process whereby decomposing plant parts exude natural chemicals that have the short-term effect of preventing other seeds from germinating. This can be beneficial in weed control if timed properly but can also be challenging for the grower by preventing the germination of seeds purposely planted in the same field. Another potential challenge of cover cropping is that for growers with limited field capacity, it can potentially tie up land that could otherwise be producing more profitable crops.

Overall, the benefits of cover cropping can outweigh the challenges when they are properly managed and utilized to their full potential. Following are five species of cover crops that we have used at ZWHF, along with management suggestions and observations we have gleaned from growing these important plants.

## Buckwheat

Buckwheat is a fast-growing, warm-weather annual cover crop. Its broad leaves and fast growth make it a good choice for smothering weeds and protecting soils from erosion. Buckwheat will flower in as few as four weeks, and it can set seed within ten weeks of being planted. We take advantage of this rapid growth period by sowing the first stand in spring after the threat of frost passes. About a week after it starts flowering we mow it with the rotary cutter or sickle bar mower, then till it in. We can repeat this process once or twice more to smother weeds, prevent erosion, and store nutrients before we plant perennials in late summer or early fall. This is a great

choice for newly tilled fields with lots of dormant seeds waiting for the light of day to germinate.

To plant buckwheat we use a spinner-type broadcast seeder and lightly cover the seed with a disc harrow or field cultivator set about an inch deep.

*Buckwheat seeding rates:* 35 to 100 pounds per acre. Use the higher rate for broadcast seeding.

*Pros:* Fast growth, good leaf canopy, cheap seed, excellent nectar source for beneficials. Buckwheat also grows well in poor soils, which adds to its versatility.

*Cons:* Doesn't produce much biomass, doesn't fix nitrogen, can set seed and become a weed before you get a chance to mow it if not monitored.

## Oats and Red Clover Combo

This combination is our favorite cover crop because of its ability to smother weeds, produce biomass, and fix nitrogen. In addition to these benefits we can take advantage of the value-added aspect of harvesting the milky oat heads and red clover blossoms for market without noticeably diminishing the biomass or nitrogen production. Oats are an annual grain crop that produces a good amount of biomass in a short period of time. Red clover is a leguminous short-lived perennial from the Fabaceae, or pea and bean, family. Like most legumes red clover fixes nitrogen by way of rhizobium bacteria in its root nodules. This nitrogen fixation can be enhanced by inoculating the seed with rhizobium bacteria before planting it. To inoculate the seed we buy fresh inoculum powder and add it to clover seed we have predampened by misting it with a spray bottle. We shake the seed/inoculum mix in a plastic bag or bucket to coat the seed evenly with the inoculum; then we sow the seed immediately after.

Planting oats and red clover together requires a two-step approach. Since the oat seed is larger we plant that first by broadcasting it with a spinner-type seeder. We then lightly cover the oat seed with a disc harrow. After the seed is covered we broadcast the red clover seed on the soil surface using the spinner, but we don't cover that seed since it is so small. Using a seed drill (precision tractor-mounted seeder) is recommended when available to conserve seed and ensure a more uniform and dense stand. As the oats germinate and start to grow, they provide cover for the slower germinating and growing red clover. These two plants grow well together and benefit from each other's company in the field. Red clover can also be "frost seeded," which means to broadcast the seed on the frozen ground in early spring.

*Red clover seeding rates*: 7 to 18 pounds per acre. Use the higher rate when broadcast seeding.

*Oats seeding rate*: Approximately 100 pounds per acre when broadcast seeded.

*Red clover pros*: Marketable harvest of blossoms for medicinal use, attracts pollinators, fixes nitrogen, deep taproots lessen compaction, good biomass producer, good smother crop, good nutrient storage, lives for an average of two years.

*Red clover cons*: To achieve its maximum nitrogen-fixing benefit, red clover should be inoculated and grown for at least two seasons, which can tie up field space. Red clover seed is relatively expensive compared to other cover crop seed; it is not killed by mowing and therefore requires some tillage effort to incorporate; the nitrogen formed in nodules on the plant's roots is soluble, like most any other nitrogen fertilizer, so it is important to time the planting of another crop to be able to use that nitrogen before it is leached through the ground. Red clover also has the potential to harbor harmful insects and disease, though this is uncommon from our experience.

*Oats pros*: Marketable harvest of milky oat heads for medicinal use. Good biomass producing, allelopathic, weed smothering, and nutrient storage properties make oats an excellent choice for building soil fertility and lessening weed pressure. Oats can also be used as a source of straw for mulching beds.

*Oats cons*: Allelopathic qualities can limit direct-seeding germination in fields oats have recently grown in. Oats have a tendency to "lodge" or to tip over, making them harder to harvest and/or chop before tillage. Oats also are reported to harbor injurious pests and disease organisms. Oats are an annual so will not overwinter for multiseason cover cropping.

## Sorghum × Sudangrass (Sudax)

This cover crop is hands down our first choice for weed smothering, capturing nutrients, and biomass

**Figure 9-12. Sudax.** Photograph courtesy of Dee Horst-Landis

production. Sudax is a warm-season annual hybrid that puts on a tremendous amount of growth in a short period of time and can grow up to ten feet tall if not mowed. This rapid, dense growth makes it the ideal cover crop for rehabilitating fields with heavy weed pressure and compacted soils. Mowing this crop at least twice during its growth period when it reaches approximately three feet high is recommended to help its roots penetrate compacted soils more deeply and prevent the plants from going to seed. Mowing also tends to encourage increased vegetative growth. Sudax will only germinate in soils that are above 60°F and will winter-kill during the first hard frost, leaving a dense cover of residue to protect soil from erosion and add organic matter as it breaks down. Because of its strong allelopathic properties and its propensity to tie up nitrogen before it fully breaks down, it is not a good candidate for a quick summer cover to be directly followed by a new planting. Sudax prefers well-drained soils with a pH between 6 and 7 but will grow fairly well even in poor soils with low fertility.

To plant Sudax we wait until the soil warms above 60°F and broadcast-seed it densely. We then lightly cover it with the disc harrow. Ideally this seed would be planted with a seed drill to conserve seed and ensure a more uniform stand. This crop is very drought tolerant and will often go into temporary dormancy when drought conditions persist, only to reemerge vigorously after the rains return.

*Sudax seeding rates:* 10 to 35 pounds per acre. Use the higher rate for broadcast seeding.

*Pros:* Incredible biomass production, weed smothering, allelopathy, nutrient storage, and root penetration make this an ideal summer cover crop.

*Cons:* Sudax needs to be planted as early as possible when the soil warms to obtain its full benefit because it isn't frost hardy. It also requires mowing to keep it from becoming unmanageable in size and going to seed. Sudax's strong allelopathic properties can prevent a fall crop being seeded. Some reports show this crop has a tendency to attract nematodes, while others dispute this claim. It also doesn't fix nitrogen.

## Winter Rye

Also known as cereal, rye is an excellent choice for a late summer or early fall cover crop that can overwinter to provide soil protection and emerge to continue growing the following spring. Winter rye grows fast in cool weather and establishes an extensive fibrous root system. This, coupled with its biomass production, makes it a great choice for smothering weeds, as well as for storing nutrients. Its allelopathic effects can be beneficial in preventing weed seed germination around transplanting time but could definitely limit the germination of direct-seeded crops. Winter rye will tolerate temperatures as low as −28°F, making it the most cold hardy of cover crops. It can be difficult to kill and incorporate unless you wait until it flowers to mow and then till it under.

To plant winter rye we prepare a clean seedbed anytime from midsummer on and direct-seed with a broadcast spreader. We then lightly cover the seed with the disc harrows. Drilling is recommended for more uniform stands.

*Winter rye seeding rates:* 60 to 200 pounds per acre. Higher rates should be used for later plantings to ensure a dense cover.

*Pros:* Winter hardy, excellent biomass producer, incredibly elaborate root system mines and stores nutrients, allelopathy can prevent weed seeds from germinating, performs well in poor soils.

*Cons:* Can be tough to incorporate if not mowed after it begins flowering, allelopathic properties can hinder germination of direct-seeded crops, can potentially harbor harmful insects and disease, doesn't fix nitrogen, doesn't perform well on poorly drained soils.

While we have discussed many ways of growing medicinals and cover crops using mechanized equipment, these are merely methods that have proven successful on our farm with the fifty-plus species we produce. With literally hundreds of species of medicinals sought after in today's marketplace, we encourage growers to choose crops and methods that are well suited for their particular farm size, environment, marketplace, and desire—and to adapt their own production systems for success. Although

**Figure 9-13.** Winter rye.

growing methods and species in production vary from farm to farm, one thing all growers can be certain of is that weed-control methods will need to be implemented to keep unmarketable species of plants from interfering with the growth and processing of marketable crops. In the next chapter we will discuss some weed-control methods that have worked for us and for other growers.

**Figure 10-1.** Field of dreams.

# CHAPTER 10

# Weed Control

I have spent many hours wondering what farming would be like if I were able to wave a magic wand and keep the plants we don't want growing in our fields from getting in the way of the plants we are trying to grow. Oh, if only life were that simple. Unfortunately weeds are something we have to live with, and while the time and expense of keeping them at bay can chip away rapidly at the bottom line, there are some effective ways that medicinal herb farmers can deal with weed challenges while keeping the cost of doing so at a minimum.

"A delphinium in the wheat field is just a weed." This adage aptly describes the fact that a weed is merely a plant growing where it doesn't belong. It's tough to discriminate sometimes, especially when most of the medicinal herbs we grow are considered weeds by many. Our employees often laugh when we tell them to "go weed the dandelions." They know we aren't telling them to pull the dandelions as weeds; rather, we are telling them to pull the weeds growing within the dandelion beds. Sometimes the dandelions become weeds when they are growing within other herbs, but we have the benefit of being able to collect them in a basket as we pull them, to then dry, process, and use them to make medicine. I suppose these could be classified as "useful weeds." It all gets a bit confusing and even comical at times, but the fact is that we are growing commercial crops, and with that production comes the inevitable fact that weeds must be controlled or our jobs become much more difficult.

## Why Weed?

Whether it is one person cultivating with a tractor or a dozen employees hoeing, the expenses involved with weeding can have a significant impact on the farm's bottom line. As farmers we have to be as efficient as possible with our methods to maintain the fine balance between good weed control and good cost of production management, and it's rarely easy to strike this balance. So why do we spend so much time and money weeding our crops when many of the species of plants we grow do just fine in the wild without being weeded?

The answer is somewhat complex, but it boils down to two important reasons. One reason we weed is that we are trying to make a living by growing herbs, and to accomplish that we have to do everything we can to maximize production while keeping our costs down. Sure, we could probably just plant the herbs, let the weeds grow around them, and charge higher prices to make up for the decreased yields and increased labor costs due to weed competition, but would people be willing to pay much higher prices to make up for our losses? Perhaps some people would, but most probably would not.

The other primary reason is that if we let the weeds run free, they start to grow closely together around and within the crops. This is great in the wild setting, where tight-knit, diverse communities of plants thrive, but on the commercial herb farm it becomes extremely difficult to separate the herbs

**Figure 10-2.** Spring hoeing. Photograph courtesy of Bethany Bond

from the weeds when we harvest and process them. Most people don't want to see goldenrod in their skullcap or ragweed in their peppermint. When we put a label on our bags of herbs stating the species of plants contained within that bag, we have to be certain that our customers are getting what they expect and that there are no "surprises" that could come back to haunt either the buyer or the seller. Adulteration of herbs, both purposeful and accidental, is a serious problem in the medicinal herb trade, and we have to do all we can to avoid it.

Adulteration is synonymous with contamination. In the past there have been some relatively high-profile examples of herbs being purposely adulterated with similar-looking plant parts to increase the profits reaped by shady producers. Fortunately, Good Manufacturing Practices and increased awareness levels from buyers and brokers are helping to prevent this from happening. Accidental adulteration can also occur when plants are harvested from fields with mixed species growing in them. Even small amounts of benign or nontoxic plants found accidentally mixed in with herb species can pose problems. Growers need to be particularly vigilant and cautious to prevent what could become a ruined sales opportunity and tarnished customer relationship or perhaps even much worse.

Weed control, especially using tractors, can become a somewhat controversial topic that we have found ourselves engaged in with some people who don't seem to understand why we spend so much energy and time "killing plants." Every season here on our farm we host farm tours consisting of diverse groups of interested (and interesting) folks from all walks of life. Often a lively discussion will ensue as a (usually young and naive) plant lover will ask why we don't just let all living beings live and grow in harmony among the crops.

Figure 10-3. Melanie and Hannah racking anise hyssop and removing rogue weeds.

One example in particular comes to mind several years back when a young woman who was staring at the ground intensely, walking barefoot very slowly and carefully, when asked, "Are you looking for something?" replied, "No, I'm just trying not to hurt any plants." She wasn't walking in the beds where herbs were growing; rather, she was walking in the grass outside the beds. Needless to say, that example seems a bit extreme, but it serves as an example of some people's feelings about killing plants, especially when they hear that we do it with a tractor. Yes, these are often the same people who drove gasoline-powered cars to get to our farm for the tour, but the tractor is sometimes viewed as an instrument of environmental destruction. "Why don't you use draft animals to farm instead of that tractor?" they often ask, to which I may reply, "Why don't you trade that car you drove here in for a horse?" Actually, I hold people who are farming with draft animals in the

Figure 10-4. Mechanical weeding with Farmall Cub.

highest regard, and if we could do what we do on the scale we are doing it without a tractor that would be great, but we can't.

# Weeding Methods

Now that we have discussed why we weed, let's talk about how we weed. There are two ways that we approach weeding. Our primary approach is to try to prevent the weeds from growing in the first place, and our secondary approach is to control those that have survived our first approach and are already growing. We have several methods to choose from in trying to prevent weeds from growing. There are inherently gazillions of dormant weed seeds lurking just below the soil surface from years of having been deposited there from many nearby sources. These seeds are just waiting for the right set of circumstances to germinate and grow; namely, being exposed to daylight during tillage to stimulate germination, then finding a niche to exploit.

The first thing we can do to minimize this is to try to prevent more seeds from being deposited in our fields. Although we can't always prevent seeds from being blown or carried into the fields from afar, we can certainly attempt to manage those in close proximity. One way we do this is by keeping borders around the fields mowed. We do this with a rotary cutter or sickle bar mower mounted on the tractor when we see the wild grasses and plants starting to flower. This mowing is done three to four times per season, and we try to maintain a border at least three tractor widths (approximately eighteen feet) around each field. Beyond that distance we mow once or twice a season, which also helps considerably. After we have done what we can to keep the local weeds from depositing their seeds in our fields, we have to try to prevent the weed seeds that are already in the soil from germinating and growing.

The preferred method we use to limit weed seed germination before we plant is the "stale seedbed" technique. Weed seeds primarily germinate within the top two inches of the soil profile. If we can encourage as many of the seeds as possible within

**Figure 10-5.** A new crop of weed seeds ready to be borne into the wind.

that two inches to germinate early, before we plant crops, then eradicate them as they germinate, we can start out by planting in cleaner or stale beds. Whereas many conventional farms use toxic preemergent herbicides to create stale seedbeds, organic farmers must rely on timing and tools. If we can till fields early enough in the spring to be able to delay planting by two or three weeks, we can get a head start by allowing weeds to germinate, then eradicating them as they grow.

After the beds are tilled, the next step we use is "flame weeding" the beds with our backpack flame weeder. Some farms use tractor-mounted flamers, which are very efficient, but for us our backpack flamer works fine for now.

Weeds rapidly succumb to flaming, especially when they are small. This method works because the heat from the propane expands and bursts the plant's liquid-filled cells, not by actually burning the plants, which isn't necessary. It is amazing how little heat it

**Figure 10-6.** Richard Wiswall discussing the art and science of flame weeding.

takes to kill weeds when they are young. We can walk very fast down the length of each bed holding the propane torch in the middle of it and the heat deflects off the middle of the bed sideways enough to kill weeds on the whole bed width. Flame weeding efficiency is limited by the size of the weeds being flamed and is best performed before or shortly after weeds have developed their first set of "true leaves." Grasses are more resistant to flame weeding than broad-leaved weed species.

## Blind Cultivation

Another effective method that we have found for creating stale weed beds before planting as well as preventing weed set in planted beds is performing "blind cultivation" with the tine weeder. This implement consists of tightly spaced rows of small-diameter spring steel tines that "tickle" the soil surface and kill tiny weeds as they are germinating. Blind cultivation means shallowly cultivating weeds

**Figure 10-7.** Newly emerging weeds ready to be cultivated.

that are unseen below the soil surface just as they are entering the "white root" stage of early germination. This approach requires us to very carefully monitor what is going on below the soil surface.

One way we do this is by using a technique we heard about through Eliot Coleman's work, which is to place a sheet of Plexiglas on the soil surface and check it once a day. When we begin to see the first little cotyledons (seed leaves) pressing against the Plexiglas, we know it's time to blind cultivate. The Plexiglas warms the soil up faster than the surrounding surface and gives the weed seeds underneath it a jump-start in germinating. When we see weeds starting to emerge through the soil surface under the Plexiglas we know that the seeds in the rest of the field are probably just starting to enter the white root stage.

The key thing to remember about tine weeding is that if you can see the weeds from the tractor seat it's probably too late to use the tine weeder, because at that point the weeds will have become too large to be killed by the tines. Then you have to consider a more aggressive approach. The beauty of the tine weeder is that its thin tines aren't aggressive enough to dislodge transplants or seedlings in the rows as long as they have started rooting. Unlike most cultivating tools that weed around the plants, the tine weeder indiscriminately weeds everything in its path, so again, timing is everything to disrupt weeds while not pulling plants out of the ground.

This tool can even be used to weed direct-seeded beds that haven't germinated yet. Take burdock, for example; we can direct-seed this crop one inch deep and blind-cultivate the bed it is seeded in during the

**Figure 10-8.** Jeff tine weeding the garlic bed.

ten to fourteen days it takes to germinate. The tine weeder only disturbs the top half inch or so of soil so the seed doesn't become dislodged. After it starts to germinate we wait until the burdock plants have developed the first set of true leaves; then we can cultivate right over the plants without hurting them a bit. The tine weeder is an excellent alternative to flame weeding to establish stale seedbeds before we plant into them. The two methods can also be used in tandem if need be.

## Plastic Mulch

Another method we use to prevent weed set, as we discussed previously, is by applying plastic mulch, which will be replaced with biodegradeable plastic or "bioplastic" mulch once it becomes approved for certified-organic farms. Bioplastic mulch is made from non-GMO cornstarch and slowly decomposes at the end of the season it is applied in. Mulches are very effective at preventing weeds from growing but sometimes the holes we punch in the mulch to place plants in become little windows of opportunity for weed seeds to see the light of day and grow. We try to plant sturdy transplants that will grow fast to shade out the weeds, but if weeds do pop up within the plants, we go through by hand and pull them while they are still small.

As much as we try to prevent weeds from germinating and growing in the first place, our efforts are never 100 percent successful. Creating stale seedbeds and tine weeding works really well for preventing most of the weed seeds from germinating and growing, but it never gets all of them. Perennial weeds, especially grasses that have survived our early tillage, are much more tenacious and can become established and grow quickly, especially during rainy spells when we can't bring tractors into the fields. In the early part of the season, with the majority of our efforts focused on seeding and transplanting, sometimes weeding takes a backseat to getting crops growing, and before we know it we are often playing catch-up. This is when our weeding approach goes from prevention to eradication, and we have to resort to different methods and equipment to prevent

the weeds from getting out of hand. Whereas many conventional farms rely on chemical herbicides to control weeds, organic farmers choose the nontoxic approach to weeding.

There are probably nearly as many weeding methods and tools as there are types of weeds, and the options can sometimes seem a bit overwhelming to the new farmer. There are many factors to consider when deciding on the methods and equipment you should use, including what type of tractor (if any) will be used, how much hand weeding may need to be done (if any), what type of soils you are weeding, what type of crops will be grown, what the bed and row spacing will be, what your budget is. After you have entered these factors into the equation, the next thing we recommend doing is to visit some farms nearby or talk to other growers about methods and equipment they are using. Go and check out some of their cultivating equipment, and get a sense for what seems to fit well into your farm plan. Seeing some of the tools that we purchased in action before we purchased them gave us a better sense of confidence as to what we were getting into. Some equipment is easier to use and requires less maintenance than others and some requires more precise driving skills to avoid damaging plants. If these things seem intimidating, there are usually more simple options available.

There are three primary approaches when it comes to controlling weeds that are already growing. One is the mechanized approach, in which a tractor pulls or pushes an implement that cultivates or flames weeds. Another approach is to hand-weed or flame-weed using handheld tools, and the third and less commonly used approach on an organic farm is to use organically acceptable herbicides. As with any type of tillage we do, weeding should only be done when soils are dry enough to lessen the damage caused by soil compaction. It is also highly recommended to weed during dry weather for the simple fact that the weeds will have a better chance of desiccating and dying when they become uprooted. Weeding during or right before rainy weather often results in many of the weeds regrowing because of the moist conditions around their roots.

## Sweeps

For the mechanized cultivation approach, the simplest and most common option is using a cultivator with sweeps. Sweeps are basically small shovel-like blades or knives that attach to flexible steel shanks, mounted on a three-point-hitch toolbar or walk-behind tractor frame. There are many different configurations to choose from, but sweeps all do basically the same thing, which is to cut shallowly through the soil as they move, uprooting, slicing, and burying weeds.

During the early years on our herb farm when we were planting everything in single rows we relied on a one-row cultivator mounted with S-shaped spring shanks with sweeps. This cultivator was attached to the three-point hitch and pulled over the row being straddled by the tractor. This single tool was very affordable and did a good job cultivating. The challenge with this tool was that because it was rear mounted, visibility was limited during cultivation, and we would often find ourselves dislodging crops from the ground, then having to tuck them back in. When we expanded and moved onto multirow raised beds, we bought a bigger cultivator with adjustable sweeps to weed multiple rows, and we also bought an offset cultivating tractor with sweeps, which greatly improved our visibility and efficiency with weeding. Let's discuss some pros and cons of using sweeps to cultivate.

*Sweeps Pros*

- They are simple, affordable, durable, and easy to maintain.
- They work well for eradicating small and medium-size weeds.
- They can be used when plants get large without damaging the plants.
- They can be used to hill soil up against plants.
- They are shallow running and gentle on the soil.
- They can be used in rocky soils.
- They can handle moderate amounts of crop residue without clogging.
- They can be used for preemergent weeding as well as postemergent.

*Sweeps Cons*

- They are challenging to use for precision weeding close to crops without damaging plants.
- They require slower ground speed because of their tendency to throw soil sideways at higher speeds. This can bury tiny plants.

## Basket Weeders

Basket weeders are an excellent choice for precisely cultivating multiple rows. They consist of two gangs of ground-driven rolling wire baskets mounted on axles supported by a square frame. Baskets can be customized to work within a wide range of bed widths and row spacing. Before we started using baskets to weed we had to do much more hand weeding where the sweeps would miss some weeds close to the plants. After seeing basket weeders in action on our friend's farm we purchased one and midmounted it to our cultivating tractor. This tool allows us to cultivate more precisely within an inch or two of the crops and has saved countless hours of hand weeding.

Basket weeders work very well in the early stages of both weed and crop growth. Since they don't tend to throw soil sideways, they can be used pre- or postemergence, even when crops are very small. When weeds get too large the baskets can tend to clog, and when plants get too large the baskets can damage them so they are best used for small weeds near small plants. Our basket weeder tends to start damaging plants as soon as they get above ten inches tall. This gives us plenty of time to get the weeds under control while the crops are growing.

*Basket Weeder Pros*

- They are able to perform very precise weeding.
- They don't tend to throw soil sideways, which could otherwise bury or damage plants.
- They work well for both preemergent and postemergent weeding.
- They are shallow running and gentle on the soil.

*Basket Weeder Cons*

- They are relatively expensive to purchase.
- They don't work as well on rocky soils.

- They don't work well with crop residue on the soil surface.
- They can damage plants above eight to ten inches tall.
- They don't work well with large, well-established weeds.

In addition to sweeps, tine weeders, and basket weeders there are dozens of other tools that can be used for mechanical cultivation. Many farms have custom-designed setups using toolbars mounted with a variety of cultivating implements suited to their own row spacing and weeding needs. Another option some farmers choose is to use a walk-behind rototiller to weed the paths between plants, then hand-hoe closer to the plants where the rototiller misses. We recommend the book *Steel in the Field*, published by the Sustainable Agriculture Network, for a more in-depth discussion on weed management tools.

## Hand Weeding

After we have gotten all the weeds we can with the mechanized approach it becomes inevitable that we will have to do some hand weeding. Since we are growing so many species of perennials, often during the growing season these plants get too large to machine-cultivate, and we are forced to do some touch-up hand weeding. Some farms rely on hand weeding with hoes or bare hands alone, which is a viable option for small-scale operations that don't

**Figure 10-9.** Jeff using a walk-behind rototiller to shallowly cultivate weeds in the vegetable garden.

use machinery. Some of our vegetable farming friends have gotten so efficient in their mechanized cultivation that they don't do any hand weeding at all. As we mentioned previously the labor costs associated with weeding can add up fast, especially when it comes to hand weeding, so we have to be as efficient as possible with our timing, methods, and tools.

We also have to try to get to the weeds while they are small, because the larger they get, the more time and effort it takes to eradicate them. One option we have is to use our bare hands to pull weeds, which can be effective in loose soils with relatively small weeds. One very important thing we continue to emphasize with our employees, especially those that are inexperienced in weeding, is that when we pull larger weeds we need to be removing the whole plant, including the roots, or the weed will just reemerge, often with increased vigor.

If we find that bare-handed weeding is not efficient or our backs become sore, we reach into the toolshed for our hoes. Hoeing is usually much faster and more efficient than bare-handed weeding because we are standing upright and moving faster along the row. We like to use long-handled hoes with thin blades for small weeds and broader-bladed hoes for larger weeds. We keep our hoes sharpened and try to disturb the soil as little as possible so we aren't exposing more weed seeds to the light. It is much more efficient to hoe weeds early when they are small rather than waiting until they start to become larger and well established.

## Flame Weeders

We have already mentioned using flame weeders to create stale seedbeds, but flaming can also be done when crops are large enough to withstand heat yet the weeds are small enough to perish. Either tractor-mounted or handheld flamers work well as long as a cautious approach is used. Some plants withstand flaming better than others, so it is good to do some trials on any plants that may be marginal in their ability to withstand heat. Flaming with a back-pack-mounted flamer has been our preferred method of dealing with small weeds that we miss during earlier passes with the tine weeder or sweeps. Some organic farms are also finding success using organically acceptable herbicides, such as those made with citrus-based compounds or acetic acid.

If all this talk of controlling weeds (both the undesirable and the desirable types) sounds overwhelming remember that, as with many aspects of the farm operation, taking a proactive approach and nipping weeds in the bud before they get out of hand is the key to keeping costs down and yields high. Along with weeds, pests and diseases compose the balance of the living organisms that make our lives as farmers more challenging. In the next chapter we will discuss how people, plants, pests, and diseases can coexist without compromising our sanity or profitability.

# CHAPTER 11

# Pest and Disease Prevention and Control

*The living and holistic biosystem that is nature cannot be dissected or resolved into its parts. Once broken down, it dies. Or rather, those who break off a piece of nature lay hold of something that is dead, and, unaware that what they are examining is no longer what they think it to be, claim to understand nature. . . . Because [man] starts off with misconceptions about nature and takes the wrong approach to understanding it, regardless of how rational his thinking, everything winds up all wrong.*

—MASANOBO FUKUOKA

Every year as the days become colder the nights grow longer, and we begin our winter "hibernation"—our family goes into prevention mode. We know cold and flu season is nigh, and we can recall from past winters how horrible it feels to get sick. We also know that with some extra precautions we can build our defenses against the bacteria and viruses that cause illness. Most importantly we nourish our bodies with good food and plant medicine. We wash our hands when we are out in public before we eat, and we try to get plenty of fresh air, exercise, and rest. We do all these things to strengthen our immune systems and defend ourselves so that if we do come into contact with disease pathogens, our bodies are strong enough to resist becoming compromised.

If all this prevention doesn't work as well as planned and we start to get that scratchy throat or runny nose, we delve into our home apothecary and start using such herbs as garlic, elderberry, echinacea, astragalus,

**Figure 11-1.** Garlic: a supreme antimicrobial. Photograph courtesy of Bethany Bond

shiitake, cayenne, and always more garlic. Usually our herbal allies do the job, but if for some reason our immune systems can't handle the onslaught and we get full-on sick, we have to treat the symptoms acutely. This is when we can turn to the stronger

**Figure 11-2.** Healthy bed of anise hyssop.

diaphoretic herbs, such as yarrow; the mucilaginous herbs, such as marshmallow; the alkaloid-rich herbs, such as goldenseal; and, of course, more garlic.

There is always the possibility that something like the flu could develop into a bacterial infection such as pneumonia, for which, God forbid, we could possibly have to turn to allopathic medicine in a chronic or life-threatening situation. Okay, that's a little dramatic, but I assume you get the point. We start out in prevention mode, which usually works great as long as we are diligent; then we gradually and thoughtfully respond to any increased threats and symptoms as they develop. If prevention doesn't work, we raise the defenses a bit. If that doesn't work and an acute situation develops, we turn to our stronger allies for help. As a last resort, we may even turn to allopathic medicine. All of this could be thought of as a direct parallel to the way we approach the health of our plants.

Before we bought our farm, one of the first things we did was to research thoroughly all the species of medicinals we were going to attempt to grow. We wanted to learn as much as possible about what type of growing conditions these plants liked and what types of pests and diseases we needed to be aware of. Luckily we didn't change our minds right then and there and decide to pursue different careers, because the amount and variety of pests and diseases listed with almost every one of the species of plants that we researched was staggering. It's a wonder there are any plants left on the face of the earth at all with so many

bugs and diseases out to destroy them, as one might imagine after reading some of the lists of plant pests.

We soon realized that much of the information we read was overstated and merely represented *possible*, not necessarily *probable*, pest and disease scenarios. Finally, just to be sure before we committed to our new venture, we spoke to other herb growers, who assured us that unless we were planting large mono-cultured acreages of herbs, we should see minimal pest and disease pressure growing these types of plants. Now, after our fifteenth year farming this land, we can say for sure that, although we have definitely experienced some pest and disease damage, it has been minimal and has affected us very little from an economic standpoint.

In writing this chapter on pest and disease prevention and control, we could have listed all the pest and disease issues that are possible with each of the more than fifty species of herbs that we grow, but if we were to do that, all we would basically be doing is repeating what we have seen reported by others. Instead, we are choosing to write here about our own experiences and methods in this realm so that we can give you, the reader, an accurate representation of what one can realistically expect on a small, diverse, organic medicinal herb farm in the far northeast region of the United States. For more information on some *possible* pest and disease scenarios with each species of herbs we grow, especially those occurring in different bioregions from ours, you may wish to refer to the plant profiles at the end of the book, as well as to the numerous other references available.

## Prevention

Although most species of medicinal herbs we grow and use are "wild" plants (meaning they haven't been hybridized or genetically modified) that have developed resistance to harm from pests and disease through thousands of years of evolution, they are certainly not immune to these threats. When we grow these plants for commerce in genetically homogenous stands, these "artificial" growing conditions limit the effectiveness of this resistance by providing ideal conditions to host opportunistic insects, mammals, birds, and pathogenic organisms looking for safe haven, nourishment, and a place to procreate.

To minimize the challenges that these organisms can pose to our crops, and consequently our livelihoods, we need to utilize a multifaceted management approach that relies in part on attempting to replicate the many ways in which plants growing in the wild have succeeded in living with these challenges for generations without our help. Our role in helping to replicate nature's blueprint for success requires that we assume the role of intermediary between the natural environment in the wild and the managed environment that we create on our farm. By expanding our awareness of the biological and energetic processes that contribute to healthy plants, then putting that knowledge to work, we become integral participants in the web of life rather than mere spectators.

For the rest of this chapter we will simply refer to insects, birds, and mammals collectively as "pests." Specific diseases will be addressed primarily in the second half of the chapter. At ZWHF we have adopted a "plant-positive" approach to managing pests, versus the "pest-negative" approach that is predominant in modern conventional agriculture. This plant-positive philosophy was aptly described in Maine vegetable farmer and author Eliot Coleman's book *The New Organic Grower*: "Plant-positive in contrast to the present approach which is pest-negative. It makes sense. Since there are two factors involved, pests and plants, there are two courses of action: to focus on killing the pest, or to focus on strengthening the plant; to treat the symptoms or to correct the cause. Since the former appears to be a flawed strategy, we might be wise to try the latter."

The plant-positive approach requires us to grow crops that are healthy enough to coexist with pests without becoming negatively affected by the pest's natural behaviors. As idealistic as it may sound to some, plants and pests certainly can coexist harmoniously without a great deal of compromise, especially when we provide the right conditions for them to do so. They have successfully coexisted in the wild

without our intervention in the past, and they can continue to do so with our assistance in the managed setting. The most important thing we can do is to grow strong, healthy plants that are better able to tolerate serving as host to herbivorous pests without succumbing to or being severely damaged by their natural behaviors.

Healthy plants grown in healthy soils are far less susceptible to pest damage than unhealthy plants are. For example, numerous scientific studies have demonstrated that plants that are stressed from over-fertilization are much more attractive to injurious insects than those growing in soils with well-balanced nutrient levels. This fact lends credence to the argument many proponents of organic agriculture have been making for years, that increased insect pressure on commercial farms during the past century is likely related in part to increases in the application of soluble fertilizers to enhance plant growth rates.

On the other hand, many scientific studies have also shown that plant stress caused by inadequate nutrient levels, both on the macronutrient and

**Figure 11-3.** Medicinal plants have a resilience and a vitality that help them naturally resist pests. Photograph courtesy of Kate Clearlight

**Figure 11-4.** On the farm we use a plant-positive approach to pest control.

micronutrient scale, can also increase the likelihood of pest damage in plants. Striking a balance by providing optimum fertility levels without over- or underdoing it is best accomplished by maintaining soils that contain ample amounts of biological activity and mineral availability, then adjusting nutrient levels from there if needed based on personal observations, soil tests, and specific plant needs, as we discussed in chapter 9.

Our plant-positive approach requires us to remain open to the fact that there may be times when we have to resort to using pest-negative methods in an emergency situation if we see the potential for economic losses. With over fifty species of medicinals in production on our farm, we can be almost guaranteed that each and every season at least one

species of herb will be challenged by, or possibly even succumb to, pest pressure. This amount of diversity allows us the confidence that challenges are to be expected. Although it can be devastating to see crops struggling or even dying, we can often view this as an opportunity to examine what might have led to the imbalance that rendered the plant vulnerable in the first place, attempt to minimize the damage by possibly using pest-negative methods in an emergency situation, then adapt our methods to avoid recurrences.

Many commercial growers may view our plant-positive approach as idealistic, naive, careless, or perhaps even dumb. We would counter those viewpoints by arguing that we are a very diverse small farm growing mostly pest-resistant species, and

**Figure 11-5.** Some plants act as natural insect traps. Here is tobacco covered in aphids. Photograph courtesy of Bethany Bond

we are capable of managing our crops in a way that doesn't predispose them to massive losses from pests. For fifteen years now we have been farming this way, and until we see a convincing economic necessity to implement more pest-negative methods, we don't feel the need to fix something that isn't broken.

# The Importance of Diversity

We only own ten acres of land, but within this ten acres is a great deal of diversity, not only among the flora and fauna but also among the landscape itself. There is a small stream that runs year-round through the back corner of the land adjacent to our spring-fed pond, which supports a small wetland habitat. We have some woods that are dominated by thin, acidic soils supporting coniferous evergreens and some that are primarily deciduous hardwoods supported by humus-rich topsoils. Most of this boreal woodland is steeply sloping, while the open land that predominates the farm is more level, gently undulating, and interspersed with small groups of trees that serve as shade and windbreak. The neighboring land that we lease could be considered an extension of our land in that its terrain is very similar to that of our own land.

From a biological perspective this collective landscape supports an incredible array of both native and introduced plant and tree species and is inhabited and frequented by countless numbers of insects, reptiles, amphibians, birds, and mammals. Of the more than fifty different species of herbs that we commercially grow and sell on this land, roughly half of them are also found growing nearby in the wild.

When we explore the woods and meadows surrounding our own land, we see a similar amount of diversity but on a much larger scale. There is a vast array of plant and tree species, as well as countless buzzing insects, singing birds, creepy crawlers, and the occasional (nondomesticated) mammal. What we don't see in this wild environment are plants being decimated by any of these critters that may otherwise be referred to as "pests." Sure, there is definitely some munching going on here and there, but we have never noticed anything that we would consider out of balance.

The reason is that there is a great deal of natural balance within this ecosystem, and there seem to be plenty of flora to support the fauna's food and habitat requirements. The only thing that really stands out is that the so-called invasive plant species that we humans introduced (then hypocritically labeled "invasive"), such as Japanese knotweed, barberry, and purple loosestrife, are rapidly tilting the balance by displacing native plants. The fact that flora and fauna are harmoniously coexisting in this environment with a natural system of checks and balances, such as predator/prey, host/parasite, fungi/substrate, and so on, is further evidence of the fact that as growers of "wild" plants, we would be wise to attempt to replicate this success on our own farms and gardens.

# Companion Planting and Beneficials

The term "companion planting" simply means growing different species of plants in close proximity for the benefit that they can provide each other. There is plenty of debate as to whether some of the old-time companion planting recommendations are based on fact or myth, but the fact is that there are many obvious benefits that certain trees, plants, fungi, and bacteria can offer one another. We can take advantage of these benefits by thoughtfully planning how and where to plant species that can assist one another in mutually beneficial ways. As medicinal herb growers

we have an incredible array of species to choose from when considering the benefits that companion planting can offer our farms and gardens.

Historically speaking, the first example of companion planting that comes to mind for most people is the "three sisters" combination of corn, beans, and squash that was commonplace in many indigenous cultures. In this rather simple but elegant system, corn provides a trellis for the beans. Beans climb up the corn and provide fertilization to the plants by way of nitrogen fixation in their roots. Squash provides ground cover for weed control and moisture retention. All three of these edible plants mutually benefit from one another's presence in this system and also benefit the end user (the human) by potentially increasing yields and lessening the effort required to produce these crops.

# Insectaries

There are many ways that different species of plants can benefit one another. One example is physical protection from the elements. Tall, sturdy plants can protect smaller, more sensitive plants from excessive wind, rain, and sunlight. Dense groundcover-type plants can serve as natural mulches by helping to smother weeds and assisting with soil moisture retention among other plants.

Probably the most obvious benefit of companion planting we have seen in our experience is the value of insectary plants. Insectaries are plants that are purposely planted with the goal of increasing plant pollination, deterring potentially harmful pests, and attracting beneficial insects that prey on these pests. This interaction keeps insect populations in check naturally and can help growers avoid having to resort to using insecticides that are often indiscriminate in their actions by killing both the benevolent and the malevolent insects. Although it is difficult to quantify this benefit in terms of numbers, we are absolutely convinced that part of the reason we see a lack of harmful insect pressure on our crops is due to the fact that many of the species of medicinal herbs we grow are very attractive to beneficial insects.

It just so happens that some of the most popular medicinal herbs we use, such as angelica, anise hyssop, blue vervain, calendula, chamomile, elder, meadowsweet, motherwort, peppermint, valerian, and yarrow, are some of the most effective insectary plants known. When these plants are blooming, we see an incredible diversity of predatory insects—such as wasps, ladybugs, flies, lacewings, and spiders—sipping nectar from their flowers in between snacking on potentially damaging insects such as leafhoppers, aphids, mites, and caterpillars. By planting these insectary herbs near plants that may be susceptible to damage by harmful insects, we not only help to create diversity and aesthetic beauty in our plantings, but we also help to minimize the amount of harmful insect pressure. This in turn prevents us from having to resort to more costly methods of controlling pests.

Figure 11-6. Angelica, a beautiful insectary plant.

## Pattern Disruption

Pattern disruption is another method of using plants to deter herbivorous insects. This method is most useful in monoculture systems, where plants are more susceptible to damage from insects that find increased success in their destructive ways by the convenience of being able to hop, crawl, or fly easily from one plant to the next. When we disrupt the continuity of a monocrop system by encouraging diversity in our plantings, we can potentially slow the insect feeding frenzies and mating orgies that can quickly get out of hand in these systems by confusing the insects and interrupting their progress. Pattern disruption can easily be integrated into herb crop rotations by simply staggering rows or blocks of different species of plants (especially insectary plants) within larger monoculture beds.

## Physical Barriers

Another method of preventing herbivores from eating our crops or spreading disease is by using physical barriers such as synthetic row covers and fencing. Row covers made of spunbonded polyester or polypropylene are porous, lightweight fabrics that can either be "floating," meaning held up by the crops themselves, or be held up above plants by various types of mechanical supports.

Row covers are extremely effective at preventing insect damage when applied and maintained properly. To be effective they should be installed at or shortly after planting and be securely closed on all sides to avoid insect infiltration. If row covers are installed on ground that harbors pest eggs or larvae, the covers can actually do more harm than good by trapping insects and providing them a nice cozy home. Careful monitoring and site selection can help avoid this possibility.

Another possible challenge with using row covers is that they limit the amount of sunlight that a crop underneath them can receive, depending on the thickness of the material used. This can potentially reduce yields of sun- and heat-loving crops, but on the other hand, crops that prefer cool or shady conditions can certainly benefit from this protection. Row covers are also useful for season extension by providing a few degrees of extra protection on frosty nights. This protection varies depending on what type of row cover is used.

Fencing is the most dependable way of protecting plants from being eaten by herbivorous mammals such as deer and woodchucks. Deer often torment farmers because of their ability to graze crops down rapidly, especially where shrinking habitat and natural food sources force them into our fields and gardens. Good fences make good neighbors, and it is comforting to think that there are ways that deer can be good neighbors without our having to resort to violence to keep them at bay.

At ZWHF we have a healthy population of deer in the woods and meadows surrounding our farm, but we have seen very little damage from deer browsing our crops. It seems here that they are mostly attracted to our echinacea plants in early fall. Perhaps they are boosting their immune systems for the long winter ahead, and who could blame them for that? During the growing season they seem to prefer the red clover and oat cover crops that we plant, which is fine by us because there is enough of that for everyone to share. Most of the species of plants that are grown for herbal medicine fortunately are unattractive to deer because of their intense aromatic qualities and other potent, unpalatable compounds produced within the plants. That said, we are not immune to deer damage, and we have heard from some herb growers that once deer get a taste for something they like and become habituated, it requires installing fencing to prevent economic losses.

There are countless methods of installing deer fencing as well as materials to construct it with. The size of the area you are trying to protect and the labor and materials costs associated with installing enough fencing to surround a given area with an impenetrable barrier are the primary considerations to take into account with deer fencing. Secondly, the amount of maintenance you are willing to perform to maintain the fence's effectiveness and the durability of materials used are also important factors in planning fences. Deer can reportedly jump as high as eight feet vertically, but they have a hard time judging depth and don't like to have to leap over horizontal obstacles.

Some growers install long-lasting vertical fencing using sturdy wooden posts sunk into the ground and supporting mesh fence material made of steel or heavy-duty plastic. Electric wires can also be used, either alone or in combination with this type of fence. Many vegetable growers in our area who prefer a lower-cost, more portable type of fencing material are finding incredible success using eight-foot-tall, thin, flexible fiberglass fence posts pushed into the ground at a forty-five-degree angle supporting four or more strands of either electric fencing or high-strength monofilament fishing line, or a combination of both. This type of fence is installed at an angle sloping up and away from the crops that are being protected. Deer have poor eyesight, so when they approach this fence, they can feel the wires but may not necessarily be able to see them. When they try to lower their bodies to slide under the wires, they are met with even more wires closer to the ground, which really confuses them. They won't jump over this fence because it requires a horizontal leap that they aren't capable of making. This type of fencing could be ideal for those farming several acres of crop species that may not be attractive to deer, interspersed with a few more "attractive" species. This simple and inexpensive fencing could be temporarily installed around the more attractive crops without having to protect large tracts of land.

Solar-powered electric fence chargers work great for locating electric fencing away from electrical services. There are some organically acceptable chemicals made of garlic, hot peppers, and bitter compounds that are supposedly effective after being sprayed on crops because of their strong odor and taste, but this method seems impractical for treating large amounts of plants, and it is likely that your customers would find these products as offensive on their herbs as the deer do.

For woodchucks or groundhogs, their vertical leap is considerably less than the deer's eight-foot leap, so a mesh fencing only needs to be at least three feet high. Ideally, it should also be buried at least a foot deep, because woodchucks are incredibly adept at tunneling. Fencing for woodchucks is probably the least effective method of controlling these varmints. Have-a-heart traps are our first line of defense for varmints, and they are remarkably effective. Since woodchucks primarily feed during daylight hours, we bait the traps with apple slices or other tasty treats and don't have to worry about more nocturnal creatures such as skunks getting into the traps. After we catch woodchucks, we drive at least a couple of miles down the road and "rehome" them into a vacant meadow or woodlot, where they will meet with far less resistance (it would be wise to consult local wildlife regulations before rehoming wild animals).

Dogs are also very effective at deterring or eradicating woodchucks and seem to relish the opportunity to go after the furry little varmints. For those who prefer a less violent approach to varmint control you could always just leave a couple of pounds per woodchuck of vegetables outside their burrows once a day to keep them satiated. We'll try not to mention the shotgun or rifle method.

Birds can also become a nuisance if we don't use preventive measures to protect our crops. Whether they are munching newly planted seeds and seedlings or gorging on elderberries, as much as we love our winged friends, sometimes we need to persuade them to dine elsewhere. There are a wide variety of techniques and devices used to repel birds. Growers need to be proactive with bird control methods rather than waiting until after the birds start doing damage. You should also be prepared to use more than one method, as birds can quickly become habituated. Birds are incredibly smart and persistent, and as soon as they see a chink in the armor, they will exploit any opportunity to grab and go.

There are three primary methods of repelling birds (other than shooting them, which we do not recommend and can be illegal). The first and probably the most common is using visual repellents such as reflective tape, inflatable "scare-eye" balloons, and fake birds of prey that frighten or disorient flying birds. Another method is to use auditory repellents such as recorded bird distress calls, bird of prey calls, and loud cannons. Protective netting is another option that, although effective, can be very

**Figure 11-7.** Jeff is pictured here with our most beloved pest control, Barkley. He kept deer and other four-legged pests at bay. Photograph courtesy of Robyn Straza; taken by Orah Moore

labor intensive and costly to install properly and vulnerable to even the smallest opening.

There is also a "conflict-free" approach with birds and mammals that growers may consider, which is to grow enough for everyone to share. This would probably only last as long as the resident populations kept the secret to themselves and didn't start inviting unwanted guests.

# Probiotic Inoculants

We have discussed protecting plants from physical damage caused by herbivorous pests, but what type of measures can we use to prevent disease pathogens from infecting our crops? We recently attended a workshop on holistic plant health presented by our friend Michael Phillips. The information we gleaned from that workshop as well as from his fascinating book titled *The Holistic Orchard* has helped revolutionize our approach to preventive health care for our plants.

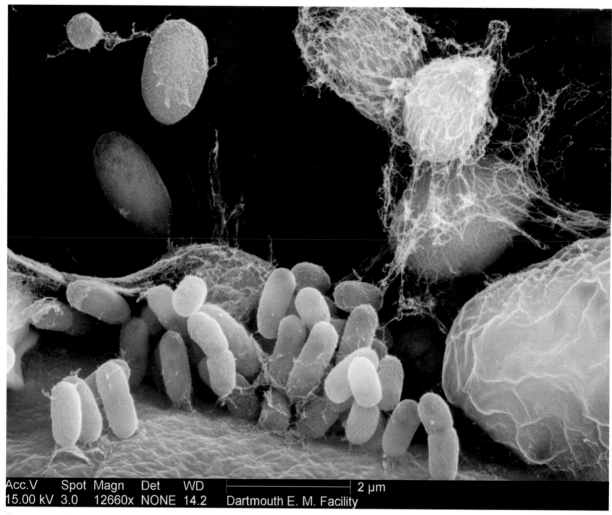

| Acc.V | Spot | Magn | Det | WD | | 2 µm |
|---|---|---|---|---|---|---|
| 15.00 kV | 3.0 | 12660x | NONE | 14.2 | Dartmouth E. M. Facility | |

**Figure 11-8.** A microscopic view of the "arboreal food web" that is found on a leaf surface containing fungi, bacteria, and disease spores. Photograph courtesy of Louisa Howard, Dartmouth College Electron Microscope Facility

Our first line of defense against plant pathogens lies in the subterranean realm, where our primarily bacterially dominated soils can benefit greatly from the addition of fungally dominated compost made from copious amounts of hardwood leaves. This soil inoculation helps to support the growth of mycorrhizal and saprophytic fungi in the "rhizosphere" or root zone, which in turn play an incredibly complex and important role in increasing nutrient and water availability to the plant's root system. These fungi serve as antibiotics that not only protect the roots from pathogenic organisms but also encourage production of secondary plant metabolites called phytoalexins that help stimulate an induced systemic resistance response within the plant.

In addition to supporting the soil's fungal population, we focus our probiotic inoculations on the plant's leaves. Through foliar application of probiotic sprays, we can help to inoculate our plants with beneficial bacteria and fungi that, like soil inoculants, not only provide nutrients to the plant but also serve to occupy a niche that may otherwise be exploited by disease pathogens. Like human skin, plant's leaves have pores, and although these pores serve biologically different functions in humans from what they do in plants, there is an interesting similarity. They both serve as a direct route through which nutrients can be absorbed and assimilated, which is an efficient conduit, but this can also render pores vulnerable to pathogenic contamination.

Plants have two types of pores, called *stomata* and *transcuticular* pores. Stomata are the larger pores on both the upper and lower surface of a plant's leaves and mainly function as passageways for the absorption of carbon dioxide and oxygen as well as transpiration or loss of water. Stomata open and close in response to daylight and environmental conditions. Transcuticular pores are much smaller than stomata and are constantly open.

Both types of pores are able to absorb nutrients (primarily micronutrients) through foliar feeding and can also benefit from being inoculated with beneficial microbes to defend against pathogens. Fungal colonization of a plant's leaf surface helps to stimulate metabolic defense compounds called phytoalexins,

which serve as an immune response to resist invading pathogens. Probiotic foliar sprays also help to occupy an otherwise open and vulnerable niche on these pores by colonizing their openings with beneficial bacteria and fungi, which can leave pathogens little room to find a host. We use two products for our foliar sprays, raw neem oil from Ahimsa Organics (www.neemresource.com) and a combination kelp/fish spray called Neptune's Harvest (www.neptunesharvest.com).

## Neem Oil

Pure neem oil is a cold-pressed oil from the seeds of the *Azadirachta indica* tree, which is native to the Indian subcontinent and has been introduced throughout the tropics. Although there are quite a number of popular neem-based agricultural products on the market, many of these are mere extracts of the pure oil that have been diluted with added compounds to increase the product's shelf life and ease in its application. Pure raw neem oil, on the other hand, is readily available and contains an incredible array of beneficial compounds that stimulate production of phytoalexins. These terpenoid and isoflavanoid

## Induced Systemic Resistance

Induced systemic resistance is a biological phenomenon in plants that scientists and laypeople are actively investigating for its role in defense of plant pathogens. When plants sense a viral, fungal, or bacterial invasion, they can respond via hormonal signals from endogenous (naturally occurring) salicylin to stimulate a systemic resistance to the invading microbe. This resistance can be manifested by an oxidative burst,[1] which can lead to cell death.[2] This can cause the pathogen to be "trapped" in these dead cells and can prevent it from spreading into uninfected tissues. Other responses include mutations in the cell wall that can block pathogenic invasion and the de novo production of antimicrobial compounds such as phytoalexins.[3]

Figure 11-9. Raw neem oil with backpack sprayer and soap.

compounds serve as systemic immune enhancers in plants in much the same way herbs such as echinacea work in our own bodies.

Using plant extracts to maintain health in plants being grown to make herbal extracts used for maintaining health in humans is but another wonderful example of the magic in plant medicine. Essential amino acids, vitamin E, and trace amounts of nitrogen, phosphorous, potassium, zinc, copper, iron, magnesium, and manganese found in neem oil all serve as nutrient boosters, while fatty acids provide a source of food for the colonization of beneficial microbes. In addition to these nutritive, immune-enhancing, and probiotic properties, neem is also a very effective insecticide that acts both as a feeding deterrent and a hormone disrupter to herbivorous insects that incidentally ingest it while feeding on plants it is applied to. One of the coolest things about neem's insecticidal properties is that it is benign to most, if not all, beneficial insects such as bees, parasitic wasps, lady beetles, and beneficial nematodes. There is conflicting information about neem's toxicity to bees. Some reports demonstrate that neem is non-toxic and is even an excellent choice for controlling mites that harm bees when applied inside honeybee hives. Other reports demonstrate a mild toxicity to bees that come in contact with heavy concentrations of neem while foraging or by having it sprayed onto their bodies. We definitely recommend playing it safe with bees by avoiding spraying crops that are attractive to bees, especially when those crops are flowering. Growers can also time their neem applications to occur in the early morning or late evening and during cloudy weather when bees are less active.

All these truly amazing attributes may cause one to stop and question whether there is anything neem cannot do to maintain plant health, to which common sense would answer that its effectiveness is directly proportional to the effort given to properly manage its application. Pure neem oil has a consistency similar to butter when it is cool, so it requires a bit of planning and effort to prepare for application via sprayer. We place the container of pure oil in a warm room, in a bucket filled with warm (not hot) water or under a greenhouse bench (not in direct sunlight) for a few hours before use to liquefy it for use. Since the oil is so rich in vegetable fats, it must be emulsified with a surfactant (we use organically acceptable liquid soap) to allow its dilution in water. The recommended dilution for pure neem oil is 0.5 percent to 1 percent.

At the lower dilution ratio we add one ounce neem oil to one gallon of water using one-half to one teaspoon of liquid soap or other emulsifier. We then mix the solution thoroughly until the oil is completely emulsified, with no oil floating on top. Then we add a bit more soap as needed. Be sure to use lukewarm water in cool temperatures. For best results use the solution within eight hours of mixing with water. Poor emulsification or exceeding the 1 percent dilution ratio may result in phytotoxicity.

We apply neem thoroughly to all plant surfaces to the point of runoff with a backpack sprayer approximately once every two weeks or so during early growth on plants we feel may need a little extra defense against pathogens or herbivorous insects. Our favorite backpack sprayer is a rechargeable-battery-powered four-gallon sprayer made by Dramm. The hand-pump Solo backpack sprayers are a more affordable alternative and also work great at applying foliar sprays. For larger-scale applications tractor-mounted or -towed models are recommended for greater efficiency.

## Kelp/Liquid Fish

Another of our favorite products for beneficial microbial inoculation also happens to be an excellent foliar feeder. Neptune's Harvest is a combination of kelp and liquid fish. Sea kelp contains many natural plant growth hormones such as cytokines and gibberellins, as well as a wealth of micronutrients. Liquid fish is sometimes confused with fish *emulsion*, which it is certainly not. Fish emulsion is pasteurized, which destroys the beneficial microbes. Liquid fish is cold processed and contains fatty acids, enzymes that feed beneficial microbes, and readily available macro- and micronutrients.

Foliar sprays of kelp/fish applied to the point of runoff support healthy microbial populations on leaf surfaces as well as in the rhizosphere. We apply foliar sprays such as kelp/fish and neem oil early in the morning or during cloudy weather when stomata are open, and we try to coat both the upper and lower surfaces of the leaves, since some plants actually have more stomata on the undersides of their leaves than on the upper surfaces. We primarily focus our foliar spraying on field crops that are susceptible to fungal diseases such as powdery mildew and anthracnose, but we also try to spray every plant in the greenhouse at least twice during its early growth period to support healthy transplants.

Finding the time to apply foliar sprays has often been a challenge, with all the other items on the to-do list, so we have to mark our calendars in advance to remind us of the importance of this preventive maintenance task. We spray plants that may be susceptible to disease once shortly after transplanting them, and we aim to reapply at least two more times at one- to two-week intervals. We don't apply any foliar sprays within one week of harvest to ensure that there are no spray residues left on the plants when we harvest and process them.

# Controlling Insects

Again, there are many different types of pests and diseases that you may possibly encounter when growing medicinal herbs, but here we are only going to discuss what we have directly experienced on our farm, how we have dealt with it, and some possible options for us to consider should we find the need to utilize pest-negative methods. We have addressed many ways that we can support our crops to help them succeed without having to utilize pest-negative methods, but what do we do when our efforts fail and we start to see the potential for economic losses?

Things happen very often in farming that are usually well beyond our control. Sometimes we can do little more than stand by and watch, as in the case of hail or a late, hard frost, and sometimes we can act quickly to perform damage-control measures before things get out of control and threaten our livelihoods. The best approach when employing pest-negative methods is the one that has the least amount of potential to cause collateral damage, such as killing nontarget species.

## Red Admiral Butterfly

When Red Admiral butterfly larvae start chomping on our stinging nettle plants, we could easily just go out and spray *Bacillus thuringiensis* (Bt) on the plants, which would kill the caterpillars when they ingest it. The problem with that method is that although it would be effective, easy, and affordable, it could have negative consequences that would go beyond just killing the Red Admiral larvae. Although Bt is a relatively benign insecticide that is currently approved for use on certified-organic farms, it is potent stuff, and it can kill nontarget species. This means that not only are the Red Admiral larvae going to die when they eat it, but any other butterfly larvae who may crawl on the nettles for a sip of dew or a munch of leaf could also possibly succumb to its effects.

The other drawback to using Bt is that when we spray it on the plant's leaves, even when it rains it doesn't all rinse off. People are making medicine with the herbs we grow, and although Bt is "benign" and approved for certified-organic use, would people really want residue from its application in their nettle tea? We wouldn't, and if it's not okay for us, it's not okay to sell to our customers. Instead of spraying Bt we have learned to time the first nettle harvest to happen just after we start to see the pretty little Red Admiral butterflies fluttering by, up and down in the nettle crop, ovipositing their eggs.

The timing for this is perfect because the Red Admirals show up every year just before flowering time. If weather or other factors delay that early harvest and the larvae start to hatch and feed, we physically remove the caterpillars by hand (with gloves on to avoid nettle stings) by shaking the clusters of newly hatched caterpillars into plastic buckets, then feeding the writhing little morsels of protein to our chickens. This physical removal is very effective

**Figure 11-10.** Red Admiral butterfly. Photograph courtesy of Revital Salomon

**Figure 11-11.** Red Admiral caterpillar cocooning in a folded leaf. Photograph courtesy of Tony Wills

and not difficult. The larvae are black-colored when they are newly hatched, and it is easy to spot the clusters against the backdrop of green leaves.

After the first nettle leaf harvest the Red Admiral butterflies return to oviposit again, and this time we wait until we see the black clusters of larvae hatching. If they are really numerous and we are concerned about crop losses, we can selectively hit the larvae clusters with the backpack flame weeder. At this point their populations have been reduced enough that we are able to get a second nettle harvest before the larvae decimate the crop.

## Colorado Potato Beetles

Colorado potato beetles (CPBs) are another species of insect that has challenged us in the past and requires a very methodical approach to control without harsh insecticides. CPBs love ashwagandha plants, which are closely related to their favorites, potato plants (they are both in the nightshade family). Bt is an effective, organically acceptable pesticide for dealing

with the CPB larvae, and although we can't say that we will never use it, we haven't had to use it yet. Since ashwagandha is harvested for the roots and not the aerial portions of the plants, if we absolutely needed to we could spray Bt without its residue being passed on to our customers but fortunately we have some far more benign methods and tools as our first line of defense. Neem oil has been effective for CPBs and has been our go-to for situations where the beetle populations elevate beyond our means to control without spraying.

We try to rely primarily on handpicking the CPB eggs and larvae while their populations are still minimal, but if they start getting out of hand we spray neem oil on the plants once every two or three days until we see the populations declining. We have to carefully monitor the crop and repeat the spraying as needed for repeated hatches; the neem oil really works well at controlling them without doing collateral damage. We also rotate the ashwagandha crop as far as possible from one field to another each season to avoid the CPB larvae, which can overwinter in the soil. The only time we saw the CPBs severely damage the ashwagandha crop was in 2013, when we had our rainiest year yet on the farm and the plants were literally drowning. We didn't spray that year because it rained so often that we felt the neem oil would be diluted and rinsed off the plants, rendering it ineffective. We ultimately had to plow the ashwagandha crop down that year and call it a complete loss, but we suspect it was the lack of oxygen from saturated soils and not the CPBs that ultimately led to the crop's demise.

Another option we could possibly consider in the future to protect the ashwagandha from CPBs is to use row covers. Our only hesitation would be that with the decrease in sunlight and heat under the row cover, the heat-loving ashwagandha may not yield as high or be as potent as it would if it were grown in the full sun. For now we will just keep picking bugs off small plants, applying neem oil sprays, and moving the crop as far as we can from previous beds each year. Unfortunately there are very few if any native insects that will prey on CPBs. Certain types of ladybird beetles or ladybugs are reputed to prey on them, but they require purchase and release into the affected field, which can be a shot in the dark.

## Japanese Beetles

Japanese beetles appear to be rapidly increasing in population each year, and there is little if anything we can do to curtail their growth. It seems whenever we see them, whether they are flying in the air or munching on plants, they are mating. Whether we are envious of their sexual prowess or just fed up with their voracious appetite for things we like to grow, they are the bane of many a plant lover's existence. They munch on leaves of many species of medicinals, yet most of this munching is more aesthetically unpleasant than damaging to the plants. We find some Swiss cheese–like holes on leaves of certain plants, but it hasn't really decreased yields or plant health to any noticeable effect.

However, there are two species of plants that we are increasingly concerned about and keeping a watchful eye on for the potential of having to protect: rosebushes (which we are recently experimenting with growing commercially) and hops vines. Japanese beetles seem to be particularly attracted to these two species, or perhaps the damage is just more noticeable. Unlike in Japan, where they are native and are kept in check by several species of predatory insects, here in the United States there is only one predator that is effective at keeping their populations in check. Tiphiid wasps (*Tiphia vernalis*) parasitize Japanese beetle larvae, but these wasps are not numerous and widespread enough to serve a noticeable role in diminishing beetle populations.

Other than handpicking adult beetles off plants, the most effective organic control we have seen is the use of neem oil as a repellent. Adult Japanese beetles, as with most insects, seem to dislike either the taste or smell of the neem oil and feed less on treated than on nontreated plants. Applying neem oil as a Japanese beetle repellent is only moderately effective and requires frequent reapplications to really work.

## Klamath Beetle

The Klamath beetle is an insect that was purposely introduced here as one of the U.S. government's first attempts at using a biological control to eradicate a "nuisance species." That nuisance species just so happens to be our beloved Saint John's wort, the crop responsible for an estimated $45 million in product sales in the United States alone in 2013. Saint John's wort is also called "Klamath weed" after the Klamath River in northern California, where the imported weed was discovered growing in the early 1940s. Saint John's wort can be toxic to ruminants when taken internally in large quantities, and cattle grazing on open pasture in the West were becoming increasingly harmed by overgrazing this weed.

Scientists discovered that in its native range in Europe, Saint John's wort was the preferred food of a beetle called *Chrysolina quadrigemina*, soon to be known as the "Klamath beetle." They imported the beetle to the United States, and their efforts were very successful in controlling the spread of Saint John's wort in cow country. Now the Klamath beetle

**Figure 11-12.** Klamath beetle on Saint John's wort flower. Photograph courtesy of Norman E. Rees

has been found virtually everywhere Saint John's wort grows. Not only does this insect defoliate the plant by chewing on its leaves, but it has also become a disease vector for anthracnose (*Colletotrichum gloeosporioides*). Anthracnose has become an incredibly challenging fungal disease for Saint John's wort growers, and coupled with the Klamath beetle's voracious appetite for this herb, we have to do everything we can to prevent this insect/disease combination from affecting our harvest.

Fortunately for us here in Vermont we are on the fringes of the northern range of Klamath beetle country. What this means is that the beetles have a hard time surviving our winters, and many—but not all—of them die off during the hard freezes. Unfortunately our winters have grown milder, so we are expecting to see more of these visually attractive but harmful insects survive and consequently increase the threat to our commercial Saint John's wort crops.

There is little we can do to control Klamath beetles other than handpicking. The reason we can't resort to using organically acceptable insecticides, even as a last resort, is that we see little beetle pressure until the plants are getting ready to flower. If we were to spray the Saint John's wort at this time, residues from the pesticide (even though it may be labeled as safe for crop application right up until harvest time) would still be on the plants when we harvest, dry, and sell them. We can't wash the crop before we dry it, so we have no option other than to use prevention and handpicking. Unfortunately, there are very few if any native insects that will feed on Klamath beetles. Other than birds and carnivorous mammals such as skunks, it is up to us to try to keep their populations in check. Our backup plan, should we start seeing increased populations of Klamath beetles and consequently increased anthracnose damage, is to try protecting the crop with row covers.

Ultimately our main concern with the Klamath beetle up to this point is the fact that it has become a disease vector for anthracnose. In addition to keeping the Klamath beetle vector in check, our primary line of defense against this fungal disease is probiotic inoculations, which we have already discussed. In addition to providing the plants with beneficial microbes, we have to combine that with good sanitation practices. Seeds harvested from infected plants can harbor disease inoculum, so we buy our Saint John's wort seed from Horizon Herbs, which has been a dependable source of disease-free seed for us.

We keep our greenhouse free of weeds and debris that may harbor anthracnose inoculum, and we try to use good hygiene practices when watering seedlings, just in case any spores lie hidden somewhere in the greenhouse or in our potting medium. We have never seen Saint John's wort seedlings become infected in the greenhouse, but we know that the possibility does exist. Out in the field the disease is unfortunately fairly common and very easy to spot, as it causes rust-colored lesions on the stem and leaves of infected plants. Severely infected plants become completely rust colored, wilt, then die. We remove infected plants from the field as soon as we spot them by placing them in a sealed plastic garbage bag. We use a very slow and gentle motion when pulling the infected plants, being careful not to spread spores on to uninfected plants.

# Powdery Mildew

Powdery mildew (*Golovinomyces cichoracearum* syn. *Erysiphe cichoracearum*) is a common plant pathogen that is easily recognizable by its telltale powdery blotches appearing on affected plant leaves. Vectors of transmission for this fungal disease include wind, insects, and even human or animal contact. Unlike many other types of mildew, powdery mildew does not require moisture in the form of free water to germinate and grow, and although it thrives in warm (not hot), humid conditions, it can actually be inhibited by rainy weather or persistently wet foliage.

The powdery appearance is caused by white, threadlike mycelia that colonize the leaf surface and produce spores. As these mycelia penetrate the leaf's cuticle defense, they parasitize the epidermal cells and slowly rob the leaf of its nutrients. Leaves then become yellow or brown and chlorotic, and although powdery mildew is rarely lethal to the plant, it causes a rapid loss in the plant's vigor and growth and makes for an unmarketable product.

We had previously seen isolated amounts of powdery mildew on our skullcap plants from year to year, and we would remove and destroy the affected plants before it spread. This was of little concern to us until 2012, when we lost a quarter acre of skullcap plants to this pathogen. That summer was the driest year we had experienced on our farm to that point, yet in spite of the droughty conditions, the skullcap had been thriving thanks to solar-powered drip irrigation providing daily watering from a stream. We had to travel for ten days in early July that summer, and before we left for our trip the skullcap crop had no noticeable health issues. While we were gone, our crew had no reason to check the skullcap, which had been planted on our leased land, so it remained unseen for almost two weeks.

When we returned from our trip and went to check the skullcap to see if it was ready to harvest, we were shocked to find out that the whole field of plants was completely infected with powdery mildew. At that point it was too late to apply any foliar sprays, and the plants withered and appeared to be dying. We consulted with a plant pathologist who told us that the spores would overwinter on the infected plant debris to such a degree that future infestations were almost unavoidable, so we plowed the skullcap in and planted a cover crop of oats. Plowing the crop down was a big mistake, however, because although the spores would likely have overwintered on plant debris in the field, we could have used several biological methods to defend the plants from reinfestation.

In hindsight it's not really that surprising that the skullcap became so infected. We were growing it as a monocrop in the full sun in tightly spaced rows, we had applied no probiotic inoculants to the soil or foliage, and no one was there to monitor the plants' health when the crop was at its most vulnerable stage. Needless to say, we learned a lot from that experience, and it was one of the major breakthroughs in our newly developed approach to utilizing probiotic inoculations for preventive plant health care.

Since that fateful season of skullcap loss we have developed proactive methods to ensure that, even though we know powdery mildew spores are present in our environment, we can easily prevent our plants from playing host to their colonization. Our primary method of defense for powdery mildew is a foliar spray consisting of neem oil and cow's milk. We have already discussed how well neem oil works to protect plants from fungal diseases, but cow's milk? Who would have thought that something as simple as milk could work as well as, if not better than, chemical fungicides at preventing powdery mildew from infesting our beloved plants?

Apparently scientist Wagner Bettiol thought so, and he performed a scientific study to prove his hypothesis. Low and behold, he was correct: Milk was approximately 90 percent effective at controlling powdery mildew on a wide range of plants with varying natural resistance to the disease and with several different species of powdery mildew fungus. The study demonstrated that a solution of 10 to 20 percent cow's milk mixed in water and sprayed weekly was equally as or more effective than chemical fungicides.

There is mixed speculation as to exactly why it is that milk is so effective at controlling this disease, but what really matters is that it works. We mix neem at a 0.5 percent concentration in our backpack sprayer, add milk at the 10 percent rate, and mix the two together. We then thoroughly spray the skullcap plants to the point of runoff while they are still in the greenhouse during early morning a few days before transplant. The runoff soaks into the potting mix the transplants are growing in and helps to inoculate the rhizosphere. After the plants are transplanted into the field, we give them a week to get acclimated to their new homes, and we begin spraying the neem/

milk combo once every week or two depending on the weather. We apply the solution thoroughly on both the lower and upper surfaces of the leaves to the point of runoff, which, again, helps to inoculate the rhizosphere with beneficial microbes and provides the plants and soil with a nutrient boost.

# Aster Yellows Disease

Aster yellows disease is a systemic disease that is fairly common among plants primarily in the Asteraceae family. There are at least three hundred species of plants known to be susceptible, some of which are our most popular medicinals, such as echinacea and calendula. We are extremely proactive about controlling this insidious disease, because once a plant becomes infected, there is no known chemical or organic cure that has proven to be effective.

Aster yellows is caused by the transmission of a bacterium-like organism called a phytoplasma. The primary vector for the transmission of this phytoplasma is the aster leafhopper (*Macrosteles quadrilineatus*), which is not only extremely efficient at spreading the disease from plant to plant but also, amazingly enough, benefits from having its average life span extended by carrying the phytoplasma.

Once a plant acquires aster yellows disease it starts to exhibit symptoms within two weeks of infection. Symptoms include deformed growth, in which flowers are replaced by tufts of stunted leaves, "mini florets" erupting from flowers, yellowing or purpling leaves, green flowers, reduction in the root system, and eventually sterility. Although it rarely causes plant mortality, once a plant becomes infected its rapid deformation results in stunted growth, poor health, and an unsightly appearance.

Since there is no known cure for this disease, we have to be extremely vigilant and proactive to control its spread. We have seen aster yellows in our *Echinacea purpurea* and *Calendula officinalis* crops for years now, and fortunately we have been able to limit economic losses by our preventive methods. First we interplant rows of aster yellows–susceptible crops such as echinacea and calendula with insectary plants such as yarrow, tulsi, and anise hyssop that attract beneficials such as green lacewings, lady beetles, predatory wasps, and flies that feed on leafhoppers.

Probiotic sprays of neem oil are our second line of defense. In addition to stimulating phytoalexins that can help to induce an immune response in the plants, neem oil has proven to be an effective feeding deterrent for leafhoppers. Neem oil needs to be applied frequently to maintain this defense, since rain, dew, sunlight, and time minimize its effect. We apply neem at least once a week to the echinacea when we can.

Our third line of defense is to monitor susceptible crops and remove and destroy infected plants as soon as we see them. Leafhoppers need to ingest the phytoplasma from an infected plant to transmit it to an uninfected plant, so if we can limit the amount of infected plants, we can limit spread of the disease. Leafhoppers feeding on our crops would be of very little concern at all to us if it weren't for the fact that they are the vector of transmission for aster yellows.

As we stated in the beginning of this chapter, we believe that preventing economic losses from pests and diseases on medicinal herb farms is best accomplished using plant-positive instead of pest-negative methods. In addition to this philosophy we need to accept the fact that losses can be expected from time to time. If we produce a wide variety of commercial crops using good organic growing practices, these losses should never be great enough to completely undermine our businesses. This approach works on our farm as well as on many others like ours, and it can definitely work on your farm.

The first step is to proactively maintain a healthy, diverse environment with natural checks and balances between predator and prey, host and parasite, fungi and substrate, and pest or disease and farmer. When and if the balance shifts in favor of the pest or disease we should utilize the lowest impact tools and methods we have in our biological toolboxes. If things appear to be getting out of hand we have to act within reason to help prevent economic losses from undermining our efforts to sustain our livelihoods. Knowing we have many choices should help farmers feel confident that we can produce healthy, pest- and disease-resistant crops profitably and sustainably, while impacting the natural environment around us in a positive way.

**Figure 12-1.** Melanie harvesting chamomile.

# CHAPTER 12

# Harvest

One hot afternoon while working on a peppermint harvest, a conversation ensued about the differences between herb farming and other types of agriculture. We were dripping with sweat and a bit sore from cutting and hauling heavy tarps of herbs to the drying shed. Although it was oppressively hot, the conversation was lively, and the heady smell of peppermint kept us energized and focused. During the course of our conversation, one of our employees commented on how much she appreciated the rhythm of harvesting on an herb farm. Before coming to ZWHF, she had worked as a field manager for a large veggie farm in California.

Her comment piqued my curiosity, so I probed further. The young woman told me that on a veggie farm the harvest is always done with an eye to getting fresh produce to market. She went on to explain that because most vegetable crops are annuals the window of peak harvest is usually very short. Big crews would come in, harvest a crop, then move on to the next. And while it was exhilarating to harvest like this and very rewarding to produce food for people, she said there was little time to build relationships with the plants or truly absorb the plants' life cycle. The conversation meandered on in a different direction, but as we continued to work in the peppermint, I kept reflecting on the nature of the ebbs and flows of medicinal herb harvests.

Clearly there are some similarities between harvesting vegetables and harvesting herbs (for instance, the harvesting techniques and the equipment), but there are also significant differences. Beginning herb farmers need to educate themselves about plant vitality; the chemical constituents of plants; the methods for harvesting blossoms, leaves, berries, barks, and roots; and, most importantly, the role timing has in all of these. It is also beneficial for herb farmers to utilize hand and mechanical harvesting techniques appropriately to maximize efficiency while maintaining the highest quality.

## Plant Vitality and Chemical Constituents

As herbalists we were taught that dried herbs should look, taste, and smell like the plants do in their living form. Additionally, these dried herbs need to be full of the medicinal properties and nutrients that make them effective. For years the bulk herb market sported dried medicinals that oftentimes looked more like hay than vibrant medicine. Things certainly have changed; we are fortunate to live in a time when quality reigns supreme and when many herbal practitioners, product makers, and customers are beginning to insist on quality over cost.

One of the keys to producing high-quality herbs is to harvest the plants when they are at their peak of vitality and when the desired chemical constituents are present at significant levels. For example, herb farmers growing lobelia (*Lobelia inflata*) need to harvest the plant after it has gone to flower and has begun producing seedpods. It is at this time when the alkaloid lobeline is at its highest concentration. Lobeline helps to relax smooth muscles in the body and is one of the reasons lobelia is used in formulas to reduce spasmodic muscle contractions.

If lobelia is harvested before seedpods are formed, the lobeline concentrations will be markedly lower.

**Figure 12-2.** Lobelia in seedpod stage. Photograph courtesy of Larken Bunce

**Figure 12-3.** Dried calendula, vivid in color and highly medicinal. Photograph courtesy of Bethany Bond

Consequently the end product will be less effective. This illustrates how essential it is for herb farmers to understand the life cycle and the phytochemistry of their crops. Simply growing healthy plants is not enough. Knowing when to harvest is key.

Additionally, with new regulations and attention to quality, product makers are now using chromatographic testing for key biochemical markers. Farmers who can consistently demonstrate the ability to grow potent herbs with desired constituent levels will gain the confidence of their retail customers and wholesale buyers. Those who do not may find it harder to sell their herbs.

A few years ago we had our first introduction to how lab results can either make or break a farm in the wholesale arena. We were working with a company that makes a supplement to treat hyperthyroidism. They had ordered five hundred pounds

of fresh lemon balm from us (*Melissa officinalis*) to use in their formula. Upon receiving the herbs, they tested the lot for both cleanliness (they wanted to be sure that the lemon balm had low microbiological loads) and also for the level of rosmarinic acid, a key medicinal constituent. The lemon balm would have been rejected if there were high microbiological levels or if the rosmarinic acid levels were low.

After spending a whole day harvesting and packaging the lemon balm and hiring a driver to deliver it, we waited anxiously to hear the results. The relief was palpable when our results came back favorable for both tests. We also learned from the lab that the rosmarinic levels for our lemon balm were higher than any they had seen from several other sources of lemon balm. This high quality ensured not only that sale but also orders for future seasons. The company contracted us to grow more lemon balm and other

**Figure 12-4.** Early spring nettle emerging. Photograph courtesy of Bethany Bond

species of plants for them, and we have been working with them ever since.

On the flip side, there have also been times when lab results have not been in our favor, and we have had to work with companies to remedy it and adjust our farming techniques in response. It is a learning process for sure but one that can be aided considerably by cultivating a deep understanding of the plants, how they grow, and when optimum times of harvests are.

One final example, illustrating the importance of harvesting herbs when the correct chemical constituents are present, can be found in the case of nettle (*Urtica dioica*). Nettle is one of our top selling herbs and is a must-have crop for every medicinal herb farm. I can hear the dairy farmers laughing as they read this, but seriously, nettle is where it is at. This single herb makes up 10 percent of our total gross sales and

is a key ingredient in our most popular value-added tea blend. Conventional crop growers may despise it, but herb farmers can take nettles to the bank.

However, to be successful with nettle it is essential to know how and when to harvest it. Nettle leaf is an incredible tonic that is fortifying to the body; full of vitamins, minerals, and protein; and wonderful for the liver and kidneys—if harvested at the right time. We harvest nettle leaf just as the plant begins flowering and well before it sets seeds. If harvested during late flowering or when it goes to seed, silica levels can increase, and it may be slightly irritating to people with compromised kidneys. The proper harvest time is essential to ensure best-quality medicine.

Fortunately for us, there are rich herbal traditions that outline harvest times for medicinal herbs. When this body of historical information is coupled with the new research emerging regarding plant biochemistry, we have a powerful tool to guide our harvesting practices. To learn more about harvest recommendations for individual herbs, see part two: Herbs to Consider Growing for Market.

## Harvest Times

While each plant has its unique combination of bioactive compounds, there are general guidelines that herb farmers can use to help plan their harvests. Harvest should occur when the plants are at their most vital stage, full of energy, nutrients, and medicinal constituents. These harvest windows correspond to the seasons and the life cycle of the plants. Nutrient uptake and the synthesis of phytochemical compounds closely follow the vegetative cycle and are influenced by environmental conditions the plants experience during their growth.[1]

Therefore, in hardiness zones 3 through 6 it is recommended that roots be harvested in the spring or fall. The roots are the storehouse for vitamins, minerals, carbohydrates, sugars, and proteins. In the spring, before vegetative growth begins and nutrients are carried up into the plant, the roots are where the medicine is concentrated. Similarly, in the fall, as the aerial parts of plants die back, nutrients are shuttled back

**Figure 12-5.** Dandelion root harvest.

**Figure 12-6.** Siberian ginseng in bloom. Photograph courtesy of Larken Bunce

**Table 12-1.** General Harvest Guidelines for Non-Temperate Zones

| Plant Part | Harvest Window | Additional Considerations |
|---|---|---|
| Roots | Fall and spring | Some herbs (e.g., *Valeriana officinalis*) produce larger roots in the spring. Biennials (e.g., burdock, *Arctium lappa*) should be harvested in the fall of the first year for optimum results. |
| Barks | Fall and spring | To avoid weakening or damaging medicinal trees, remove bark from pruned branches rather than scoring the tree. |
| Leaves | Early spring through late summer | Many herbs will give multiple aerial harvests throughout a season. |
| Blossoms | Summer through early fall | If seeds/fruits are desired, be judicious in blossom harvests. No blossoms equates to no fruit. We lightly harvest the blossoms of elder (*Sambucus nigra*) and hawthorn (*Crataegus* spp.) so we get an additional berry harvest. |
| Seeds/ Fruits | Late summer to fall | Timing is essential. Birds and animal pressure can significantly impact yields. Consider using netting or time harvest to avoid animal competition. |

into the root, making it a perfect time to harvest. The spring and fall are also excellent times to harvest medicinal tree and shrub barks. The bark's inner cambium layers house the vascular system of the plant, and in the spring and fall these layers are full of "sap" that in some species of plants is highly medicinal.

In the late spring and summer, growth occurs primarily in the aerial parts of the plant; namely, the leaves, stems, and flowers. In general this is the window for aerial harvest. Harvest should be timed for dry, sunny days when the plants are clean, not dusty or speckled with soil (which can occur after a heavy rainfall). Blossoms tend to be a little trickier. Some blossoms, such as Saint John's wort (*Hypericum perforatum*), should be harvested in bud and early flowering stage. Other blossoms, such as calendula (*Calendula officinalis*), elder (*Sambucus nigra*), or arnica (*Arnica* spp.), are harvested when fully open.

As the season progresses from summer into fall, the plant switches from vegetative growth to reproduction. The plant's energy and nutrients concentrate themselves in its fruiting bodies. In response to this shift in the plants, our harvest turns from leaves to berries and seeds. This harvest rhythm, from season to season, following the plants' natural cycles, keeps us connected to the plants in a meaningful way. We are in concert with the natural world, and as a result the medicine we harvest is strong and full of vitality.

# Harvest Methods for Leaves, Blossoms, Fruits, Barks, and Roots

We have spent countless hours through the years harvesting medicinal plants and we are always on the lookout for more efficient ways to harvest them. Inspiration and innovation has come from many places: creative employees, collaboration with other herb farmers and engineers, and (most recently) postings on social media sites such as Facebook and Instagram. As the old adage states, "Necessity is the mother of invention," and employing efficient harvesting techniques is essential, not only to the production of good medicine but also for the profitability of the farm. Each part of the plant is harvested in a unique way and can be fraught with its own set of challenges and rewards.

## Leaves

Depending on the scale of the farm and the size of the crop, leaf harvest can be done in many ways. Some people like to harvest leaves by handpicking them from the stems directly into baskets in the field. While this can be meditative and "gentle" to the plants, it

**Figure 12-7.** Brian harvesting tulsi.

is time consuming and not necessarily recommended for producing commercial crops profitably. When we were first learning to farm medicinals, we spent time with our mentors Matthias and Andrea Reisen on their herb farm in New York. These two farmers are a wealth of information and helped shape many of the farming practices used at our farm.

I'll never forget the first motherwort leaf haul we did with them. Andrea handed us a tarp and a field knife and told us to "keep up." A whirling dervish with a knife, Andrea started knocking out this long row of motherwort in a truly impressive fashion. No handpicking leaves for her. She cut the entire plant back using a quick slice of the knife, leaving ten to twelve inches of leaf and stem intact. She explained that the plant would regenerate, and she would get at least one, if not two, more cuttings that season. In no time our tarps were full, and we dragged them into the barn.

This is still how we harvest the majority of our leaf crops today. We cut the entire aerial part of the plant, dry the leaves on the stem, then process the dried material to remove the leaves (see chapter 14 for postharvest processing techniques). Besides using field knives we have also experimented, with varying success, in using scythes and are experimenting with mechanizing this harvesting process using a sickle bar cutter. Regardless of what tool is used, the key to leaf harvest is cutting plants efficiently and then processing them immediately after harvest.

*Rewards:* Many medicinal herbs will give more than one harvest of aerial parts per season.

*Challenges:* Cleanliness is essential. Make sure to harvest clean, dry plants to avoid soil contamination on leaves. Mulching plants can also help minimize dirty leaves. Leaves also contain "field heat" (heat gained from solar radiation) and need to be processed immediately after harvest or may become anaerobic and begin to compost in the field. For fresh orders, cool the newly harvested leaves by spreading them out on a tarp in the shade. For larger batches we stack drying racks in the shade and place the herbs on them to cool. Let the leaves rest in the shade for up to an hour to remove field heat before packaging fresh orders. Plants that are not being shipped fresh should go directly onto drying racks (see chapter 14 for postharvest processing techniques).

## Blossoms

When we are harvesting blossoms, it is important to know if the blossoms should be picked in full bloom or in budding and early flowering stage (see part two). For some blossoms, such as Saint John's wort (*Hypericum perforatum*), we use field knives to chop the top three to four inches of the plant, harvesting the entire flowering top and some of the medicinal leaves. Some plants, such as arnica (*Arnica* spp.) and calendula (*Calendula officinalis*), create multiple flushes of flowers, flowering for weeks and, in calendula's case, months at a time.

**Figure 12-8.** Vermont Center of Integrative Herbalism students displaying calendula harvested during a field trip to ZWHF.

Calendula and arnica are picked by popping the flowers off where the flower joins the stem. We harvest these blossoms every few days until the plant stops producing flowers and starts going to seed. Our flowering trees and shrubs, elder (*Sambucus nigra*) and hawthorn (*Crataegus* spp.), are harvested by snapping the flower clusters off and drying them intact. To reach the blossoms on the tree, we use ladders and/or the tractor bucket loader, harvesting into baskets tied around our waists. One word of caution: We also harvest the hawthorn and elder berries, so we keep our flower harvest moderate, leaving many blossoms on the plants to produce fruit.

*Rewards:* Harvesting flowers is a truly rewarding experience. It's a straightforward hand harvest that is very pleasing to the senses, with beautiful blossoms, lovely aromatics, and pollinators buzzing from flower to flower.

*Challenges:* Reaching the blossoms can be strenuous to the body. Many of the plants, such as chamomile and arnica, are low growing and require a lot of bending to harvest. Similarly, calendula can be a backbreaker because of its plant height and also the incredible volume of flowers the plants put out. The taller species, such as elder and hawthorn, are difficult in that they require ladders or a tractor bucket to reach. For our field crops we are experimenting with growing some plants in raised beds to help ease the strain of harvest.

## Fruits

Anyone who has been berry picking knows the obvious considerations when harvesting medicinal fruits. Pick the fruits when fully ripe, be gentle to avoid bruising, and process or refrigerate immediately. There is a strong market for fresh hawthorn berries and elderberries. These berries can also be dried, frozen, or made directly into value-added products such as tinctures and syrups. There are also fruiting bodies, such as garlic scapes, that, while not traditionally

**Figure 12-9.** The very first blossom harvests at ZWHF in 1999: Melanie's sister and brother-in-law harvesting calendula and chamomile, and Jeff's ninety-year-old grandmother and sister harvesting red clover.

**Figure 12-10.** Lily with her elderberry harvest.

thought of as a medicinal crop, can be cut and sold for culinary purposes. We have had success marketing scapes to local restaurants and co-ops and selling them in bundles at farmers' markets.

*Rewards:* Berries are easy to pick and process, and there is great demand for medicinal berries on the market. We simply can't produce and harvest enough to keep them in stock.

*Challenges:* Like the flowers, gaining access to the berries can pose difficulties and require additional equipment. Also, animal pressure can decimate yields, so covering with netting or careful monitoring of harvest times is essential.

## Barks

Tree barks, such as slippery elm (*Ulmus rubra*), cramp bark (*Viburnum opulus*), and wild cherry (*Prunus serotina*), are an important part of herbal medicine. Additionally, barks are another way farmers can diversify their crops by utilizing the trees on their land. "Barking," or bark harvesting, is relatively simple. Using a sharp knife and gloves for protection, make a cut through the outer bark, going past the inner cambium layer but stopping at the woody

center. The medicine is found in the inner bark and cambium layers, so there is no need to cut into the wood center. Once the initial cut is made, use the knife to peel back the bark into strips. These strips can be used fresh or dried, depending on the species.

One primary consideration when harvesting bark is the health of the tree. The inner cambium is the vascular system of the tree, and if barking is done correctly the inner cambium can be harvested without doing damage. However, if done improperly, it can weaken and kill the tree. When harvesting bark, it is essential to avoid removing bark from around the entire circumference of the trunk or branches. This is known as girdling or ring barking, which will kill, or "girdle," the tree. One technique harvesters use to avoid girdling is to take bark in three-inch by five-inch "band-aid" strips spaced out over the surface of the tree.

Even though this is a sound technique and these strips will eventually heal, we prefer to collect bark in a different manner. In the spring or fall we prune branches off our medicinal trees and use a drawknife to harvest all the bark off these prunings. This method is not only more efficient, because you can harvest a lot of bark in a relatively short amount of time, it is gentler on the tree. Pruning helps the overall health of the tree and still allows for a sustainable harvest.

*Rewards:* The medicinal barks are in good demand in the market and are a wonderful way to utilize trees as another sustainable crop.

*Challenges:* The most time-consuming aspect of this harvest is pruning the trees and collecting branches. Stripping the bark from the branches is not difficult, but care should be taken when using sharp knives to avoid cutting hands.

## Roots

Root harvest is done in the beginning of the spring or in mid- to late fall, depending on the species. If harvesting in the spring, we collect the roots before aerial growth begins. In the fall we wait until the tops of the plants have died back. To lessen the impact on the soil and to make digging easier, we try to avoid harvesting in rainy weather and allow the soil to dry

Figure 12-11. Knotweed root harvest.

Figure 12-12. Jeff knocking soil off of roots before transport.

out. Prior to fall root harvest, we use our rotary or sickle bar cutter to mow down the plants and often run chisel plow tines deep in the row parallel to the roots to loosen the soil around them.

To dig the roots we employ a number of different tools. For hand digging we use a spading fork, being sure to dig deeply enough to free the entire root system. Some plants such as burdock (*Arctium lappa*), have long taproots; others, like valerian (*Valeriana officinalis*), have a network of smaller rootlets that form a dense mat. Once the root is freed from the earth we bang the root-ball against the handle of the spading fork to knock off the soil. Because we do so much work to build soil fertility, we like to leave as much of our soil in the field as possible. This also makes it easier to handle, wash, and process the roots.

Other tools we use to dig roots are mechanical and run off the tractor. The chain digger (also referred to as a potato digger) and bed lifter are two pieces of equipment that have greatly increased the efficiency of our harvesting and helped produce larger volumes of roots, while saving our backs. These tools are discussed in detail in chapter 6. Even though many

farmers may start out with limited equipment these two tools are well worth the investment, and we highly recommend that farmers who plan to do a significant volume of roots explore these options.

*Rewards:* Roots, unlike aerial parts of the plant, are more stable, and do not degrade as quickly. This allows farmers more time to harvest and process without fear that the roots will biodegrade in the process.

*Challenges:* Root harvest can be very strenuous, so for large-volume crops it is best to explore mechanical harvest. Roots need to be washed and processed before going into the drying shed, whereas leaf crops can be dried straight out of the field with little or no prep. When planning a root harvest it is essential to allow enough time to process the roots before drying unless they are to be shipped fresh.

# Hand Harvesting versus Mechanical Harvesting

There are three primary considerations for the small-scale medicinal herb farm when it comes to deciding whether to harvest crops by hand or by machine: volume of production, cost of production, and product quality. For farms with fewer than five acres in production, harvesting crops solely by hand could be considered a viable option, provided extra hands are available when needed, especially during harvest time, to help keep production flowing.

At ZWHF we hand-harvested everything for the first five years in business, with Melanie and me and one or two part-time employees and with our total planting acreage never exceeding five acres. However, when we consider the labor costs we racked up paying people to dig burdock roots with spading forks, cut nettles by hand with field knives, and so on, our costs of production would have been much lower and our profitability higher had we invested in some equipment to assist us with the harvest.

Why didn't we buy automated harvesting equipment earlier in the game? Good question, and one we have asked ourselves repeatedly. The primary answer to that question is cash flow. Early on in our venture we felt that we never really had enough extra capital to afford those "luxury items." With all the nonnegotiable costs such as seed, fertilizer, fuel, payroll, equipment maintenance, utilities, and so forth, our profit margins were slim, and we felt that we never really had that little cushion of extra money to buy a root digger or a sickle bar cutter.

Mistakenly, we had decided early on that we were never going to borrow money to finance farm expenses. Making payments on a loan would have cost considerably less than the labor we paid out digging roots by hand, but at the time Melanie and I didn't feel like borrowing money was a good idea. Our philosophy has changed since then and we have taken advantage of some low-interest agricultural loans to make improvements that have yielded quick return on investment.

Finally, we went to an agricultural auction and bought a beat-up antique ground-driven chain digger for $500. Another $200 went into buying a rear differential from a 1974 Jeep and some angle iron to weld and convert the digger to PTO driven. For $700 plus some of our own tinkering labor we now had an automated piece of harvesting equipment that literally paid for itself the very first day we used it.

We had planted a few acres of angelica, burdock, and echinacea roots for wholesale accounts, and we were daunted by the thought of how long it would have taken and how much it would have cost us in labor to dig that quantity of mature roots by hand. The chain digger was able to dig in a couple of days, with one person driving a tractor and three people picking the roots up and loading them into trucks, what would have taken six to eight days to do by hand with the same size crew. By hand we would have had to pry the roots out with spading forks, shake the excess soil off them by banging them on the fork's handle, then load them into the trucks. The chain digger automated the first two steps of the process so that all we had to do was pick the roots up off the ground and load them into trucks.

We have seen similar increases in efficiency with a sickle bar mower. In the past we cut our taller leaf crop plants by hand with field knives or scythes. The larger the plants got, the more difficult it became to hand-cut them because of the thickness of their stems. We soon figured out that a sickle bar mower mounted behind the tractor can cut a 350-foot-long bed of nettles in as little as five minutes. It takes a crew of three approximately fifteen total additional minutes to pick the plants up and load them into trucks, for a total of 20 minutes × 3 people = 60 minutes total labor, plus a bit of fuel and tractor time. By hand we were spending approximately two hours or more × 3 people = 6 hours total labor, depending on the size of the plants, harvesting the same amount of material.

The best thing about this automation, in addition to the savings in labor, is the fact that we are not compromising product quality at all with these mechanized harvesting processes. In fact, the less time it takes to get plants from harvest to processing, the higher the retention in quality. Digging roots with a chain digger and cutting herbaceous material with a sickle bar mower would probably be best defined as "semiautomatic" or "mechanically assisted" harvesting, since the material is harvested automatically, then collected and loaded by hand. This is much different from the gigantic fully automated harvesting machines, such as the combines found on large farms,

**Figure 12-13.** Chain digging angelica root in the spring.

where the entire process is fully automated. Fully mechanized harvesting often involves a compromise in product quality because of the heat caused not only by friction from aggressive mechanical processing but also by anaerobic conditions that can occur with high-volume quantities of plant material condensed during harvest and transport.

There are certain crops grown on the small-scale herb farm that are not suitable for any type of mechanical harvesting. Blossom crops such as arnica, calendula, elder, red clover, and Saint John's wort, for example, are best harvested by hand to ensure that only ripe, open blossoms are being taken without excess stem and leaf material or immature buds, something no amount of mechanization we are aware of that has been developed yet can discern.

Equipment exists in the foreign marketplace for automated chamomile harvesting, but it is exceedingly expensive to purchase, operate, and maintain and has not yet become readily available in the United States. Low-growing crops such as thyme, spilanthes, and dandelion leaf are also difficult to harvest automatically without performing elaborate field-leveling conditions to allow for tight tolerances between ground and cutting devices. Medicinal fruits such as elderberries, hawthorn berries, and schisandra berries are also not suitable for mechanical harvest. Hops cones can be harvested mechanically with hops pickers that are fairly common in hops-growing regions, but they are expensive to purchase and maintain for small-scale growers.

Ginkgo leaves are another species whose leaves are primarily harvested by hand. There is much debate as to whether it is best to harvest ginkgo leaves before they naturally drop or after. The possibility exists for vacuum-type collectors to be used to suck up fallen leaves for transport to processing.

Another challenge when considering mechanical versus hand harvesting is the amount of foreign material such as weeds present within the desired crop. Oftentimes it can be more cost effective to selectively hand-harvest leaf crops with field knives rather than attempt to pick foreign material out of already harvested mixed species. This is another reason we try to be extremely proactive with our weeding practices, so we don't have to spend unnecessary time removing contaminants from crops during harvest and processing.

# Wild-Harvesting Medicinal Herbs

*In God's wildness lies the hope of the world—the great fresh unblighted, unredeemed wilderness. The galling harness of civilization drops off, and wounds heal ere we are aware.*

—JOHN MUIR[2]

In the wilderness lies our medicine, the medicine that is found in the untamed energetics of nature that feed and inspire our spirits, and in the medicinal plants that heal our bodies. The profound vitality that is present in wild plants has always been a treasured aspect of herbal medicine. For thousands of years people have journeyed into the wilderness to hunt and gather what was needed for their apothecaries. Wild harvesting, also known as wildcrafting, has a rich tradition and is also an important part of herbalism today.

There is nothing like hiking into the creeks and riverbeds of Vermont on a summer afternoon in search of elderberries, or the feeling of joy when a journey into a hemlock forest yields exquisite reishi mushrooms (*Ganoderma tsugae*) full of immune-enhancing medicine. As a rule, herbalists like to be out in the wild edges of the world, often with collecting baskets and digging sticks in hand. In fact, one of Melanie's fondest childhood memories is of her step-mother Rosemary suddenly pulling their car over midtrip, to see if a "secret stash" of wild plants was ready to harvest. They'd pile out of the car, slip on boots, and head out in search of herbal delights.

Because of variable and sometimes harsh growing conditions, coupled with the forces of natural selection at work in plant populations, wild plants are often some of the most potent medicinals to be found. Many wild-harvested plants command incredible prices on the market, and gatherers can feel compelled for many reasons to reach into Mother Nature's larder for their harvest. For many species of plants, when they are harvested properly, wildcrafting is a sustainable practice. However, for some species this practice is not acceptable. For these plants, high demand coupled with other factors such as habitat destruction, low reproductive rates, and highly specific growing conditions has had a devastating effect on plant populations, and wild harvesting threatens the very existence of many of these species.

It is essential that herb farmers and wildcrafters working with plants today become well educated about the plants they harvest and understand them in their larger environmental context. It is important to know which herbs can regenerate themselves fully a short time after withstanding a wild harvest and which ones cannot. It is also highly beneficial to the herb farmer and our plant allies to understand which threatened plants can be cultivated and to establish plantings of these species on our farms to provide alternatives to wild-harvested plants.

Cultivated sources of prized medicinals can ease the pressure facing our wild populations but only if we have enough farmers committed to growing them. Balancing the desire to wild-harvest with our responsibility to be good stewards of the plants is something many herb farmers and herbalists grapple with. Fortunately, there is a growing body of knowledge coming out of the dynamic work of such organizations as United Plant Savers (UpS), the Convention on International Trade in Endangered Species (CITES), and the International Standard for Sustainable Wild Collection of Medicinal and Aromatic Plants (ISSC-MAP) that can help inform our practices.

## United Plant Savers

In the year 2000, Dr. Richard Liebmann, ND, wrote, "As we enter the new millennium, a twenty-year, sixteen-organization study reports that thirty-four thousand plant species—12 percent of plants worldwide and 29 percent of plants in the United States—have become so rare that they could easily disappear."[3] Thirteen years later, in September 2013, Susan Leopold, executive director of United Plant Savers, and rangers from the Great Smoky Mountains National Park made national news when they brought to light the destructive poaching of American ginseng in the

Figure 12-14. ZWHF is a proud member of United Plant Savers.

Figure 12-15. American ginseng seed ripening. Photograph courtesy of Larken Bunce

southern Appalachian Mountains. Wild American ginseng is a supreme adaptogen and has a staggering market price of upwards of $800 a pound, making it one of the most sought-after medicinals in this country and abroad. This high demand, coupled with habitat destruction and ginseng's slow rate of maturation and sexual viability are the major reasons, Leopold and many scientists assert, that the very survival of this species is in question.

These are two examples that clearly illustrate the increasing need for the sustained and rigorous examination of wild-harvest and conservation practices in the herbal community. Both examples come from work generated in large part by the organization United Plant Savers. UpS was founded in 1994 by Rosemary Gladstar and a group of dedicated herbalists who were deeply concerned about the fate of medicinal plants in the wild. Gladstar writes of her early observations,

*For the first couple of years, I wandered through our woodlands (in Vermont) in a state of happy anticipation . . . I was ever on the lookout for those mysterious and oh so famous eastern woodland medicinals: ginseng, goldenseal, bloodroot, black cohosh . . . But after several seasons passing with nary a sighting, I began to doubt that any of these native medicinals remained . . . I began to realize that many of the oldest plants of the eastern deciduous forest, including many important medicinal plants, had either completely disappeared or were in short supply.[4]*

To reverse this trend, UpS has worked for the past twenty years with community partners to protect native medicinal plants of the United States and Canada and their native habitat while ensuring an abundant renewable supply of medicinal plants for

**Table 12-2.** Medicinal Plants to Avoid Harvesting in Wild Populations (from United Plant Savers)

| "At Risk" Plants | "To Watch" Plants |
|---|---|
| American ginseng (*Panax quinquefolius*) | Arnica (*Arnica* spp.) |
| Black cohosh (*Cimicifuga racemosa*, syn. *Actaea racemosa*) | Butterfly weed (*Asclepias tuberosa*) |
| Bloodroot (*Sanguinaria canadensis*) | Cascara sagrada (*Frangula purshiana* [*Rhamnus*]) |
| Blue cohosh (*Caulophyllum thalictroides*) | Chaparro (*Castela emoryi*) |
| Echinacea (*Echinacea purpurea*) | Elephant tree (*Bursera microphylla*) |
| Eyebright (*Euphrasia* spp.) | Gentian (*Gentiana* spp.) |
| False unicorn root (*Chamaelirium luteum*) | Goldthread (*Coptis* spp.) |
| Goldenseal (*Hydrastis canadensis*) | Kava kava (*Piper methysticum*) (Hawaii only) |
| Lady's slipper orchid (*Cypripedium* spp.) | Lobelia (*Lobelia inflata*) |
| Lomatium (*Lomatium dissectum*) | Maidenhair fern (*Adiantum pedatum*) |
| Osha (*Ligusticum porteri, L.* spp.) | Mayapple (*Podophyllum peltatum*) |
| Peyote (*Lophophora williamsii*) | Oregon grape (*Mahonia* spp.) |
| Sandalwood (*Santalum* spp.) (Hawaii only) | Partridgeberry (*Mitchella repens*) |
| Slippery elm (*Ulmus rubra*) | Pinkroot (*Spigelia marilandica*) |
| Sundew (*Drosera* spp.) | Pipsissewa (*Chimaphila umbellata*) |
| Trillium, Beth root (*Trillium* spp.) | Spikenard (*Aralia racemosa, A. californica*) |
| True unicorn (*Aletris farinosa*) | Stoneroot (*Collinsonia canadensis*) |
| Venus flytrap (*Dionaea muscipula*) | Stream orchid (*Epipactis gigantea*) |
| Virginia snakeroot (*Aristolochia serpentaria*) | Turkey corn (*Dicentra canadensis*) |
| Wild yam (*Dioscorea villosa, D.* spp.) | White sage (*Salvia apiana*) |
| | Wild indigo (*Baptisia tinctoria*) |
| | Yerba mansa (*Anemopsis californica*) |

generations to come.[5] They have strived to accomplish this in a variety of ways, including the creation of the following:

- "At Risk" and "To Watch" lists of threatened medicinals
- Nursery and bulk herb directories
- Educational and "Take Action" guides
- A 366-acre botanical sanctuary in Ohio
- A Botanical Sanctuary Network Guide (a step-by-step guide to medicinal plant and habitat stewardship)
- Seed and root giveaways
- Seminars and conferences
- *The Journal of Medicinal Plant Conservation* and UpS Bulletins[6]

Their work includes political activism, educational outreach, and conservation, harnessing the creativity and collaboration of wildcrafters, growers, researchers, manufacturers, and consumers of medicinal plants. UpS is a true grassroots organization that is making significant and sustained changes in the herbal community. It is also an important resource for herb farmers. First and foremost are the "At Risk" and "To Watch" lists in table 12-2. These outline plants that farmers should commit to cultivating and avoid harvesting in the wild. Also helpful to herb farmers is the UpS's Cultivator's Corner, which provides information on cultivating threatened medicinal herbs. These plants are often tricky to cultivate because of the plants' life cycles and the specificity of growing conditions. Breakthroughs in growing techniques are communicated in this resource and also in the UpS journal. Other important resources UpS provides on their website are their plant profiles and Plant Conservation Links and Resources.

ZWHF has been a member of UpS for fifteen years and has conducted research into the cultivation practices for threatened medicinal plants found here in Vermont as well as outside our region. We

Figure 12-16. Goldenseal in a woodland bed.

## "At Risk" and "To Watch" Plants Cultivated at ZWHF

Arnica (*Arnica* spp.)
American ginseng (*Panax quinquifolius*)
Black cohosh (*Cimicifuga racemosa*,
    syn. *Actaea racemosa*)
Bloodroot (*Sanguinaria canadensis*)
Butterfly weed (*Asclepias tuberosa*)
Echinacea (*Echinacea purpurea*)
Gentian (*Gentiana* spp.)
Goldenseal (*Hydrastis canadensis*)
Lobelia (*Lobelia inflata*)
Mayapple (*Podophyllum peltatum*)

sustainably and profitably grow seven of the "At Risk" and "To Watch" herbs and are experimenting with growing others. This allows us to help preserve native plant populations while still providing people with much-needed medicine. As new herb farmers enter the field or existing farmers diversify into the realm of medicinal herbs, consider the threatened plants that grow in your bioregion and explore ways to cultivate them for your community. UpS has a wealth of information; it is also a wonderful way to network with other growers, and the directories can be a great place to promote work and market crops.

## Methods for Sustainable and Ethical Wild Harvesting

While we do not wildcraft any threatened plants at all, we do harvest many other medicinals from their natural habitats. Wild harvesting is something we truly enjoy. It has a rhythm and feel to it that is considerably different from harvesting a cultivated crop (although we like that, too). Perhaps it is the hunter/gatherer instinct. Maybe it is the sense of adventure and discovery. While all plants are a gift from the earth, finding a beautiful stand of mullein in bloom or wild skullcap plants nestled at the edge of a field feels like treasure.

A key aspect to any successful harvest is setting the right intent and being mindful. Gifted herbalist, author, and scientist Ryan Drum captured this beautifully when he wrote, "I believe that the wildcrafter's mindset, clarity of healing intent, and process rigor affect the eventual herbal healing outcome in human health. Herbal healing is more than a few known dominant physiologically active molecular species and is best served by continuous conscious healing intent from harvest to individual use."[7]

Before we start collecting we set our intentions that the plants will be harvested in a way that is healing and beneficial for the medicine being made, the plant population, and the earth. We give thanks to the plants and work hard to be as gentle to the surrounding ecosystem as possible. We harvest with care to leave enough of the wild plant stands to ensure the plant's health and ability to regenerate, while also leaving enough for the animals and other creatures that may feed or browse off it or use it for habitat. Often herbalists talk of the Rule of Thirds: Harvest a third for yourself, leave a third for the plant population, and leave a third for the planet.

In many ways, however, this is an oversimplification, as harvesting a third of many plants can be much too much, but the general philosophy

## Herbs and Analogs

After introduction to the UpS list of "At Risk" and "To Watch" plants, students always ask, "Well, what can we use in its place?" Oftentimes the choice is simple: Choose a cultivated species rather than one harvested from the wild. When cultivated species are not available, then it is best to find a plant analog. An analog is something having an analogy or similarity to something else. For our purposes, this indicates parallels in function or end results between two or more medicinal herbs.

In most instances, it is important and necessary to use a variety of analogs for the "At Risk" or "To Watch" herb because an analog generally satisfies only some of the therapeutic actions of a particular plant species and does not demonstrate all medicinal actions of that plant. It is sometimes difficult to find replacements for our tried-and-true herb friends, but it also can be very satisfying and will expand your expertise, while helping to replant our future.

Whenever possible, use what grows around you. Oftentimes those herbs hold the most potential for helping you to heal. Why not choose alien (non-native) plants for food and medicine, leaving the more fragile native plant species to flourish. Many alien plants are extremely powerful medicinals and will be a welcomed addition to your medicine chest.

See the Appendix A for an alphabetical list of some of the United Plant Savers "At Risk" and "To Watch" lists, accompanied by suggested analogs

JANE BOTHWELL,
MARCH 2000, REVISED 2014

**Figure 12-17.** Elder flower fairy.

## Medicinal Plants Wild Harvested at ZWHF

Blackberry root (*Rubus* spp.)
Elder flower and berries (*Sambucus nigra*)
Hawthorn berries (*Craetagus* spp.)
Linden leaf and flower (*Tilia americana*)
Raspberry leaf (*Rubus* spp.)
Red clover blossoms (*Trifolium pratense*)
Saint John's wort leaf and flower
   (*Hypericum perforatum*)
Wild cherry bark (*Prunus virginiana*)

speaks to the mind-set of not overharvesting. When wild harvesting it is important to understand the pressures and inputs placed on the land. Are other people harvesting this same spot? Is this area clean of pollution? Are these plants in a reproductive state, and if so, is it problematic to harvest? How much do I really need or want to harvest, and can the plant regrow and the population remain strong if I do? Are there ways to harvest, replant, or care for the land to help ensure plant regeneration?

**Figure 12-18.** Melanie planting Solomon's seal.

As farmers work with medicinal plants and make careful observations of ecosystems over time, it often becomes clear what works for certain species and what does not. For example, red clover seems to thrive with heavy picking, producing more flushes of blossoms. This is not so with elder or hawthorn blossoms. If we pick these blossoms, the plants do not produce more until the following season, and we have also significantly reduced the plants' ability to produce medicinal berries and have removed an important food source from the habitat for a season. Therefore, it is important to become familiar with each species' life cycle (see part two for plant-by-plant profiles) and begin by harvesting small amounts of plants you have questions or concerns about.

Another key consideration is time of transit from harvest site to drying or processing facility. Harvests can soon turn to compost if not processed quickly. For some plants, especially blossoms such as red clover, you may need to make more than one excursion to avoid degradation of the harvest while in transit.

## General Guidelines for Wild Harvesting

**Avoid wild-harvesting plants on the United Plant Savers' "At Risk" or "To Watch" lists.** Instead, preserve wild populations by working to conserve ecosystems and to cultivate these species.

**Limit wild harvest to plants that are "wholly renewable."** Many plants grow prolifically and are not at risk of being negatively impacted by wild harvests. Often thought of as our "weedy species," these plants, like the mints and burdocks of this world, are ripe for the picking.

**Understand bioregional plentitude and local abundance.** Just because bloodroot or goldenseal may grow well and abundantly where you live or in the ecosystems around your farm does not mean that it should be harvested from the wild. These plants are harvested by the tons annually, and their habitats are under great pressure. Often these plants take years to reproduce and should only be used from a cultivated source.

**Be mindful of human impact on fragile ecosystems.** Anyone who has walked through a calcareous

## Growers Needed! Wild American Ginseng on the Brink of Extinction

In the dappled shade of the hardwoods, a king of the forest is tucked quietly in the leaf litter and humus-rich soil; its branches of green leaflets are resplendent and bright. This stand of wild American ginseng (*Panax quinquefolius*) is healthy and mature, sporting plants from multiple generations, many with fruiting bodies of bright red berries. Some plants are very old, some are seedlings, and others are juvenile plants with single- and double-pronged growth. Each plant is a treasure of botanical diversity and highly sought-after medicine and is also an integral part of the ecosystem.

Historically, wild American ginseng was a prized remedy in the healing traditions of Native Americans, folkloric herbalists, and Chinese medicine practitioners and was used as an exemplary tonic for longevity, adrenal health, and fertility. Today it is still one of the most popular and well-known herbs in America and Asia, with exports of twenty to thirty tons on average annually.[8] But woven through all this rich history and botanical wealth is a problem—a really big, complex, and nuanced problem. How herbalists, botanists, wildcrafters, forest managers, and farmers address this problem will determine whether this extraordinary plant will survive or will tragically disappear into legend.

### Habitat and Growth Cycle

Wild American ginseng is a long-lived woodland perennial that grows in deciduous forests in the mountains of the eastern United States and Canada. In the United States its range extends from northern Maine south to Georgia and as far west as Minnesota. It is a slow-growing species, taking at least five to nine years to reach reproductive maturity, and reproduces exclusively by seed. Repopulation is also hindered by low germination rates; on average wild American ginseng generates approximately one viable plant for every thirty mature seeds. In addition, these plants face many environmental pressures, including increased deer populations that browse on the herbaceous part of the plant, loss of habitat from development and mountaintop removal, the encroachment of invasive plant species, and overharvesting.[9]

### The Problem

Wild American ginseng represents a "perfect storm" in plant conservation. Ginseng is slow to grow and slow to reproduce, is in high demand, commands an amazing market price, and has been historically wildcrafted in large quantities since the eighteenth century; as a result generations of people have been making some part of their living through "digging 'sang." Additionally, ginseng has been touted as "one of the most important of the non-timber forest products collected" in U.S. forests.[10] Not surprisingly, harvesting occurs on both private and public land, with permits being issued for harvest on state-owned land and U.S. national forests. And while it is illegal to harvest in national parks, poaching is occurring at such an alarming rate that it made national news this year.

With all these factors in place it is highly questionable whether it is possible to sustainably wild-harvest American ginseng at this point in time. Botanists Janet Rock and Gary Kauffman and ecologist Nora Murdock write in their 2012 paper, "Harvesting of Medicinal Plants in the Southern Appalachian Mountains,"

*Individual ginseng plants are very long-lived (plants over a century old have been documented), but are slow to reach maturity and reproduce in the wild, which intensifies the impact of heavy or repeated harvesting on populations. For populations that are already dropping to dangerously low numbers in the wild as a result of heavy harvesting, the effects of severe consecutive drought years (like 2007 and 2008), added to harvesting impacts, could result in annihilation of the species in some areas.*[11]

But despite these scientific findings, there are still resources in the industry that promote wild harvest. These resources contend that harvesting in the proper season, the "dig some, leave some" approach, and planting seeds after harvest will mitigate the removal of natural stands of ginseng. While these practices have merit, they do not off-set the enormous negative pressures facing wild populations. The research is becoming increasingly clear that the proper "season" for the wild harvest of American ginseng is now closed, and it is time to explore sustainable options.

## Possible Solutions and Next Steps

Increased efforts in conservation, stricter monitoring of state and national land, and rigorous enforcement of anti-poaching laws should continue to be key focus areas. However, as long as there is significant demand and high prices for ginseng, there will be incentive for people to look to the woods for their income. It is time to change this paradigm and help people look to the woods to grow ginseng rather than wildcraft it. Andy Hankins, extension specialist for alternative agriculture at Virginia State University, has helped farmers successfully grow wild-simulated ginseng in a sustainable and profitable way. From his decades of work, Hankins stresses the importance of cultivating ginseng in the woods rather than in fields under shade cloth.

In the 1990s America, Canada, and China culti-vated millions of pounds of shade-grown ginseng, but it has not been as successful or lucrative as promised. Not only are there incredible difficulties associated with shade cultivation (high setup costs, debilitating disease pressure, and dependence on toxic fungicides), field-cultivated ginseng fetches a low market price ($6 to $18 per pound) and is less sought after than its woodland brother.

Fortunately, wild-simulated American ginseng, because it is grown in its natural habitat, sidesteps these issues and is a viable alternative to wild har-vested ginseng. The bottom line? Hankins writes the following about the market:

*In selling dried roots of wild simulated ginseng . . . it is hard to find any product that is easier to sell. In Virginia, there are 45 certified ginseng buyers spread out across the state. These buyers are regulated by the Virginia Department of Agriculture and Consumer Services—Office of Plant Protection. A list of the certified buyers can be obtained from that office. All that a grower has to do is drive to the buyer's house or store or service station, carry the roots in, watch as they are weighed and accept payment if he agrees with the price that is offered. If the grower does not like the price that is offered, he can take his roots to the next buyer down the road. A grower who has a large volume of roots to sell often will allow buyers to make bids on his roots to get the highest price. Some growers sell directly to large herb companies who buy ginseng for export to Asia. In a few states, ginseng auctions have been organized to help both the buyers and the sellers. Current price information is easy to obtain from several sources. Marketing wild simulated American ginseng roots is easy because market demand is very strong for this scarce commodity.[12]*

Hankins's conservative estimates show average cost of production on a half acre of wild-simulated American ginseng at $3,768.00. The possible net income on a half acre of wild-simulated ginseng, assuming root yields of about fifty pounds dried and a conservative price of $260 per pound of dried roots, is around $9,000.[13]

Certainly there are challenges facing growers of wild-simulated ginseng. They will have to grapple with site selection, slow maturation of plants, and the possible need for security from poachers. That said, a strong case is being made for the widespread imple-mentation of wild-simulated cultivation methods. Wild-simulated cultivation has the potential to take pressure off the wild ginseng populations while pro-viding people access to this powerful herbal medicine.

bog knows what I'm talking about—where every footfall can create pools of water and disruption in the fragile groundcover. Even if the plants you plan to harvest are hardy, prolific growers, sometimes the habitat can be susceptible to damage or erosion if not approached thoughtfully. Heed the camper's and hiker's adage: Leave the place better than you found it. Take care to fill in holes, replace leaf litter, and minimize human impact.

**Replant.** There are some plants—angelica and valerian, for example—that can be easily regenerated by spreading the seeds after harvest. This is especially helpful when harvesting roots, which ends the life cycle of the individual plant. Take a page out of Johnny Appleseed's book; sow a little as you go, and help ensure that there are more plants in the future.

**Be sure of proper plant identification.** Proper plant identification really goes without saying, but it's not uncommon to see a new harvester bring in misidentified plants. For example, in our region a common mistake by inexperienced wildcrafters is harvesting red elderberries (*Sambucus racemosa*), which are toxic, instead of black elder (*Sambucus nigra* or *S. canadensis*), the delicious

medicinal treasures. Not only is this potentially dangerous, it is also wasteful and disheartening to compost plants that can't be used. Plant species need to be properly identified before harvest, and there are many good field guides available for keying out medicinal plants. Some of our favorites include: *Newcomb's Wildflower Guide* by Lawrence Newcomb, *Peterson Field Guide to Medicinal Plants and Herbs of Eastern and Central North America* by Steven Foster and James A. Duke, and *Medicinal Plants of the Mountain West* by Michael Moore.

**Use good harvesting techniques.** Familiarize yourself with the plant's "harvesting window"—the time of year when the plant contains the highest vitality and strongest medicinal properties. Also explore the tools and techniques used for harvesting different parts of the plant (the roots, leaves, berries, barks, and fruits). Depending on what you are collecting, you will at the very least need a good digging fork, sharp pruners, a knife, gloves, and baskets or bins. Having a good "toolbox" will make wild harvesting a whole lot easier and help ensure that you are harvesting plants efficiently.

# CHAPTER 13
# Geo-Authentic Botanicals

So far in this book we have tried to encourage more people to consider growing medicinal herbs. I hope we have demonstrated that there exists here in the United States a viable market for high-quality, certified-organic, locally produced botanicals, yet most of the raw materials we use in our domestically produced herbal products continue to be imported from foreign countries. We have also discussed the fact that almost every species of medicinal herbs widely used in commerce could possibly be grown here in the United States, whether they are grown in the harsh alpine tundra of Alaska, the lush tropical volcanic soils of Hawaii, the temperate rocky coast of Maine, or the fertile plains in between. Within the boundaries of our country lies an incredible amount of environmental diversity, enough to closely replicate just about every condition on Earth where our incredible healing plants are found growing in their native and preferred habitats. However, the question we really need to ask ourselves is, just because we *could* possibly grow every species of medicinal plants used in commerce here in the United States, does this mean we *should* attempt to grow them all commercially?

## The Concept of Terroir in Medicinal Herb Production

"Terroir" is a term commonly used to describe the unique characteristics that certain foods and beverages (namely, cheese, coffee, tea, and wine) develop as a result of being grown or produced in certain regions of the world. At the heart of the concept of terroir lies the assumption that distinct climate, geography, and soil types and their interactions impart unique gastronomic qualities to plants or animals grown within a particular region, qualities unable to be replicated elsewhere, even when given seemingly identical conditions.

**Figure 13-1. Peruvian woman at market.** Photograph © 2014 Steven Foster

**Figure 13-2.** Wine growing in the Chianti region. Photograph courtesy of Francesco Sgroi

**Figure 13-3.** Yellow ginger (*Zingiber officinale*). Photograph © 2014 Steven Foster

The concept of terroir is thought to have originated in ancient Greece, where amphorae (clay vessels) containing wine were stamped with seals marking the regions where the wines originated. Certain regions soon became favored over others for the quality of the wines produced therein, and wines began bearing designations on their seals based primarily on the location where they were produced and secondly on the type of grape they originated from or the vintner who produced them. The French later developed the concept of terroir into a set of wine laws, or "appellations," that in time became a model adopted by other countries for other products such as cheeses that were deemed worthy of bearing unique geographical indications.

Chianti wine is an example that many people are familiar with. The grape varietal that Chianti is made from is called Sangiovese and is a widely adaptable grape frequently grown in other regions, including here in the United States, but only wine made from Sangiovese grapes that are grown in Tuscany is able to bear the Chianti name because of laws that only allow specific wines to bear geographical indications.

Recently there is a growing trend in the international trade of medicinal and aromatic plants to designate provenance in regard to specific species or varieties deemed worthy of distinction and protection. This designation is referred to as "geographical indication (GI) botanicals." A GI botanical is named after a geographical area, indicating that it is produced within a particular area, and its quality and characteristics depend on natural, historical, and cultural factors. An example of a GI botanical is Luoping Yellow Ginger (*Zingiber officinale*). This specific cultivar is grown in Luoping County in the eastern Yunnan province of China, an area often referred to as the "home of ginger."

The concepts of terroir and provenance regarding their influence on the qualities of medicinal plants is an important and intriguing one to consider. Whereas the term terroir is primarily ascribed to *gastronomic* qualities imparted by plants grown in specific regions, one has to wonder how much influence regional variations in geography, geology, and climatology have on the *medicinal* attributes of plants growing in a given region.

When discussing the influences that specific geographical and environmental factors have on plants, we should consider what happens to a particular species when we remove it from its native habitat and either attempt to replicate that habitat artificially and grow the plant similarly or even grow the plant in entirely different environmental conditions from those that are found in its native habitat. The cultivated version may grow, look, smell, and taste similar to its wild counterparts, but when we delve into the molecular and even energetic realms within that plant, what differences if any do we find?

We can certainly isolate, analyze, and quantify the active medicinal compounds found within these plants, but do those analyses reveal the entire picture? What about the more qualitative organoleptic methods of analysis? Isn't the way a plant tastes or feels in our body as important as what the laboratory reveals, or possibly even more important? Every human being has biological mechanisms that are relatively similar yet totally unique, and the effect that a plant has on one person is often subtly and sometimes even profoundly different for another person. Opinions vary greatly and are often subject to a certain degree of bias.

For example, we feel that the ashwagandha (*Withania somnifera*) extract we make from plants that we grow here on our farm in Vermont tastes and feels more medicinally active than any other ashwagandha extract we have ever used. My guess is that an herb farming counterpart of ours or perhaps an ayurvedic practitioner from India, where ashwagandha grows wild, would likely disagree with our opinion and maybe even scoff at the idea that high-quality ashwagandha can be produced in such an extremely different environment from that in its native region. Common sense should tell us all that the ashwagandha from India is likely to be more potent, in that it produces greater concentrations of secondary metabolites, which form the basis for most of its medicinal activity because of the harsh conditions it undergoes growing in such an arid climate. Plants growing in those subtropical regions also have much more time to mature than they do here in Vermont, they receive more direct solar radiation because of their proximity to the equator, and they grow in soils that are minerally and biologically different from ours.

We do everything we can on our farm to mimic the way ashwagandha grows in its native habitat. We start the seeds indoors on heat mats very early in the spring to extend their growing season, and we plant them in well-drained soils in the full sun. We withhold irrigation during dry spells, and we wait to harvest them as long as possible after they have matured and gone to seed. Yes, we admit that it is definitely a stretch to say that the way we grow ashwagandha here in Vermont comes even close to mimicking the environment in its native habitat, but we as well as many of our customers agree that our ashwagandha makes potent, high-quality medicine. Not only does this ashwagandha feel good physically in our bodies, but on the conscious level it also feels good to know who grew, harvested, and processed

**Figure 13-4.** Ashwagandha harvested from ZWHF ready for post-harvest processing.

those plants how and where and that they didn't need to travel thousands of miles to end up in our teas and tincture bottles.

So if we remove a plant from its native or preferred habitat and grow it in a different environment, what molecular and energetic changes are possibly taking place within that plant? Here in Vermont, where black cohosh (*Cimicifuga racemosa*, syn. *Actaea racemosa*) is a native (and threatened) woodland plant that is most often found growing in the shade in fungally dominant soils, we primarily grow this species in the full sun in bacterially dominant soils because it seems to prefer growing out in the open with more sunlight. When I say "prefer" I mean that from our somewhat limited and merely visual perspective, the plants grown in full sun seem to exhibit more vigor and grow larger than those that we grow in the full shade in our woodland beds.

However, rapid growth does not necessarily equate to potent, high-quality medicine . . . or does it? The majority of black cohosh root that we sell is harvested from plants grown in rows in the full sun, and we receive very positive feedback from our customers regarding the quality of the medicine coming from those roots. We should also note that what we consider full sun here in Vermont is much different from full sun in other parts of the country. Vermont runs pretty much neck and neck with the state of Washington as the rainiest, cloudiest states in the United States, so "full sun" here is actually more akin to partial or even full shade elsewhere. Black cohosh would probably not be able to tolerate growing in full sun in most other regions. This is an example of merely flexing the concept of growing plants outside their preferred habitats, not necessarily turning it upside down.

# Medicinal Potency in "Native" Plants

Many herbalists and consumers of herbal products agree that some of the highest quality and most potent plant medicine is extracted from plants harvested from the wild in their native or preferred habitat. Although that consensus is likely valid in most cases from a quality standpoint, the challenge with that belief lies within the fact that wild plants coming from their native and preferred habitats often cannot provide the herbal product industry with the quantities of many of the most popular wild-harvested botanicals necessary to meet consumer demand. We have seen the results of our folly in assuming that natural regeneration will compensate for our lust for harvesting prized botanicals from wild stands. Look at what we have done to wild American ginseng (*Panax quinquefolius*), for example (see sidebar, page 182). This species of plant is literally teetering on the brink of extinction in the wild.

The scarcity of most of these highly sought-after botanicals still found growing in the wild is one of the reasons commercial cultivation of herbs, especially those that are rare and threatened, is so important. Although as herbalists we would love to use mature American ginseng roots wild-harvested from the Appalachian foothills because of their exceedingly high quality, we seek alternatives because to use those wild plants would be totally immoral and almost certainly illegal. Luckily, wild-simulated production of ginseng is helping to providing a viable, ecologically responsible, and high-quality alternative to wild-harvested American ginseng.

So now the question becomes, how loosely can we interpret and apply the principle of "wild simulated"? What happens if we were to establish wild-simulated American ginseng production in eastern Asia in an attempt to decrease the amount of wild-harvested American ginseng exported to Asian markets? The environmental similarities between forests in eastern North America and those in eastern Asia are remarkable, so remarkable in fact that botanists and plant geographers have dubbed this phenomenon

**Figure 13-5.** American ginseng (*Panax quinquefolius*) being sold in Asian markets. Photograph © 2014 Steven Foster

the "disjunct eastern North American–eastern Asian range." "The similarities of the forests of Japan, central China, and the southern Appalachians in appearance as well as in ecological associations are in many instances so great that a sense of déjà vu is experienced by botanists from one of the regions visiting the other."[1]

## American versus Asian Ginseng

In spite of the phytogeographical similarities between these two regions on opposite sides of the earth, the difference in the medicinal qualities of the plants found within these similar ecosystems is often very distinct. While from a morphological viewpoint *Panax quinquefolius* (American ginseng) and *Panax ginseng* (Asian ginseng) are relatively indistinguishable, from a medicinal and energetic perspective the differences between them are remarkable. They share similarities, in that they are both adaptogenic tonic herbs useful for helping us adapt to stress and counter fatigue, but the way they work is very different. Whereas American ginseng is considered a cooling, internally moisturizing tonic herb useful in strengthening qi (vital essence) and nourishing the yin or feminine aspects of our constitutions, Asian ginseng is constitutionally more warming and drying and is indicated for yang-deficient aspects.

Chemically, although they both share the same medicinally active compounds known as ginsenosides, the concentrations of several distinct types of ginsenosides found within both species vary enough as to provide the two species with their own very unique actions. These two morphologically almost identical yet chemically rather unique species provide us with an excellent example of the influence that geographical and environmental factors can have on the medicinal activity of botanicals. While we could certainly grow American ginseng in Asia and Asian ginseng in North America, we would find that they wouldn't bear the signature qualities that make each of them unique to their home ranges not only because of their slight genetic differences but also due, in part, to complex variables in their native habitats that play a significant role in influencing their chemistry, as subtle as those variables in habitat may seem to be.

**Figure 13-6.** Child of maca farmers at Caracancha Farm, Junin, Peru, taking a break atop sun-drying maca roots.

Photograph courtesy of Ed Smith

At the beginning of this chapter we posed the question of whether we should attempt to commercially produce every single species of medicinal herbs found in the world that are used in commerce here in the United States, even if we could. According to Josef Brinckmann, who is arguably one of the world's leading authorities on the global medicinal herbal products trade, the answer is a resounding "no." Josef explains:

*The concept of provenance is understood and important to people who enjoy good wine. For me, it is not so different for medicinal plants. I prefer herbal products that contain "geo-authentic" botanicals. Personally, I prefer native American plants like echinacea from Canada or the United States but I prefer native Asian plants like northern schisandra (Schisandra chinensis) from where it grows naturally in northeastern China and/or neighboring Russian far east. I don't think that I would ever prefer maca (Lepidium meyenii) to be grown in the Rocky Mountains but would prefer it from the high altitude Peruvian farmers of the Junin Plateau of the Andes, where it has been traditionally grown by indigenous farmers since pre-Colombian times. Recently, European researchers tried to cultivate maca in the Czech Republic (field and greenhouse). Fortunately for the Peruvian maca farmers the experiment failed and they won't have a new competition for their new crop.[2]*

Brinckmann's mention of the Peruvian maca farmers provides us with another intriguing consideration for our discussion of whether we should attempt to grow all of the botanicals we use in this country, including many that are considered "exotic." Would it be fair for us to take a plant such as maca, whose cultivation in its native environment supports an entire culture of farmers who depend on its commercial export for their economic as well as perhaps cultural stability, and possibly "rob" them of that stability by growing "their" crop on "our" soils? As Brinckmann mentioned, luckily for the maca farmers, successful large scale commercial cultivation of this crop outside its native range seems unlikely, at least for the time being. But how many more "agri-cultures" are at risk of becoming destabilized or already have become destabilized because of our desire to support our own agricultural economy and herb demand?

# Social and Ecological Standards

Fair trade certification practices have improved economic opportunities for disadvantaged communities by providing them with greater access to world markets. Many businesses involved with the global importation of goods take pride in promoting their efforts at participating in this practice. Recently, two separate international organizations sharing common interests were formed to establish standards to help protect both the environmental aspects of native medicinal plant commerce and the cultures that rely on cultivating or wild harvesting these plants. The International Standard for Sustainable Wild Collection of Medicinal and Aromatic Plants (ISSC-MAP) is primarily an *ecological sustainability* standard with supporting elements of economic and social sustainability. FairWild Standard is primarily a *social* standard with supporting elements of ecological and economical sustainability. Josef Brinckmann, who is a founding member of the FairWild board, says:

*There are an estimated 50,000 to 70,000 MAP (medicinal and aromatic plants) species used in traditional systems of medicine but only about 3,000 occur in global trade. About two-thirds of these are wild collected and are likely to continue to be wild collected in the future for various agronomic and economic reasons. Most species that are easy to cultivate, about a thousand species, are already in cultivation, though many of these also continue to be wild-harvested if still available in their native habitats. For other species, cultivation would be costly in research and technical input, as well as land and agricultural inputs.*

*Some medicinal plants take several years to reach maturity, for example, many require over four years and some over ten years before a first harvest is possible. A farmer would need to conduct a crop rotation over a few decades before it would be possible to determine whether a sustainable system, which is feasible and profitable, is possible.[3]*

Brinckmann's point is well made, in that even if we were to attempt to produce all three thousand species of herbs used widely in commerce here in the United States, our efforts would likely be met with many challenges and failures because of the very specific environmental conditions and long growth periods required to produce many of these species. He goes on to explain that:

*Notable exceptions are medicinal plants that have been cultivated for a very long time and are no longer known to even occur in the wild, for example the Chinese medicinal plant ginger (Zingiber officinale) rhizome, which is now cultivated in tropical and sub-tropical areas throughout the world [Asian, South American, African and even North American (Jamaica and Hawaii) farmers have developed unique varieties of ginger over time]—or kava-kava (Piper methysticum) rhizome, no longer occurring in the wild, with scores of distinct cultivars that have been bred by different tribal groups in Fiji, Vanuatu, Samoa, Tonga, and Hawaii.*

*A good example of an important medicinal plant, whose global market demand would not likely ever be handled through cultivated sources, and would also not be a good candidate for cultivation experiments in the United States would be Ural licorice (Glycyrrhiza uralensis) root. Ural licorice root is among the highest volume and most widely traded medicinal plants in the world. The commercial supply of Ural licorice is collected from wild populations in sandy lands, dry riverbanks and grasslands on hills in just a few Asian countries, namely Afghanistan, Kazakhstan, Kyrgyzstan, Mongolia, Pakistan, Tajikistan, and parts of the People's Republic of China (PRC), particularly the PRC's northwestern frontier areas (e.g. of Inner Mongolia Autonomous Region, Ningxia Hui Autonomous Region, Qinghai Province, Xinjiang Uyghur Autonomous Region; and Gansu Province). Like many subterranean plant parts, medicinal quality of licorice comes from mature roots, rhizomes and stolons with a minimum age of six years. Even if an American farmer could mimic the optimal ecosystem conditions and environmental stressors in order for licorice root to develop a composition, quality and strength comparable to wild mature roots, they would need to set up a crop rotation extending out many years with the first return on investment occurring at the sixth year when the first harvest would be possible. Farmers generally can't afford to speculate so far into the future.[4]*

## Social Implications of Growing Botanicals Outside Their Native Range

Now that we have posed many questions concerning the commercial production of botanicals outside their native ranges, how do we, as commercial herb farmers, make wise choices regarding which species to grow and which species would best be left to grow in their native habitats by the cultures that have traditionally grown them? Should we focus our efforts solely on species that are either native to, or have naturalized in, our own bioregions, or should we attempt to grow some of the "exotic" species that are in high demand yet must be grown outside their native environment?

In our search for answers to this important question, let's begin by discussing the social implications of growing botanicals outside their native ranges. Take maca, for example. The Peruvian maca farmers are a

**Figure 13-7.** Ancient Inca agricultural terraces. Photograph courtesy of Jorge Láscar

good example of an agrarian society supported by the commercial cultivation of a native medicinal plant. In 1989 the United States National Research Council labeled maca one of the "lost crops of the Incas," and the plant was declared in danger of extinction.[5]

Shortly thereafter, in the 1990s, the purported aphrodisiac and fertility-enhancing effects of maca were discovered and widely promoted, and the "maca boom" took off. Commercial maca production began in earnest and not only provided economic and social stimuli to a unique Andean culture but also helped save an incredible medicinal plant from possible extinction. Successful commercial maca production outside its native range seems unlikely, but what if it had been discovered that maca could viably be commercially produced in other high-altitude areas, such as the Rocky Mountains of the United States, for example?

What would have happened to the Peruvian maca farmers who had established small individual plantations averaging approximately four hectares each (enough to support their families) if demand for export of their unique crop had plummeted because maca was being commercially produced elsewhere? Thankfully, their agrarian lifestyle appears to have long-term security based on the increasing demand for maca root, but their story is unfortunately the exception rather than the rule. Many other cultures in developing nations have experienced devastating social, economic, and environmental decline as a direct result of the global trade of medicinal and aromatic plants, and we as herbalists and consumers of herbal products are primarily responsible for these declines.

So what about small-scale commercial herb growers and wildcrafters here in the United States? Although we live a relatively safe and comfortable lifestyle in comparison to many of our counterparts in developing nations, as commercial producers in the midst of a economic recession, we face many of the same challenges that they face: increasing costs of production, cropland and native plant habitats being polluted or lost to development, competition from big business, political instability, climate change, and on and on. Is supporting domestic (U.S.) herb farmers—some of whom, like the Peruvian maca farmers, depend on agriculture for our economic as well as our cultural stability—a better choice than supporting foreign farmers by sourcing imported instead of domestic herbs?

The answers to these and many other questions concerning the social implications of the commercial production of botanicals outside their native ranges are extremely complex and contain countless variables. Although the likelihood that a small, diverse organic medicinal herb farm in the United States will produce a crop or crops that could negatively affect growers or wildcrafters in another country is fairly remote, we should still make well-informed and well-thought-out choices when considering which if any exotic crop species to cultivate commercially on our own farms—and the social impacts those choices may possibly have on others.

# Ecological Implications of Growing Crops Outside Native Ranges

From an ecological standpoint the choices we make regarding growing crops outside their native ranges can have a tremendous influence on our environment, both negative and positive. On the positive side we can help to supply the medicinal herb market with cultivated alternatives to species that are threatened with extinction from overharvest and habitat loss, even if these species are not native to our own bioregions. We can also help reduce global carbon emissions by providing a local alternative to botanicals that are commonly shipped from hundreds or thousands of miles away. We can provide certified-organic alternatives to botanicals produced using chemicals and other unsustainable agricultural practices, and we can increase the diversity on our farms and landscape by growing a tremendous variety of interesting plants that are useful to humans as well as to the wild inhabitants of our local ecosystems.

We also have to consider the negative implications that introducing and growing exotic species of plants can have on our environment. Nonnative plant species often behave differently when they grow in environments other than those they have naturally adapted to. A plant that may play an integral role in a diverse environment in one region can possibly dominate the landscape and force native plants out of another region. We have certainly seen the impacts this can have with species such as Japanese knotweed that have dramatically altered the landscape by being imported out of their home ranges and into new and vastly different ecosystems.

When asked about commercially growing exotic species on herb farms in the United States, Josef Brinckmann brings up some examples of possible negative ecological, social, and economic impacts this practice can have:

*If we aim to supplant the current market demand for sustainable managed wild plants with cultivated plants, introduced far away from the natural range, the incentives for local people to preserve and protect the world's last remaining*

**Figure 13-8.** Live plant sales at ZWHF often include nonnative species such as gotu kola, ashwagandha, rhodiola, and schisandra. Photograph courtesy of Bethany Bond.

*areas of biodiversity could be lost. There may be a higher risk of loss of biodiversity through development and encroachment if the local, rural and indigenous communities lose their source of income presently earned from export trade of wild medicinal plants and plant products. Other less sustainable economic uses for the meadows and forests could prevail in the absence of a market for the wild medicinal plants.*[6]

While Brinckmann brings up an excellent example of a possible negative economic impact that commercially growing exotic species of medicinals outside their native ranges can have on a global scale, what about the positive economic impacts this practice can have here on small organic herb farms in the United States? At ZWHF we grow and wildcraft over fifty species of medicinal herbs. Of those fifty or so species, only six species are considered native to our bioregion. Thirty-seven of these fifty plus species are imports that have naturalized here in the United States in and outside our bioregion, and seventeen of the species that we grow could be considered exotic, meaning they definitely aren't native and probably haven't yet naturalized here in the United States.

Many of these exotic species are widely cultivated in the United States and elsewhere outside their native ranges as medicinals, and some are even sold in nurseries and garden centers as ornamentals. Some of the best-selling species that we grow are exotics. Species such as ashwagandha, astragalus, codonopsis, garlic, ginkgo, lavender, licorice, red sage, rhodiola, schisandra, Siberian ginseng, spilanthes, and tulsi are highly sought after, some in the form of dried bulk herbs and others as live potted-plant specimens for people to grow in their own farms and gardens. If we were to forego growing these exotic species our sales volume and profitability would certainly decrease. Ashwagandha and tulsi in particular are two of our top ten species in terms of sales volume and would be hard to remove from our offerings. Sure, we would probably do okay and be able to survive as a business without these exotics, but they are an important economic facet of our diverse enterprise.

# Herb Quality: The Most Important Factor?

The quality and medicinal activity of the herbs that we grow and sell are also important factors for growers to consider when deciding whether to grow crops outside their native and preferred habitats. After all, this is why we grow medicinals in the first place, to produce high-quality products that help to increase and maintain the health and well-being of the customers who support our efforts through their purchases. In this book we discuss ways to grow healthy plants that produce potent medicinal activity and the importance of maintaining that medicinal activity through our postharvest processing and storage methods, but how do we quantify our efforts? How do we factor quality and potency into our decisions regarding whether to grow exotic species, and which if any of the thousands of exotic medicinal species can we choose from?

There are some obvious considerations, such as which species are adaptable enough to be grown outside their native and preferred habitats and will grow well in our own bioregions. Just because they are adaptable enough *to be able to grow* on our farms, this doesn't necessarily mean they *will* grow well and produce high-quality medicine. We have experimented with growing several species of exotic medicinals at ZWHF and have had great success with some and limited to no success with others.

Licorice (*Glycyrrhiza glabra*) is a good example of a species of plant we tried to field-grow during the first two years we were in business. Although the plants survived the winters and appeared to be able to grow here, we weren't seeing a lot of vigor and vitality in those plants in the field. Instead of removing licorice from our offerings altogether, we decided that we would only sell live potted nursery plants of this species and give other people the chance to grow it, either for medicine or simply as a specimen plant. This was a great option for us and one we have utilized with many other species of exotic plants that can be grown here but are not necessarily suitable for field production. Just because a species of plant may not be suitable for producing potent plant medicine by growing in our particular bioregions, that doesn't mean that they don't have commercial value.

Plant lovers love to grow interesting plants, especially those that are rare or hard to find in regular nurseries and garden centers. There are a few species of plants that we only sell as potted nursery plants (licorice, passionflower, gotu kola) because they are not viable for commercial production here in Vermont. Live nursery plants are a great option for growers to consider marketing as value-added products.

While licorice was obviously not suitable for growing and marketing as a bulk medicinal, there are several other species of exotics that we have had great success producing high-quality potent plant medicine from. We mentioned ashwagandha and tulsi as being economically important species that we grow and sell. The reason for this economic success is that we are able to produce high-quality, potent plant medicine by growing these plants in the field and marketing them as bulk medicinals as well as potted nursery plants.

How do we know that these plants produce potent medicine when commercially grown here in Vermont, thousands of miles away from their native subtropical habitat? We know this because when we started experimenting with these species during the first years we were in business, the first thing we noticed was how well they grew here. The plants exhibited excellent vitality and produced bountiful yields. As we mentioned previously, just because a plant grows well, grows fast, or grows large, that doesn't necessarily equate to potent medicinal activity, but it often does give us a pretty good indication that good things are likely to be happening on the molecular and energetic levels. In addition to observing the vigor and health of the plants, we made extracts with the dried material we produced after harvest and observed how the plant medicine felt in our bodies.

One can gain an incredible perspective as to the medicinal activities of plants by the frequent consumption of their extracts. Drinking tulsi tea and ashwagandha tea or tincture almost daily for a period of time and feeling their effects in our own bodies gave us a pretty good indication that those plants had produced some potent medicine. In addition to our own observations, we relied on the feedback from our customers. Their positive feedback gave us a less biased confirmation that we had made a good choice in deciding to produce these crops commercially. Our feelings about tulsi were confirmed when a wholesale buyer who purchases a large quantity of our crop tested it for essential oil content and other medicinal compounds and reported to us that the tests showed exceptional medicinal activity. Our biggest challenge at this point, after fifteen years of growing these two exotic species on our farm, is producing enough of them to keep up with the demand, and that is a challenge we can accept and attempt to overcome.

**Figure 13-9. Licorice (*Glycyrrhiza glabra*).** Photograph © 2014 Steven Foster

# Growing Naturalized Species of Botanicals

Growing exotic species of botanicals in Vermont thousands of miles away from their native ranges is a good example of greatly pushing some phytogeographical boundaries. Let's now shrink those boundaries a bit to discuss growing plants that are native to or have naturalized here on our own continent, yet are not found growing in the wild in our own specific bioregions.

Echinacea is a good example of a plant species native to the prairies of the United States. There are nine known species of echinacea, three of which (*E. angustifolia*, *E. pallida*, and *E. purpurea*) are primarily used medicinally. All three of the "medicinal echinaceas" can be grown where we farm and live in Vermont. They are hardy perennials that can survive our harsh winters by going dormant and emerging again in springtime. *Echinacea purpurea* is the most adaptable, easiest to grow, and most marketable species and is one of our top-selling herbs by volume. We attempted to grow *E. pallida* and *E. angustifolia* here, and although they survived and grew here, they didn't thrive and exhibit good vitality the way *E. purpurea* does, so we stopped growing those two species and have continued to grow *E. purpurea*.

The environmental conditions found on the prairies of the Midwest, where echinacea is native to, and the environmental conditions found here in the green mountains of Vermont are vastly different. The soils are different, the weather is different, the amount of solar radiation is different, yet in our experience as well as that of our customers, the *Echinacea purpurea* we grow commercially here produces high-quality, potent plant medicine. *Arnica montana*

**Tennessee purple coneflower**

*Echinacea tennesseensis*
Asteraceae

This is one of the rarest wildflowers in the U.S. Now known only from five natural populations within a 14-mile radius in central Tennessee, it was the second plant listed as endangered by the U.S. Fish and Wildlife Service in June 1979. One site was destroyed by the development of the Nashville Superspeedway, which opened in 2001.

**Figure 13-10.** *Echinacea tennesseensis* is an example of plants that only grow in specific bioregions. Only five natural populations exist in central Tennessee.

and *Arnica chamissonis* are two more examples where we have brought plants that are native to the alpine zone of the western mountains and have commercially grown them successfully here without any obvious compromise of their medicinal activity.

# Determining Efficacy or Medicinal Potency

Note that in that last sentence we used the phrase "obvious compromise." Full disclosure is warranted here, in that most of what we have stated in this chapter in terms of the quality of herbs we produce on our farm is based on qualitative analysis. This means that our family, friends, employees, and customers have reported their opinions to us about their experiences using our herbs. These opinions are merely based on organoleptic analysis. This means that these people think our herbs look, smell, taste, and feel good according to their senses, and they feel that these herbs can help to improve or maintain their health.

As previously mentioned, there is definitely some degree of bias involved with most of these analyses, especially from those who know us. We feel that these opinions, as unscientific as they may be, are as important as and possibly even more important than what the laboratory reveals when we have the specific medicinally active compounds analyzed and quantified. The fact that many prominent herbalists and herbal product manufacturers continue to purchase products from us year after year and express their pleasure in doing so irregardless of any laboratory analysis does help to substantiate these opinions. More importantly, it reinforces our feeling that we are choosing the right species to grow in our environment and are growing and processing them well regardless of where they are native.

In a hypothetical side-by-side comparison through scientific analysis, ashwagandha grown in India could very likely contain higher levels of withanoloides and other active medicinal compounds than

ashwagandha grown in Hyde Park, Vermont. How important is the degree of concentration of active compounds as demonstrated through quantitative scientific analysis? The answer to that question is highly dependent upon whom it is asked to and would likely vary according to the species of plant in question, the mode of its administration, and the therapeutic indication for its use. Another important point to consider is the fact that, although we know that plants contain bioactive compounds that are capable of promoting health and well-being in humans as well as in other living organisms, we rarely know exactly how these mechanisms work and what role individual compounds play in determining quality and action.

Ultimately for the commercial herb farmer, it is up to us to make well-informed decisions regarding which species, native or nonnative, we choose to grow and sell, and it is up to our customers to make well-informed decisions regarding who and where to purchase their herbs from. When given the choice between purchasing a pound of certified-organic, geo-authentic ashwagandha root from a large corporate herb company that purchased that root from an (unknown to the consumer) herb farm somewhere thousands of miles away or purchasing a pound of certified-organic, non-geo-authentic ashwagandha root from a small, family-owned herb farm within fifty miles of their home, the consumer would be wise to consider several factors, in addition to the concentration of bioactive compounds in that pound of ashwagandha root. These factors include the following: What types of soil conservation and labor practices were practiced on those two "competing" farms? How and when was that ashwagandha root harvested, dried, processed, packaged, and stored? Were the people that provided the labor to produce that produce treated and paid fairly? Who directly profits from the sale of that product? What is the carbon footprint of that product, and how well does the product make that consumer feel when using it? It's good to know that we all have choices.

# CHAPTER 14

# Postharvest Processing

We have been teaching classes to and consulting for aspiring medicinal herb growers for years, and we often begin our discussions by asking the audience or our clients some background questions to get a sense of where they are in terms of goals and limitations. When we ask them what they feel are the biggest challenges or limiting factors with starting their own farms their answers often have a very common thread. These potential farmers aren't usually concerned about the difficulties they may face growing and harvesting the plants; rather they usually anticipate that their biggest challenges will be drying, processing, and marketing what they grow.

Although this is a valid concern, we have good news! The drying and processing parts are not as difficult as most might think. Actually, when it comes down to a comparative analysis of time, energy, and risk involved with herb production, growing and harvesting the plants is generally much more costly, difficult, and risky than drying and processing them. Yes, marketing can certainly be challenging, and we will discuss that in chapter 17, but for now let's focus on how to process medicinal herbs after they are harvested.

Now that you have already performed the difficult tasks of prepping fields, planting, weeding, growing, and harvesting, it's time to prepare your plants for the final step in the process, which is to sell them. Before you can sell the herbs you need to make them ready for sale, and to do this there are a number of choices to consider. Will your herbs be sold fresh or dried? If they will be sold fresh how will you wash, pack, and ship them? If they will be dried, how and where will your drying take place? How will you process

the herbs after they are dried? Where will you store dried herbs after they are processed? How will they be packaged for sale? Let's examine some options.

## Fresh Herbs

There is a solid market for freshly harvested herbs because some herbal product manufacturers and retail customers prefer to make their herbal products from fresh plant material. Approximately 10 percent of the material we harvest at ZWHF is sold fresh. Fresh herb sales are great because this saves us the energy of having to dehydrate and process the plants. The main challenge with selling fresh herbs is

**Figure 14-1.** Processing Japanese knotweed postharvest.
Photograph courtesy of Bethany Bond

**Figure 14-2.** Fresh anise hyssop straight from the field.

that they need to be chilled, packaged, shipped, and received before their quality starts to deteriorate. As of the writing of this book, we are not mandated to wash fresh herbs before sale but with increasing GMP regulations, we anticipate the possibility of having to triple wash herbs much the same way vegetable producers are now having to do within a short period of time.

At ZWHF we require anyone wanting fresh herbs to preorder and prepay for these products ahead of time so we can harvest and ship the day they are harvested without potential delays in making the sale or tracking down shipping information. When customers preorder the fresh herbs early in the season we give them an estimated harvest date and tell them that we will call approximately forty-eight hours before harvest to let them know they are coming. This notification call gives them enough time to prepare for their fresh herb arrival and to avoid having the herbs sit in the post office or on their doorstep rotting. There are four primary plant parts that we harvest and offer for fresh sale: leaves, blossoms, roots, and barks. Each one of these plant parts gets a slightly different treatment before it is packed and shipped.

## Leaves

Leaves that are harvested for fresh sale must be cooled and packed, before we ship them. First we harvest the plants at the peak of potency and bring them into the shade to cool on drying racks. From there they are placed in paper grocery bags or wrapped in paper, taped shut, properly labeled, and put into their final packaging, which is either a cardboard box or, for small retail orders, a padded "bubble mailer." We never use plastic to package fresh herbs because it can promote bacterial growth by inhibiting the plant parts from "breathing" in transit. The packages are then placed into a refrigerator or cooler to chill until they are picked up by the shipping company.

We give customers two options when it comes to shipping. If they choose overnight shipping, we guarantee that the herbs will arrive in good condition; if they don't we provide a full refund or replacement. If the customer chooses two-day or longer shipping, which is less expensive, they assume the risk of the herbs' possibly spoiling, and we don't offer a guarantee. Most choose the overnight method, but it can be expensive, and often the shipping costs much more than the cost of the herbs in the package.

Packing fresh herbs requires a thoughtful approach to ensure that you retain as much quality as possible. We have experimented with using ice packs to preserve freshness in shipping and have found mixed success. The challenge with ice packs is that they can add significant weight to the package, thereby increasing the shipping costs. They can also melt and leak in transit, which is why we only use water-filled recycled plastic water bottles double-bagged in ziplock plastic bags and then wrapped with paper. Nontoxic gel packs are okay as long as they are sealed and wrapped in paper to prevent contaminating herbs if they burst.

With large boxes of herbs figuring out how to distribute ice packs evenly and efficiently can be challenging. The best approach with placing large volumes of herbs in large boxes is to separate the mass into equal parts and place each of these portions into its own paper bag with its own ice pack. We use the smallest recycled plastic water bottles we can find for ice packs. Customers decide whether or not they want to pay for the added shipping expense of the ice packs.

## Blossoms

Blossoms sold fresh, such as calendula, Saint John's wort, red clover, arnica, and milky oat heads, are harvested at the peak of potency as the flowers open. Since these flowers are only open for a day at the most, they are not prone to bacterial contamination from prolonged exposure in the field and therefore are not washed. Instead, they are harvested, bagged in paper grocery bags, labeled, packaged, refrigerated, and shipped. As with shipping leaves, customers may choose overnight shipping and/or ice packs with our quality guarantee or two-day-or-more shipping at their own risk. Blossoms are very susceptible to spoilage in shipping if it takes longer than one day for the package to arrive, so we highly recommend that our customers choose overnight shipping.

## Roots

Roots are dug primarily during their dormancy in either early spring or late fall. After digging the roots by hand with spading forks or with a tractor-mounted root digger, we shake any excess soil off the roots, place them in bulb crates, and bring them to the washroom. We then use pruning shears or knives to cut any stems growing out of the roots. For larger multistemmed roots we chop them into quarters or smaller using field knives. We leave smaller taproots such as dandelion whole. For some of the larger, denser roots we use rubber mallets with the field knives to assist with chopping them. The gnarly, multiple stems of these large roots can become packed with soil and small stones, making them difficult to wash if we don't quarter them.

After quartering the roots we have two options for washing them. If it is a small harvest of roots—less than fifty pounds fresh weight—we will often wash them by placing them on stainless steel mesh screens and spraying them with a high-pressure hose. If this is a large harvest of roots—over fifty pounds or so—we will use the barrel washer. Barrel washers are great for washing large volumes of roots efficiently and thoroughly. Another option that we used during the early years of our business was an old-fashioned Maytag wringer/washer with the wringer arm removed. These

**Figure 14-3.** Fresh elderberries being processed.

### Guidelines for Shipping Fresh Herbs

- Have clear communication with customer as to arrival date/perishability.
- Cool and/or chill herbs before packing to remove field heat.
- Use paper packaging—no plastic.
- Add proper labeling, including species, lot number, certified-organic, Certificate of Analysis (COA) when necessary, etc. (COA are discussed in chapter 17)
- Use ice packs when requested.
- Seal boxes well to prevent debris from entering box.
- E-mail tracking number to customer.
- Ask customer to report any damage or spoilage within twelve hours of receipt of package.

open-basin agitator-type washers work great for small to medium-size operations. Old wringer/washers aren't difficult to find in yard sales, flea markets, and old barns and houses. After the roots are washed we air-dry them on stainless steel mesh drying racks for an hour or two, then bag them in new paper grocery bags and label, package, chill, and ship them. We offer to mill fresh roots for an additional fee depending on the order size, but most fresh roots are

**Figure 14-4.** Root washing station. Photograph courtesy of Bethany Bond

shipped either whole or quartered. Roots have a lower perishability factor than leaves or blossoms, so we generally ship them to arrive in two to three days.

### Barks

Barks are the easiest plant part to ship fresh because all we need to do after harvesting them is cut them into small enough pieces to fit into paper bags. We do give people the option of having them milled finer for a small additional fee. After we package and label the barks they are ready for shipping. Barks are very nonperishable, so we send these the least expensive way possible.

# Drying Herbs

As small to medium-size farms strive to gain a foothold with customers in the medicinal herb market they

must rely on product quality to differentiate themselves from low-priced, imported, mass-produced herbs. Some of these herbs are field-dried out in the open with poorly controlled conditions, much the same way we see hay being dried on our local farm fields. Others are processed in extreme conditions to increase the speed and volume of throughput, often at the expense of quality. To set ourselves apart from (or ahead of) high-volume, low-quality producers and suppliers, we need to demonstrate compelling reasons that our customers should pay more for our herbs than they would for lower-priced, mass-produced herbs.

Here is an opportunity for herb farmers to let the product sell itself. It is difficult to describe with words how fantastic our dried herbs are in a way that can match the complete organoleptic experience. Organoleptic refers to using one's organs or senses

**Figure 14-5.** Drying freshly harvested herbs for storage and sale.

to analyze something. Sight, smell, taste, sound, and touch all can have an enormously persuasive effect on our customers' opinion of the quality of our products. The tactile experience of handling freshly dried herbs that look and smell almost identical to their former living selves can be a very convincing marketing tool.

To obtain the quality that sets our dried herbs above the rest, we need to dehydrate them in a way that attempts to retain everything that was in the fresh, live plant while it was growing in the field, with the exception of one thing . . . the water. Our teacher and mother, Rosemary Gladstar, taught us at an early age how important it is that the dried herbs we use to make teas and herbal products look, taste, and smell as close as possible to the living versions of these plants before they were harvested, and we take this recommendation to heart.

# The Basics of Drying

This technical section was contributed by Alexander Otto.

Dehydration is one of the oldest methods of preserving food and other organic perishables. By removing moisture from plants, then storing them in airtight packaging, we deprive bacteria, yeasts, and mold of the ability to grow and multiply, as well as slowing or stopping internal enzyme-assisted degradation and browning.[1] Degradation by these internal mechanisms is a vast subject, the details of which are beyond the scope of what we can cover here. Storage will be discussed further at the end of this chapter.

In the case of dried-leaf product forms produced by small to medium-size growers, drying also prepares the plants for the subsequent postharvest processing step that separates the leaves from the stems and converts them to the required size range, known as the industry standard "cut and sift" grade, typically a quarter inch or less, with as little dust formation as possible.

## Goals of Drying

Herbal dehydration by small to medium-size growers is motivated by five factors:

1. Removing enough moisture to meet the levels required for each specific plant product to ensure long-term storage without spoilage
2. Retaining the highest amount of bioactive compounds possible.
3. Reducing the weight of the plants, making them easier to handle
4. Preparing the plants for straightforward, economical, and high-quality postdrying separation of the product components, typically leaves from stems
5. Meeting the above goals with a reliable, affordable approach that produces a high-quality product

Achieving these goals requires consideration of physical, chemical, weather, and plant-specific factors. If methods and equipment are not adequately established, then the process can damage even the highest-quality fresh plants and be so costly as to undermine the profitability and viability of the operation. Setting up a customized approach for the specific requirements of each grower can significantly benefit from affordable expertise that is currently developing in the United States and that is becoming increasingly available.

## Choosing from the Various Approaches to Drying

Without quality, efficiency, and economic constraints, removing moisture from freshly harvested plants is simple and can be accomplished by many methods. Vast quantities of hay, firewood, and other dried agricultural products are produced by field-assisted weather combinations of relatively dry air, wind, and sunshine. However, in our case the main goal of dehydrating herbs is to remove moisture from the plants while retaining as much of the bioactive medicinal compounds as possible in a high-quality

dried product form suitable for long-term storage and human consumption.

Reliably and economically producing high-quality dried herb products requires a well-designed and well-executed approach. Many methods have been used to dry herbs, ranging from simply hanging plant bundles on rafters with natural ventilation to quite complex and expensive (but well controlled) continuous belt-feed systems with high-output heaters, which are best suited to large-scale operations. In other cases freeze-drying techniques have been utilized, but they require elaborate and expensive equipment.

On the organic medicinal herb farm we have to choose the most affordable and efficient methods we can come up with for our dehydration systems and scale of operation. This may mean retrofitting or putting additions on old buildings, or constructing new structures, while utilizing the best and most affordable equipment available. We also have to identify the appropriate drying conditions for each type of plant and adapt our approach to changing weather and plant conditions.

## Hot-Air Batch Drying

In this section our focus will be on a widely employed set of methods for drying herbs known as "hot-air batch drying" with the main steps illustrated in figure 14-6. Hot-air batch dryers are recommended for small to medium-size herb farms because they dry an entire "batch" of herbs at one time when they are harvested at peak potency—versus systems that are much more expensive to set up, require very high volume, and are more complex. Before getting into drying system design and operation, it is important to become familiar with our drying-related goals, plant characteristics vital to drying, and essential components of the drying process. Since we have invested a great deal of effort and expense into growing high-quality plants, we need to ensure we can convert these plants into high-quality value-added product forms.

The following sections describe the role of factors that are key to successfully setting up and drying medicinal herb plants, as well as offering vital guidelines. However, the specific approach for establishing a highly efficient, high-quality drying setup may very likely require additional technical analysis. Expertise is becoming increasingly affordable and available on a consulting basis to help achieve this setup relatively quickly and effectively without the grower having to spend a great deal of time and effort figuring out and applying all the technical pieces.

## How Dry Must the Finished Product Be?

Herb plant components that are going to be stored and sold need to be sufficiently dry to ensure suitability for long-term storage without degradation. The moisture content (MC) of a plant may be expressed either relative to a "wet" mass basis composed of the initial water mass plus remaining "dry" mass of material in the freshly harvested plant ($MC_{wb}$), or relative to just its dry mass basis ($MC_{db}$). Since our plants are freshly harvested, then dried to remove some of the moisture, it is usually easier for us to use $MC_{wb}$ for process and quality control, whereas scientific studies often express results in terms of dry basis. This is not a problem since dry-basis moisture content data can be readily converted to wet-basis data by the expressions below.

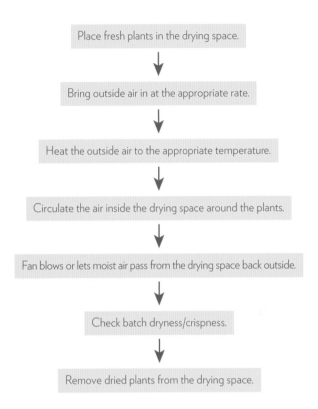

**Figure 14-6.** Steps in the hot-air batch-drying approach. Some steps are optional depending on weather conditions.

**Table 14-1.** Final Moisture Content for Common Medicinal and Herbal Plant Components

| Species | Common Name | Plant Component | % MC_fwb |
|---|---|---|---|
| *Arnica montana* L. | Arnica, wolf's bane | Flower | 10 |
| *Calendula officinalis* L. | Calendula, pot marigold | Flower | 12 |
| *Chamomilla recutita* L. | German chamomile | Flower | 12 |
| *Coriandrum sativum* L. | Coriander | Seed | 10 |
| *Foeniculum vulgare* Mill. | Fennel | Seed | 8 |
| *Hypericum perforatum* | Saint John's wort | Plant | 10 |
| *Levisticum officinale* Koch | Lovage | Leaves | 12 |
| *Malva silvestris* L. | Common mallow | Leaves and flowers | 12 |
| *Melissa officinalis* L. | Lemon balm | Leaves | 10 |
| *Mentha piperita* L. | Peppermint | Leaves | 11 |
| *Plantago lanceolata* L. | Plantain | Plant | 10 |
| *Verbascum phlomoides* L. | Mullein | Plant | 12 |

a) $\%MC_{wb} = 100MC_{db}/(MC_{db}+1)$,

    if $MC_{db}$ is just plant (water mass)/(dry mass) and,

b) $\%MC_{wb} = MC_{db}/(MC_{db}+100)$

    if $MC_{db}$ is expressed also in percentages.

Final moisture content ($MC_{fwb}$) guidelines are available for some, but far from all, significantly produced herb species.[2] For example, threshold $MC_{fwb}$ levels in table 14-1 are published in several reference documents. Based on these levels, as a general rule a 10 percent or less final wet-basis moisture content is acceptable for aboveground plant components as well as roots. Seeds typically require 8 percent or less MC.

As will be described later, the last little bit of water removal by drying takes the most energy and time, so processing to 12 percent compared to 10 percent $MC_{fwb}$ would be very helpful. However, presently, affordable tools are unavailable to measure that content accurately.

## How Do We Determine Moisture Content during or after Drying?

Unfortunately, tools for inexpensively, easily, and accurately measuring moisture in our plants during or after drying are currently unavailable. Suitable tools are, however, presently under development, and within a few years they will likely be available for use in our operations. For now we must rely on indirect assessments of dryness combined with setup of a system that, when working as intended, reliably dries our specific plant components to at or below the 10 percent wet-mass-based target moisture level.

Several tried-and-true qualitative measures of dryness are very useful in the meantime to monitor drying of our plants and guide us toward high-quality output. Among these, tactile evaluation of the change in leaf compliance from flexible to brittle and crispy stands out as the most direct and common method.

In some form or another this approach is easily administered because it can be rapidly applied without any equipment. It consists of bending and rubbing a selected sample of the target plant component between the fingers or hands, to assess if the component is limp and flexible or crispy and brittle,

indicated by a rustling sound, as well as its breaking apart into pieces. It can also be applied as the plants are being processed in their drying environment. As with all tests, *representative sampling* is vital to get an accurate and comprehensive assessment of dryness everywhere in the plant batch. For example, testing at several positions through the thickness of the plant bundles as well as testing bundles at different center and edge locations in the entire drying batch will provide a clearer picture of plant dryness at the extremes of slower and faster drying regions. This helps ensure that no unacceptably wet regions remain when the decision is made to pull the batch from the dryer for further processing.

A second more accurate and quantitative weight loss method requires some basic equipment; for example, a scale that can measure up to about five pounds with at least a one-ounce accuracy and perforated baskets or trays of known mass for use in holding the plant samples to be weighed. Immediately after harvest several samples from the harvest are placed in these trays to a similar depth as in the drying operation and weighed for initial wet mass. The samples in the trays or baskets are then included in the drying operation at strategic locations in the drying area while still in the trays. Weighing these as drying progresses provides direct data on moisture loss that can be compared to expected loss from prior runs, or from published data on moisture content for that specific plant type.

If weight loss testing is supplemented by the tactile, flexible-to-crispy transition assessment, then the grower can determine with a few batches weight-based moisture content that corresponds to the onset of brittleness for their plants. At this point the instances of weighing can be greatly reduced, with monitoring relying much more on the rapid tactile method.

But what are the actual moisture contents of leaves of different plant types when they become crispy? As a general rule, at 10 percent or less moisture, leaves are quite crispy and easily crumbled. Flexible-to-crispy transition occurs at different moisture contents for different plants—and this data is unavailable for many herb plants. It is, however, likely to be more available in the near future, since this area is addressed by at least one small company that is currently developing affordable small-to-medium-scale automated methods for removing the dried leaves from the stems at just the right time to produce the highest quality products.

This information is vital because, after drying, the process relies on crispiness to separate leaves from stems and to break up the leaves into standard product sizes. It is also important because for lower cost and higher product quality drying the plant component at or just below the acceptable 10 percent moisture content is best, as further drying can potentially remove more of the vital chemical components that provide it with value, as well as costing more in terms of energy and tying up your drying facilities during critical harvest times.

## How Much Moisture Do Plant Components Contain When They Are Harvested?

Water is a growing plant's primary constituent. $MC_{wb}$ is almost universally between 50 and 90 percent, and with most leaves and flowers it is in the 75 to 90 percent range, as illustrated in table 14-2. The remaining 10 to 25 percent of the mass provides the structure we see and the chemical components we value. Specific fresh-plant moisture contents are published for some common herb plants. With

ongoing literature analysis and additional tests this important information will gradually become much more readily available, although, as shown in table 14-2, it is made much more complex by variability of growing condition and plant attributes.

## Drying Mechanism End Point

In any drying environment a harvested plant will lose moisture by evaporation, with the water leaving by desorption through the plant surfaces, until eventually it reaches its equilibrium moisture content (EMC) in that environment. Our task in drying a plant component is to put it into a drying environment in which its EMC is at or below our acceptable moisture content (AMC) target 10 percent level, thereby providing the driving force for water evaporation to reach 10 percent.

By drying in our facilities we can get close to EMC. But we would have to wait a very long time for full convergence of our plant component onto EMC. Instead we can get to a steady state moisture content (SMC) that is close to EMC and below AMC, and where further moisture loss is substantially stopped at our drying condition.

## Relative Humidity

Relative humidity (RH in percent) describes the gaseous water content of air relative to saturation at

**Table 14-2.** Variable Water Content of Fresh Herb Plants[3]

| Species | Common Name | Component | Condition | Fresh %MC$_{wb}$ |
|---------|-------------|-----------|-----------|------------------|
| *Urtica dioica* L. | Nettles | Leaves | NA | 81 |
| *Echinacea purpurea* | Echinacea | Leaves | 2-year growth | 75 |
| | | Flowers | 2-year growth | 80 |
| | | Stems | 2-year growth | 72 |
| | | Leaves | 5-year growth | 79 |
| | | Flowers | 5-year growth | 71 |
| | | Stems | 5-year growth | 62 |
| *Calendula officinalis* | Calendula | Flowers | Just opened | 86 |
| *Trifolium pratense* | Red clover | Flowers | First ones | 82 |

RH equal to 100 percent with respect to pure liquid water. The commonly used chemical term, "activity of water," describes the driving force for it to go from one state, and in our case also one place, to another, thereby decreasing its activity relative to its pure liquid state, where its activity is defined as the number one. Since water is present in fresh plants as a relatively dilute liquid, its activity there is only slightly less than one. However, in air, gaseous water's activity can go almost full range from zero to one because its activity there is given by RH as a fraction, where $a_w$ = RH/100.

At RH equal to 100 percent, then air's $a_w$ = 1, and it is in equilibrium with pure liquid water. Therefore, this air cannot take in any more gaseous water, and plants cannot be dried in it by water evaporation. At lower RH levels the gaseous water's activity is much less than the liquid water's activity in the plant, driving evaporation and drying. However, as drying proceeds, water activity in the plant is decreased by decreasing purity and an increasing fraction of more chemically bound water, slowing drying until it halts when the activity of the water in the air and plant become equal. In this case the plant has approached its EMC, and conditions toward the end of the drying step must be set up to ensure that EMC is at or below the target 10 percent AMC. However, in earlier stages of drying, a higher RH in the drying environment will work quite effectively.

Relative humidity also provides a measure of the amount of water that can evaporate into that air. For example, at RH equal to 10 percent, the air is very dry, only containing 10 percent of the water it could contain if it were saturated at RH equal to 100 percent. Plants can only air-dry until the air in close proximity to them increases in RH close to equilibrium with the plant water. To avoid slower drying from this kind of proximal saturation, the moisture-enriched air around the plants must be removed. It is most easily swept away by fan-induced forced air convection or just natural air convection and replaced with drier, lower RH air, as described in more detail later.

## Heating the Air and Plants

The evaporation of water also requires energy. This energy can come from many sources, including (1) relatively dry air around the plants, with zero fuel cost but also with low energy density and drying rate, (2) added energy from inexpensive but only occasionally available sunlight, and (3) lots of added energy from reliable but expensive burning of fuels such as wood, gas, or oil. We will almost always need to get the energy for drying from a combination of these sources, with as much as possible coming from the (relatively) free dry air provided by optimal weather conditions and sunlight. If all the energy of drying comes from combustion, however, a quarter to a half gallon of fuel oil or four to eight pounds of dried firewood on average are required to produce one pound of acceptably dried plant product (assuming an initial benchmark $MC_{wb}$ of 80 percent).

## The Role of Temperature

As air's temperature increases, its water capacity increases. If saturated air (RH 100 percent) at 72°F is heated to 105°F, its RH drops to about 30 percent, even though its water content is unchanged. Further, the water capacity of air at 105°F and RH of 32 percent is three times its water capacity at 72°F with the same RH. For drying speed everything improves as temperature is increased, and large-scale operations may apply much higher temperatures to reduce their drying times and investment in equipment. However, key chemical constituents in the herbs, such as vitamins and essential oils, can break down or evaporate if temperatures increase too much.

Chemical, physical, and visual state degradation with increasing temperature and other conditions is unique to each plant and plant component and involves analysis of many quantitative and qualitative factors that put it beyond the scope of what we can cover here. However, as a general rule there is very little risk with any plant up to about 100°F[4] and some risk with some plants in the 105 to 110°F range, and above 110°F some plants will still retain their quality, while others will lose quality very quickly, often in ways that are not apparent to the senses but that do alter internal chemistry in a pro-foundly negative way. Here is a brief summary of higher temperature drying trade-offs:

- Drying time is greatly reduced because of (a) faster water evaporation kinetics and (b) lower RH levels from heating outside air with fixed moisture content to a higher temperature.
- Quality can be degraded above critical tempera-tures by (a) evaporation of essential oils and other key chemicals, (b) molecular breakdown of vital temperature-sensitive chemicals, and (c) changes in color, appearance, form, and cohesion.
- The cost of increased energy and fuel used to heat air is offset by the need for much less air because of the rapid increase in air's moisture capacity with increasing temperature. This is

evident in figure 14-7, showing the amount of air required to dry one pound of typical fresh plant at different RH levels and temperatures.

## How Much Air is Required for Drying?

As seen in figure 14-7, the volume of air needed to dry one pound is relatively large. Although still large, the volume decreases rapidly as air tempera-ture increases and also as RH decreases. The data in figure 14-7 is also used to estimate the in/out fan capacity required for drying in a reasonable time in a structure of a certain size at the process RH levels.

To put this in context, if outside air is heated to 100°F, where it has an RH equal to 50 percent, a ten-foot by ten-foot drying shed can only dry two pounds of fresh plant before its air becomes loaded with water. But about 250 times as much plant

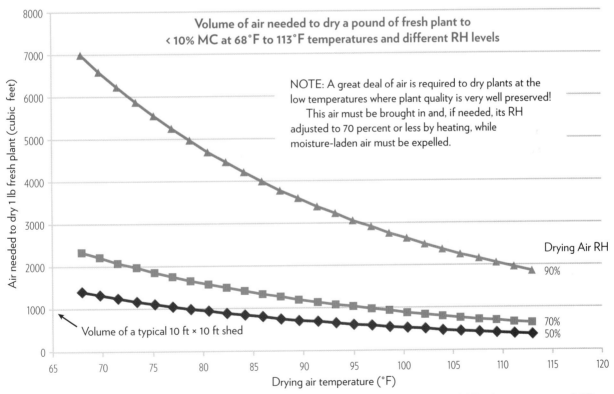

**Figure 14-7.** Illustration of the volume of air required to dry one pound of plant to 10 percent MC$_{wb}$ from 80 percent MC$_{wb}$ for different RH levels and drying air temperatures. Air at the low temperatures employed to make the highest quality product cannot hold much water. A five-gallon bucket of water-vapor-loaded air contains the equivalent of about half a teaspoonful of liquid water. Image courtesy of Alex Otto

(about five hundred pounds in this case) can be put into such a shed in one batch for drying. As a result, the air in the shed must be replaced with outside air between 250 and 1,000 times to dry the plants at or below a 10 percent AMC level.

## Equilibrium Moisture Content

Scientists experimentally determine gaseous water adsorption (water pickup) and desorption (drying) into and out of plant components. Their data is typically presented as sorption isotherms (graphs) describing the steady state moisture content of the plant at constant temperature and increasing RH level. To obtain these data a plant sample of known starting state is set inside a chamber with RH and temperature fixed at the selected test level. Its weight gain or loss is then monitored over time until it ceases to change. Repeating the experiment at different RH levels and temperatures with different plant starting-moisture levels identifies each plant's unique steady state moisture content at one temperature and a range of RH levels, and from these the plant's EMC levels can be estimated.

Although theoretically the steady state moisture contents found by desorption and adsorption should be equal, mechanisms within most plants prevent full convergence. For herb drying processes, however, these data, where available, provide very useful guidance on the connections among temperature, RH, and EMC for each plant type, allowing practical selection of drying conditions, where the EMC of the plant component is below the target 10 percent AMC level.

An example of isotherm data in figure 14-8 illustrates the trends for MC: (A) is data for adsorption and desorption at 77°F (25°C), while (B) illustrates that EMC decreases at fixed RH as temperature is increased. This again increases the appeal of higher-temperature drying; however, as noted previously, higher temperatures can negatively affect product quality.

## Using Desorption (Drying) Data

Desorption data can be very usefully combined with data on the decrease in RH as air is heated up to determine temperatures to which outside air with a certain RH and temperature need to be heated to dry a specific plant type to the desired 10 percent AMC. As an example, figure 14-9 illustrates how the conditions for drying lemon balm and peppermint leaves with heated air on a dry (50 percent RH) and rainy (100 percent RH) day with outside air at 70°F can be estimated from RH, versus temperature data for air and the sorption data for these plant leaves. The intersection points of the heated, initially 100 percent RH and 50 percent RH air data with the plant 10 percent EMC RH versus temperature data is the lowest temperature where, in principle, drying to 10 percent MC is possible. In practice, higher temperatures are needed to get to 10 percent in reasonable time, with manageable air inflow and air heat-up. The following illustrate some very important drying-related points.

- Each plant has a unique EMC response. In this case the RH of lemon balm at 10 percent EMC is almost twice that of peppermint, making peppermint a more challenging and energy-intensive plant to dry to an $MC_{wb}$ of approximately 10 percent.
- Lemon balm can be dried to 10 percent MC even on a rainy 70°F day by heating it to the high-quality-producing 90 to 100°F range, while peppermint can only be fully dried in such conditions by heating it well above 105°F.
- This kind of analysis can produce very useful guidelines for each grower's specific plant types, quality retention goals, drying setup, and climate conditions, provided sorption data is available.

## Determining the Steady State Moisture End Point

The method is not as difficult as it may seem. The previously described weight loss tracking method to monitor extent of drying of your plant batches is based on techniques used to determine sorption and EMC in the laboratory. Its application to farm conditions for obtaining initial and final moisture content at your specific conditions for your crops is described below.

If your product is dried leaf, for example, you will need three sample drying trays, loaded with whole

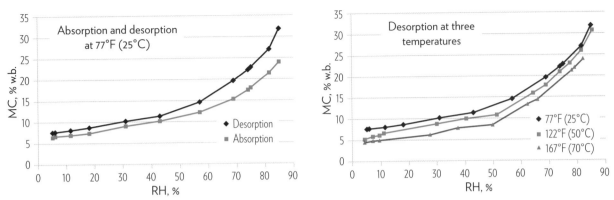

**Figure 14-8.** Sorption isotherms of *Artemisia dracunculus* (tarragon). (A) Data at 77°F (25°C) shows typical, increasing steady state moisture content as RH is increased, and the small but significant gap between desorption (drying) and adsorption (water pickup). The EMC of this plant is somewhere between these two curves at each RH level. (B) Desorption at 77°F (25°C), 122°F (50°C) and 167°F (70°C) shows the decrease in steady state moisture content as temperature is increased. Since we are concerned with drying, desorption isotherms are best suited to assist our drying setup and process design.[5] Image courtesy of Alex Otto

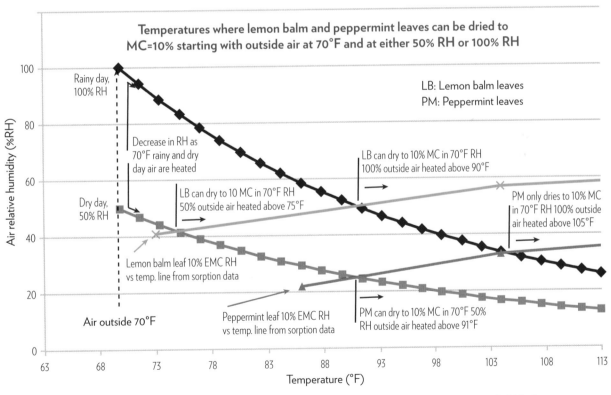

**Figure 14-9.** Example combining sorption data with a 10 percent AMC target and the decreases in air's RH when it is heated, to identify conditions where 10 percent AMC can be attained. As seen, these conditions can be very different for each plant type, with some plants presenting much greater challenges to full drying with humid or rainy outside conditions.

Image courtesy of Alex Otto

plants (leaves and stems), and will monitor their weights from the start of drying through a standard drying run by following these steps:

1. Load the drying space with a standard batch of plants as in a regular run. But take a representative selection (about six pounds) of fresh plants and divide into two samples, one about two pounds, the other about four pounds.

2. Load preweighed empty trays: For the four-pound batch cut off and separate all leaves from stems; put the leaves in one and the stems in the other empty tray to a similar depth as what you have in your full-scale drying racks. Put the remaining whole fresh plant sample onto the third tray.

3. Weigh all three trays containing the samples (use a scale with at least a one-ounce accuracy).

4. Place trays into the drying space containing the batch, preferably within half an hour of the time the full batch went in and less than two hours after harvest.

5. Dry according to your standard operation, and weigh the three trays initially after about every four hours' added time, as weight loss is likely to be most rapid initially. Also, record RH and temperature in your drying space, as well as outside, at each weighing. Place the trays back into their original drying site after each weighing. Later, weighing can be spaced out to every eight or twelve hours. Each weighing is also a good time to check the crispiness of your plants on the drying racks.

6. Record the gross and tray-subtracted weights, and subtract these weights from the initial plant weights. Divide the change in weight by the initial weight, and multiply by 100 to obtain percentages. To calculate weight loss rate between each weighing, take the percent of weight lost just in each time increment and divide by the time increment.

7. You will very likely be approaching steady state when suddenly the weight loss rate is less than about a tenth of what it was after the second four-hour weighing. The RH and temperature when weight loss has stopped is the steady state condition for weight loss.

8. Graphing the data as percentage weight lost versus cumulative drying time will help provide a good visual of what is happening. Plotting rate versus time will help clearly show where drying is fast approaching steady state (as rate is approaching zero).

9. To determine remaining and initial moisture content:

   a. *Remove almost all remaining moisture:* Place the three samples dried to the steady state condition in an oven at about 200°F for about five hours and weigh, while ensuring no fragments are lost. Thicker plant components such as some stems may need longer times.

   b. *Your freshly harvested moisture content $MC_{hwb}$:* Starting weight minus this final weight divided by the starting weight all multiplied by 100 estimates the $MC_{hwb}$ value for your crop components.

   c. *Your steady state moisture content $SMS_{wb}$:* The weight at steady state minus this final 200°F baked weight divided by the initial weight multiplied by 100 gives you $SMC_{wb}$ at the specific RH and temperature of the batch. As with other areas of drying, affordable consultants are available to help determine these values.

You only need to apply the above sequence a few times to get a good picture of how dry your plants get at the common conditions you can apply in your particular drying facility, and from that, what changes you may need to make to improve dryness without degrading quality.

## Using Fans in Combination with Heating

Fans are recommended for high-quality drying because they (1) enable controlled introduction of large quantities of drier, lower RH air from outside, and (2) they expel the moisture-laden air inside the drying space and offer much more effective forced-air circulation to sweep away moisture-laden air from next to and between the drying plants, compared to natural convection.

With a fan blowing air out, the incoming air can be readily passed through a region heated by, for example, gas, oil, wood, or sunshine—the direct heat, however, should not contact the surface of the drying plants. If temperature and RH of outside and inside air is monitored, fans and heaters can be adjusted quickly to ensure drying stays on track.

*Configuration*

The heat source may be placed in line with the incoming air to get it to temperature before it enters and mixes into the drying-space air, or in a simpler form it may be placed inside the drying space in the center, edge, or corner regions of the drying space. But placing the heat in the drying space is more likely to introduce hot spots and reduces control over the drying conditions, while making it more challenging to, for example, load wood into the woodstove because of the temperature in the drying shed. Either arrangement can work, as long as outside and inside conditions are monitored and inside conditions controlled (see the "Summary of Hot-Air Batch Drying Factors" sidebar). Affordable RH-plus-temperature meters are readily available and should be strategically placed inside and outside the drying space.

## Instruments for Your Drying Facility

At least two, and preferably three, RH/temperature meters are very useful, one outside and away from the drying building, and one or two inside the drying space, with one in the vicinity of the air heating/entry and the second far away and in a representative place between the rack stacks. The meter in the incoming heated airstream will confirm for you the extent of temperature rise above outside air, and the rack-stack meter can be used to adjust the air-in and air circulation rates. If RH in these stack zones exceeds about 75 percent, drying is slowed significantly.

## Estimating Fan-Blow Capacity

The factors that matter here are batch weight, volume of your drying space, and drying conditions; namely,

### Summary of Hot-Air Batch Drying Factors

- RH outside: If it is too high, heating is needed.
- Temperature outside: Heating incoming air speeds up drying, but avoid overheating.
- RH inside: If high, bring in lower-RH air, with heating needed if outside RH is also high.
- Temperature inside: It should not typically exceed 105 to 110°F for best quality.
- Airflow in/out: Fans and vents help maintain faster, higher-quality drying conditions.
- Heating of incoming air: With fuel or solar, this increases the temperature to speed up drying.
- Air circulation: This sweeps moisture-laden air away from plants to speed up drying.
- Batch load weight: Heavier load = more water to be removed = longer drying time.
- Drying-space volume: Larger volume dries plants faster but still requires much outside air.
- The type of plant being dried: This affects the required conditions to reach target dryness.
- Plant placement: This must be loose and not too thickly layered (less than five inches) so drying air passes between plants.

the inside RH and temperature that you can maintain for different outside conditions. As an example, the method described below can be used to estimate the maximum fan capacity you will likely need.

1. Estimate your drying space volume by multiplying its length, width, and height together.
2. Next estimate the maximum fresh plant load going into the drying space; say, eight hundred pounds.
3. From figure 14-7 you see that you may need to move about one thousand cubic feet of air per pound of plant weight.
4. In the best case you may target completing drying in approximately eighteen hours.
5. With a margin of two a fan with capacity as calculated here is needed: 2 × 800 × 1000 ÷ 65,000 = 25 cubic ft / second fan.

**Table 14-3.** Drying Guidelines for Specific RH and Temperature Conditions

| Condition—RH and temperature | Actions—when both conditions are true |
|---|---|
| Outside RH >90% and Outside temp >90°F | Very difficult to dry to EMC in highest quality regime.<br>Fresh batch MC can be reduced initially with only 10°F heat.<br>Reaching AMC = 10% requires at least 20°F heat and very large amounts of outside air and internal circulation; add more heat as drying progresses. |
| Outside RH 80–100% and Outside temp <90°F | Heat needed with lots of blown-in air and circulation.<br>Add heating to raise temp by at least 15°F, 25°F if outside temp <80°F. |
| Outside RH 50–80% and Outside temp 70 to 100°F | Fresh batch MC can be reduced initially just by blown-in and circulated air.<br>To complete drying:<br>Add heating, raise inside air temp by at least 15°F.<br>Blow in and circulate lots of air.<br>Ensure RH inside stays ~50% or less with heat and/or fan. |
| Outside RH < 50% and any outside temp | Successful drying possible without applying heat, but heat still helps.<br>Turn fans as high as possible, especially during initial stage.<br>Add some heating for later stages if outside temp <75°F. |

Note: Heat may be from solar or burning fuels such as wood, oil, or gas. Inside RH must be at <50% toward end of drying to reach AMC ~10%. Best quality results if drying temperature stays below about 105°F.

For flexibility, rugged variable-speed fans are the best, and one or more similarly sized fans can also be applied to provide internal forced-air convection. Because hot air rises, one or more convection fan should be positioned to circulate air from near the ceiling of your drying space around the inner surface of your drying space and between the racks.

*This concludes Alex Otto's technical contribution.*

# How We Dry Our Herbs at ZWHF

As a general guide, if RH outside is 50 percent or less and drying temperature is greater than 75°F in the drying space, the equilibrium fresh weight based on steady state moisture content of many herbs will be between 5 and 10 percent, and the leaves will be crispy. This is the drying environment goal for the long-term preservation of dehydrated botanicals. Drying herbs to lower than 5 percent moisture content is a waste of additional time and energy and can reduce the overall quality of the finished product.

So how do we apply all of that science to our practices in simple terms? One way we can think of drying herbs is to imagine the process in terms of "sponges and buckets." The "sponge" represents the fresh plant material in the drying shed, which starts out pretty saturated at an average of around 80 percent water by weight. The first "bucket" represents the air in the drying shed, which starts out relatively empty (ambient) but quickly becomes filled with water vapor as moisture exits the plants. All the air outside the drying shed can be thought of as being in many second buckets, with each bucket able to hold a certain amount of outside water vapor depending on its relative humidity and temperature. The number of outside buckets we have to bring in and "fill" with water vapor increases very quickly as the outside humidity increases, but that number decreases as we heat the air up from its outside temperature.

Our goal in drying herbs is to gradually wring out the water from the sponge (the herbs) into the drying-space buckets, continually bring full buckets outside (remove water vapor), and bring emptier buckets (dry air) inside. Challenges abound during this process, especially when we start wringing out the sponge so fast that the inside buckets fill up before we can bring them outside and bring in emptier buckets. The process becomes even more difficult if rainy or humid conditions keep the outside buckets so full that we can only add a small amount of additional water to each bucket once we bring it inside. Raising the temperature of the outside buckets of air as they enter the warmer drying shed increases the amount of water vapor they can hold, increasing drying speed as long as the inside buckets are moved outside when they become somewhat filled with water vapor.

With the proper combination of heat, airflow, and ventilation inside the drying shed and with relatively low humidity outside, we are able to dry a full batch of leaf crops in as little as twenty-four to thirty-six hours. Blossoms take approximately thirty-six to sixty hours because of their more dense nature, and roots take on average forty-eight to seventy-two hours. During rainy weather, and periods of high humidity, especially when large volumes of wet material are being dried, these durations can become prolonged, and we often have to supplement the process with a great deal of supplemental heat. This is why we make every attempt to dehydrate our herbs during dry weather. But since ideal weather can be elusive and with so many crops coming into season simultaneously, sometimes we just have to keep the process going or we can fall behind with our harvesting and drying regime.

When the weather cooperates we take advantage of natural convection, without supplemental heating, which dramatically decreases both the time involved with drying and our drying production costs. However, during cloudy, humid, and rainy weather we have to choose forced convection and heating to dry the herbs as efficiently and quickly as possible. Supplemental heat in our drying shed at ZWHF is supplied by two wood-burning furnaces, one located on each end and in opposite corners of the drying shed. Large commercial moveable ventilation fans distribute the heated air over and through the plants on drying racks, which assists with wicking moisture out of and away from the fresh plant material.

The best way to dehydrate herbs while retaining maximum quality is to start the drying process cooler and finish it hotter. As discussed in the previous section maximum temperatures that each plant type can tolerate vary. However, as a general guideline maintaining temperatures below 105°F is recommended for maximum quality retention. Heat is a key component of the drying process, but too much heat can be detrimental not only to the quality of the herbs but to the drying process itself. If we start drying freshly harvested plants at excessively high temperatures we risk what is known as "case hardening," a situation in which the outermost plant cells dry too rapidly, forming a physical barrier that can dramatically slow the evaporation of moisture remaining in the plant and thereby compromise herb quality. By slowly increasing drying temperatures during the early stages of the dehydration process we can ensure that moisture is exiting all parts of the plants at a relatively uniform rate.

During the process of filling the drying shed with freshly harvested herbs, we begin supplying natural ventilation by rolling the sides up and opening the doors, windows, and ventilation shutters. If the wind is calm we will often turn the ventilation fans on. This airflow starts the process of gradually drawing moisture out of the fresh plants. Depending on the relative humidity levels and weather conditions forecasted for the first eighteen to twenty-four hours or so of drying, we either close the drying shed if wet conditions (>50 percent RH) are forecast or leave it open if the humidity levels look to be relatively dry (<50 percent RH). The next morning we check the forecast again. If clear sunny weather is expected we close the drying shed to take advantage of solar radiation and set electronic controls that ventilate the drying shed if it reaches the maximum temperature or humidity setting determined by the type of crop being dried.

We use both a continuous on/off ventilation system and an automated one. Continuous ventilation is provided by a small "squirrel cage" fan mounted near the peak on the drying shed's gable-end wall. This fan can run constantly when the drying shed is closed up and serves to pull water vapor out from the drying shed, increasing the inside air's ability to absorb more moisture. If the RH of the outside air is higher than the drying shed interior RH, we can manually turn off this small fan. Automated ventilation is produced by louvered shutters on one end of the drying shed and a four-foot-diameter ventilation fan on the opposite end. This shutter and fan system is controlled in relay by a combination humidistat/thermostat. Not only does this controlled ventilation cool the house down if it reaches a maximum temperature setting, it also serves to ventilate water vapor if a maximum humidity setting is reached.

Our drying shed is covered with a woven nylon fabric shade cloth that is rated at 85 percent shade. This cloth protects the herbs from the damaging effects of ultraviolet radiation, yet it allows enough sunlight in to provide solar raising of the temperature inside the drying shed to approximately 10°F to 20°F above the outside temperature. This solar gain provides us a great advantage during sunny weather because we don't usually have to supplement with artificial heat. However, during inclement weather we still need to increase the temperature inside the drying shed to keep the dehydration process moving along efficiently. This is when we fire up either one or both of our wood-burning furnaces to assist with drying. During the second day of drying we have already removed most of the moisture from the fresh plants.

During day two or three of drying, depending on the crop type, we gradually raise the temperature inside the drying shed until we reach the maximum temperature for that particular crop. There are too many variables with crop types and drying conditions to give detailed specifications here, but one thing we can say for sure is that time and observations with your own drying system and the crops you are drying will help guide you in your decision making and allow you to maximize efficiencies within your system.

Designing and building the drying shed should be considered one of the most important and well-thought-out tasks you will perform when starting your own medicinal herb farm. A well-designed facility will be centrally located to receive fresh herbs for immediate dehydration. It will be large enough or expandable to be able to handle the processing requirements for both present and future acreage. It will be designed to operate in the most energy- and labor-efficient way possible, and it will be affordably built while not sacrificing quality or structural integrity. A wide variety of drying shed sizes and designs are used on medicinal herb farms, some elaborate and some simple. Some facilities are made by retrofitting or adding onto existing buildings, and some are built new. Like any other agricultural building, the drying shed needs to be designed and constructed to handle the weather and should ideally, in colder and wetter climates, have a curtain drain installed around it to divert melting snow and rainwater away from its foundation.

# Drying Shed Design and Construction

Since the drying shed is used to dehydrate plants, we have to design and maintain this building to be as impermeable to the weather as possible. Though it need not be hermetically sealed, it should be able to be sealed up as tight as possible during rainy and humid weather, while also allowing controlled exhaust of water vapor and intake of cool, dry air through ventilation openings. Instead of building our drying shed on a cement foundation, which would have added significantly to the cost, we excavated under the footprint of the structure to a depth of approximately two feet and removed the topsoil. We then installed a curtain drain around the perimeter using perforated PVC piping, filter fabric, and crushed stone and filled the entire excavated foundation with clean sand.

After grading, leveling, and compacting the sand, we installed a vapor barrier on the ground surface using two sheets of used (expired) greenhouse poly and covered that with heavy-duty woven landscape or road fabric. This vapor barrier and landscape fabric makes an incredibly durable, dry, and easy-to-clean floor for our drying shed. Without the vapor barrier, moisture could seep up from the ground into the drying shed, making the job of dehydrating herbs

**Figure 14-10.** Building our drying shed.

more challenging. The curtain drain helps disperse the massive amounts of water that can pour off the building during heavy rain and snowmelt. For those considering using a drying shed for also growing plants in the spring we recommend "crowning" the floor so that it is slightly convex. This crown will help water flow away to the outside of the building when you are watering plants. For those using retrofitted existing structures or building new wood- or steel-framed drying sheds, this same type of floor design can be utilized, unless a cement foundation is desired in which case floor drains should be installed.

## Designing the Exterior

The exterior of the drying shed should have doors that are large enough to allow for bringing large amounts of plant material into the building on tarps, in wheeled carts, or in crates. Louvered ventilation shutters installed on the building's gable end can be electrically wired to open and close thermostatically or hydrostatically to ventilate water vapor and cool the house to prevent high temperatures from negatively affecting herb quality. Fans installed on the gable end opposite the shutters are recommended to assist with ventilation. Circulation fans inside the structure should be evenly spaced, or better yet, moveable to allow for even air distribution.

For covering greenhouse-type structures we recommend following the same procedure used for greenhouse construction by attaching two layers of greenhouse poly (not construction or household poly, which is not treated for UV protection) inflated by a small squirrel cage fan to provide structural stability to the poly itself, as well as provide insulation to the greenhouse. This is similar to the way double-paned windows provide insulation benefits to our homes. Shade cloth is then installed over the poly to provide UV protection to the plants.

## Considering Interior Space

The interior of the drying shed should be designed and built to hold as much fresh plant material as possible during drying, while also allowing ample room for people to work. Horizontal drying racks made of nylon or fiberglass window screening stapled to wooden frames support the herbs during dehydration while allowing air to circulate around the plant material. These racks can be designed to be removable or can be permanently installed, but we recommend the removable option for many reasons.

Removable racks can be easily disinfected by being removed from the drying shed and rinsed in large containers of an organically acceptable sanitizing solution such as SaniDate (we use our tractor's bucket loader to hold the sanitizing solution). Permanently mounted racks are more challenging to clean as they would need to be sprayed and rinsed in place in the drying shed, allowing excessive water and sanitizer to collect on the floor. Removable racks are also recommended because they can be easily rearranged if need be to take advantage of variable drying rates within the drying shed without having to remove the plant material from racks. For example, in our drying shed the herbs on the bottom racks tend to dry slower than those on the upper layers because of the nature of heat dispersal. When the herbs on the upper racks become dry, we can remove them from the drying shed and place the slower drying herbs in their place without having to take the herbs off the racks and redistribute them onto another rack.

These movable racks allow increased versatility by helping us ensure that entire herb batches dry uniformly and efficiently. Our drying racks measure four feet by seven feet and weigh approximately ten pounds when empty. We experimented with several different sizes of racks and found this size to be large and sturdy enough to hold plenty of plant material while also being manageable enough for one person to move. Drying racks are supported by a permanent wooden framework that allows for easy access to the racks during the process of loading, moving, or cleaning them.

## Materials for Construction

All the wood we use for the racks and supports is 1×2 spruce or pine, which is affordable and easy to work with, yet strong and durable enough to withstand heavy loads and constant use. Some of our drying racks are fifteen years old and are still solid. The screening is attached to the wood frames with staples placed approximately every three inches. Occasionally, with heavy loads on the racks, such as when drying roots, a staple may pull out, so we leave a staple gun handy in the drying shed for quick repairs. We also have a large construction-type magnet that we can wave over that rack full of roots if we notice that a staple has pulled out—to prevent a staple from ending up in someone's bag of herbs.

Several people have asked why we don't use a more natural, nontoxic material for screening. If there was a natural alternative that was durable enough to withstand daily use and yet not become wet and moldy when in contact with live plants, we would consider using it, but we haven't found a suitable alternative to fiberglass and nylon screening. We have considered stainless steel and aluminum, but stainless screening is prohibitively expensive and aluminum has a higher propensity to introduce potentially toxic aluminum compounds onto the plants.

The drying rack support frames are built to hold two vertical rows of racks side by side, with eight racks per vertical row. The top rack is seven feet off the ground, and the bottom racks are twelve inches off the ground. We leave about six inches of space in between the top of one rack and the support bar holding the next rack. This allows us to layer the herbs up to three inches thick while still allowing three inches of airspace to assist with drying. There is also a six-inch space between the two vertical rows of racks, which gives us plenty of room to slide the racks in and out for loading, unloading, and moving them. Loading fresh herbs onto racks that are higher than four feet off the floor can be challenging, so we pull the empty upper racks out one at a time and load fresh herbs onto them by placing them between two racks at a lower working height. When loading racks with fresh herbs, two people simultaneously load each rack by working opposite one another. This allows us to rack herbs more efficiently without having to extend our reach and potentially strain our backs. We usually have two teams of two people racking herbs while the rest of the crew harvests. One person drives the truck or tractor to transport freshly harvested herbs to the drying shed.

After the herbs have finished drying they become much lighter, and one person can remove all the herbs on a single rack by sliding the rack out, tilting it down, and dumping the dried herbs onto a waiting tarp or cart, by which they are taken to the process-ing room. After an entire batch of herbs is dried and the shed is empty, there tends to be a lot of debris on the racks and floor, so we roll up the sides of the drying shed and blow everything off the racks and floor with a gas-powered leaf blower. In the past we manually swept the racks and floor with brushes and brooms, which would take more than one hour of labor. Now the same process is completed much more thoroughly with one person using the leaf blower in approximately ten minutes' time.

# Drying Leaf Crops

We have found that the best way to determine when our herbs are ready for processing and storage is by using the organoleptic method of using our senses to guide us in our decision making. Leaf crops are the easiest to test using this method. Our first indicator is the air in the drying shed. When we sense that the herbs are almost finished drying, we enter the drying shed. If the air inside the building feels moist and the humidity gauge is reading over 50 percent chances are good that water vapor is still leaving the crop and more time is needed. If the air feels dry and the humidity gauge is around 50 percent or lower, we know we are getting close.

At this point the next thing we do is squeeze handfuls of herbs on the drying rack. If we hear a distinct, dry rustling sound and feel and see dried leaf crumbling off the stems, we know we are close, and we investigate further by removing several handfuls of herb off a few racks in different areas of the drying shed and bringing them to the processing bench. We then rub this material over the stainless steel mesh sieves that we use for processing. This process of rubbing herbs over sieves to remove stems is called "garbling"; we will discuss this procedure in further detail below.

If the stems are still somewhat pliable and the leaf is easily crumbling off the stems and sifting down through the steel mesh, this is the indicator that we have reached the target low-moisture content (<10 percent MC), and we are ready to begin processing. By picking up a handful of the leaf that has sifted through the sieve and squeezing it into a ball in our hands, we can get a good feel for the moisture content. If the leaf crumbles easily and is able to be crushed by hand into fine particles, it is ready to be processed and stored. If this "ball" of leaf feels pli-able and doesn't crumble or sift into fine particles, the herbs need a bit more drying time. This window of opportunity—when the main stems (not the tiny leaf stems) have retained a bit more moisture than the leaf—is the ideal time for us to process the herbs, because at this point the stems don't crumble into pieces while we are garbling, yet the leaf is fully dry and is easily garbled.

There are times when other tasks can delay us in monitoring leaf crops for dryness, and the stems become more dry and brittle. Although this can add a bit more labor involved with garbling, it is not deleterious to the drying process or the quality of the finished product. In fifteen years of drying herbs this organoleptic method of determining leaf moisture content has been a reliable guide. As long as the relative humidity in the drying shed is low, the temperature is relatively high, and the leaf passes our visual, auditory, and sensory tests, we can be confident that our MC is at or under 10 percent for optimum quality and storage. Leaf crops generally require twenty-four and forty-eight hours drying time to reach a MC of 10 percent.

## Garbling

"Garbling" is the industry term for separating herb leaf from stem. In large-scale herb production for what we call the herbal "mass market" virtually all of this processing is done mechanically with large equipment. The challenge with this type of auto-mated processing is that the intense mechanical action required to process tons of herbs in a short period of time creates friction. Friction creates heat, and heat degrades the quality of medicinal herbs by volatizing essential oils, vitamins, and other important medicinal compounds. This equipment can also be very expensive and exists primarily in the Asian, Indian, and Eastern European herb mar-kets. However, small-scale affordable equipment specifically for small- and medium-scale growers is under development.

We currently garble all our leaf crops by hand using sieves made of stainless steel wire mesh attached to wooden frames. To do this, we remove five to ten racks at a time of dried herbs from the drying shed by dumping the racks onto clean tarps placed on carts. We then bring the carts to a process-ing bench, which is located outside during pleasant

**Figure 14-11.** Garbling racks.

weather or inside during inclement weather. Manual garbling is a two-step process. For the first step we remove herb leaf from plant stems by hand-rubbing the whole dried plants over a half-inch stainless steel mesh sieve. This sieve is approximately thirty by forty inches, large enough to handle long-stemmed herbs without difficulty. Two people can work simultane-ously on an eight-foot-long bench, removing whole dried herbs from the cart and rubbing them across the sieve, which causes the leaf to crumble through the mesh while large portions of the stem stay on top and are swept by hand into a discard pile for com-posting. During this process we use brushes to sweep piles of the processed herb into large food-grade bags attached to the edge of the garbling bench with squeeze clamps.

Our primary goal at this point is to get all of the dried herbs out of the drying shed as quickly as possible to retain maximum quality and also to make room for the next harvest. After we finish the first garble and the drying shed is empty, we tightly seal all the large bags of herbs and cover them with tarps to protect them from light. When we have time, often later that day after the drying shed is refilled, we return to our bench to finish leaf processing by a second garbling step. The sieves we use for second garbling are made of quarter-inch stainless steel mesh attached to thirty-inch square wooden frames. This smaller mesh breaks the leaf down into a quar-ter-inch industry-standard grade known as "cut and sift" and also helps us to remove any remaining pieces of large stem. The wooden frames on these smaller sieves don't need to be as large as the first garbling sieves, since we are no longer dealing with long pieces of stem.

Garbling creates a lot of dust, which can be det-rimental to breathe, so we always use dust masks to protect our lungs. We also wear canvas gloves to protect our hands and hats or hairnets to keep our hair out of the herbs. It is vital to use good hygienic practices and provide sanitary conditions to avoid introducing harmful microorganisms and materials.

During garbling we also perform quality control by removing insects or other foreign objects that may have come onto the drying plants. Drying kills the insects, and they are easy to spot and remove during both stages of garbling. After we have garbled the herbs into the final cut-and-sift grade they are ready for packaging. We store them in large food-grade bags sealed tightly with heavy-duty wire twist ties; label the bags with the species name, weight, and lot number; and place the bags in either sealed cardboard boxes or gasket-edged fiber drums. From there they go into dry storage until they are sold.

## Drying Flowers

Flowers, such as those of arnica, calendula, chamomile, elder, and red clover, are dried whole and require an average of thirty-six to seventy-two hours to reach acceptable low steady state moisture content. These flowers are much more susceptible to the degradation of quality from exposure to light; therefore we dry all our flower crops in a special drying loft. Unlike the drying shed with its 85 percent shade cloth, this loft has a solid metal roof and is completely dark inside when the doors are shut.

Ventilation is provided by circulating fans, and there is no supplemental heat source in the blossom drying loft. Soffit vents provide natural airflow and ventilation, and the blossoms are dried on the same type of racks as in the main drying shed. The process for determining when blossoms have reached EMC is similar to leaf crops: We use our senses to guide us, but since blossoms are so much denser than leaves we have to delve deep into the center of the blossom to see what is revealed in the slowest drying portion of the plant.

Calendula has the largest and most dense blossoms of all our flower crops and therefore takes the longest to dry. When we think we are getting close to acceptable dryness with calendula, we search several racks for a handful of the largest blossoms we can find and rip or cut them open to determine the moisture content in the center of the blossom. It isn't difficult to determine if the blossoms are dry or not by exposing this center portion. Calendula is extremely sticky and resinous, especially before it is dehydrated. As moisture leaves the blossoms, the

resin also dries and becomes less sticky. When we cut into the center of the blossom, squeeze it, and roll it into a ball, it becomes quite obvious whether or not it is dry. Any moisture remaining in the center of the blossom will smear on your hand, which is a good indicator that more drying time is needed. The flower petals are at AMC when they make a rustling sound when shaken together, usually before the center is at AMC. Reaching AMC throughout the entire blossom is the end goal.

All of this is a process that is best learned through experience, and we recommend checking any blossoms that are put into storage bags each day at least for two to three days after packaging. Since blossoms are sold whole and not processed, we put them into food-grade poly bags immediately and seal them tightly with wire twist ties. Even after years of drying calendula we still have a mandatory procedure of rechecking each bulk package that is sealed and ready for storage. Even if just a few blossoms in a bag of calendula are put away before they are fully dried, the whole bag will reabsorb some of that moisture and take on a damp feel. Instead of making a dry, rustling sound when the blossoms are shaken and stirred by hand, they won't make a sound and will feel damp to the touch. If this happens, we empty the entire bag back out on drying racks and give it another twelve hours or more to become fully dry.

Unlike leaf crops, where we have never had a single batch spoil in storage, we did have some calendula losses during the early years. There are few lessons we have learned in farming that are as expensive as having to compost hundreds of dollars' worth of calendula blossoms that we thought were dry but really weren't. This has taught us the value of redundancy when determining whether our calendula is dry enough and ready to be put into dry storage.

Other blossoms, such as arnica, chamomile, elder, and red clover, are much less dense and take considerably less time to dry than calendula. The easiest way for us to determine acceptable final moisture content for these blossoms is to rub a handful of blossoms over our quarter-inch garbling sieve. Any moisture in the blossoms will be obvious, as it will smear on the steel mesh and feel moist to the touch. We also bend the stems, and if they snap rather than bend, this is a good indicator that they are dry.

For red clover we try to pick out handfuls of the largest blossoms and analyze them fully for this testing, and similarly to calendula we rely on redundancy by rechecking bags for two to three days after they are initially sealed to ensure that we have thoroughly dried the entire batch. Once we are certain that they are dry and ready for storage, we label them with the proper species name, weight, and lot number; seal them tightly in food-grade plastic bags with heavy-duty wire twist ties; and place them in sealed cardboard boxes or gasket-fitted fiber drums until sale.

## Drying Roots

Roots are dried in the main drying shed and are much more tolerant of exposure to light and higher temperatures of up to 110°F with some species. They also take longer to dry than leaves and blossoms. Unlike leaf and blossom crops, roots are all washed thoroughly before drying. Because of their variable nature different species of root crops are dried under different conditions and can take up to eighty-four hours to reach acceptable low steady state moisture content.

Some roots, such as dandelion and goldenseal, are dried whole because of their smaller size, then milled after they reach acceptable moisture content. Larger multistemmed roots, such as angelica, ashwagandha, astragalus, black cohosh, echinacea, and yellow

**Figure 14-12.** Drying roots in our old Quonset-style drying shed.

dock, are chopped by hand into quarters, dried, then milled. Quartering these large roots not only makes them easier to clean, it also significantly decreases the time it takes them to dry by exposing more of their surface area to the drying effects of forced and natural convection.

Roots such as burdock, comfrey, elecampane, and mallow, which are higher in moisture content and rich in mucilage (gooey!) are quartered, washed, and partially dried for thirty-six to forty-eight hours to remove at least 25 percent of their moisture. After this initial par-drying period these roots can be milled into large pieces approximately a half-inch in diameter and spread out on racks again in the drying shed to complete their drying process. The reason we par-dry these type of roots is that they take so long to dry because of their high moisture and mucilage content. If we didn't par-dry them, yeast and mold levels in the roots could possibly elevate because of the lengthy time they sit on the drying racks while still moist.

Another reason we often par-dry before we mill them is that if we were to mill them fresh, they would turn into a gooey mess while being milled and would clog the milling machinery. After we par-dry then mill these roots, we spread the milled root pieces out on drying racks and finish the drying process. After they have reached acceptable moisture content, we remove them from the drying racks and mill them again. This second milling produces the final cut-and-sift grade, which is the industry standard.

To determine acceptable moisture content for whole roots we choose a handful of the largest pieces of dried roots from several racks and snap them in half by hand. If they are still pliable and resistant to snapping, they need more time drying. If they easily snap in half with us using our fingers, they have reached acceptable moisture content, given the fact that humidity levels in the drying shed are under 50 percent and the temperature is above 80°F. We can also easily determine moisture content during milling, which we will discuss below. This unscientific method for determining acceptable moisture content is also remarkably effective.

**Figure 14-13.** Jeff milling roots.

## Milling Roots

"Milling" means grinding, and there are a number of methods and tools to choose from to grind herbs. We mill roots and barks only; leaves are garbled, and blossoms are sold whole. Since we dry our roots either quartered or whole, we need to mill them when they are partially dried or after they are completely dried. We have tried to mill fresh roots in the past and found that it makes a huge mess because of the high moisture content in the roots. There are a few species of roots we mill fresh such as ashwagandha and Japanese knotweed because of their drier, woodier nature.

For milling roots we have two machines to choose from: One is a gas-powered wood chipper, and the other is a small electric wood chipper. The gas-powered chipper is great for doing larger volume because we can literally pour buckets full of dried or partially dried roots into the hopper, and they exit the discharge shoot into a food-grade plastic fifty-five-gallon drum that is cut to fit around the discharge shoot so we don't lose any material or risk contamination from foreign material. Between batches of herbs, especially different species, we clean the insides of these machines thoroughly with an organically acceptable hydrogen peroxide–based disinfectant called SaniDate. After cleaning the machines with SaniDate, we rinse them with fresh water to remove any sanitizing solution. We use the electric chipper for smaller batches of roots or to mill things to a finer grade than the larger chipper can.

Hammer mills are the preferred type of milling machine because of their efficiency and versatility grinding roots, but they are generally much more expensive than chippers. We have not purchased a hammer mill but are currently considering buying one to improve our milling efficiency and quality. We recommend that growers new to the medicinal herb market establish clear guidelines with potential buyers for product grade and consistency. For example, there is a large market for powdered herbs, especially root powders, and the production of large volumes of powders could help justify the purchase of more expensive milling machinery such as hammer mills. For now, on our farm our chippers have met all of our milling needs. If we get custom orders for a particular uniform cut size for our roots we can use a Clipper seed cleaner with different-size screens to separate the material into distinct grades. Used Clippers can often be found for $200 to $400 and can be a great investment for the herb farm, not only for grading milled herbs but also for cleaning seed collected for sale or planting.

Milling creates a lot of dust, so we wear dust masks to protect our lungs. We also wear hats or hairnets to keep our hair out of the herbs and nonlatex gloves to help maintain sanitary conditions with the herbs. We have the option of milling outside in the open air during good weather conditions, which is ideal for dust control, or we can move inside to the processing room during inclement weather. The processing room should have a ventilation fan or dust collection system to manage the copious amounts of dust produced during milling. Another option to help keep the dust down when milling is using cotton filter bags attached to the discharge shoots of our chippers. These filter bags have to be emptied frequently but do a very effective job at decreasing the volume of airborne dust.

After milling roots we package them in food-grade plastic bags tightly shut with heavy-duty wire twist ties and label them with the species name, weight, and lot number. Roots are stored in sealed cardboard boxes or gasket-fitted fiber drums in our dry storage room until sale.

## Drying Barks

Barks, such as wild cherry bark, are our easiest crops to process. After harvesting and peeling the inner cambium from culled branches with a drawknife, the bark goes immediately into the drying shed, where it is dried for thirty-six to seventy-two hours at relatively high temperatures. It is easy to determine when barks are dry because the pieces readily snap when bent and aren't pliable or moist to the touch. When barks reach acceptable moisture content, we immediately chip them in our small electric chipper, which mills the pieces into a cut-and-sift grade. From there they are labeled with the species name, weight, and lot number; bagged tightly in food-grade plastic bags with wire twist ties; packaged in sealed cardboard boxes or gasket-fitted fiber drums; and placed in dry storage until sale.

In summarizing this chapter we hope that you, the reader, have gained useful knowledge with which to process your own herbs without feeling overwhelmed by some of the scientific methods we have provided to add depth and detail to this important facet of the medicinal herb farm. Please bear in mind that drying and processing herbs on your own farm can be as simple or as complicated as you the producer wish it to be. We started out with a tiny lean-to-style greenhouse stapled to the side of our garage and utilized screens from the windows of our house to dry our first years' harvest. There was nothing fancy or complicated about that facility at all.

Those dried herbs turned out to be of very high quality because of our care in ensuring that the environment they were drying in and the methods we were using were conducive to quality retention. It isn't difficult to produce and process exceptional-quality herbs, but it does require an adequate facility, some basic equipment, and, most importantly, good observational skills to monitor progress and make good decisions regarding how you harvest, process, and store your valuable crops. We wish you the best of luck with your postharvest processing!

# CHAPTER 15

# Herb Packaging and Storage Recommendations

Dehydration is a very effective way of extending the shelf life and preserving the quality of bulk medicinal herbs. However, the effectiveness of maintaining quality during storage is highly dependent on several factors, including time, the storage enviroment, the microbial and moisture content of the dried product being stored, and the materials and methods used in storage. Even the highest-quality dried herbs with low microbial counts placed in ideal storage conditions will inevitably undergo a gradual and irreversible degradation of quality, some more rapidly than others.

Ideally growers will package and ship herbs as soon as possible after harvest and processing and hope that customers are prepared to use or sell the dried herbs while their medicinal and nutritional compounds are still potent. Realistically, though, we often have to warehouse quantities of dried herbs on our farms to maintain inventories for year-round sales opportunities. Our customers may also need to store the herbs they purchase from us until they are used or sold. Although we cannot completely halt the inevitable but gradual diminution of potency that occurs when dried herbs are placed in storage, we can certainly minimize some of the factors that can contribute to this decline.

## Maintaining Quality in Storage

How much quality degradation occurs in what time period, you may wonder? That question is impossible to quantify accurately. There are so many variables, given the number of species of plants and the way they are processed and stored, that it is difficult to predict what type and exactly how much degradation is occurring in storage. In general, blossom crops such as arnica, calendula, chamomile, and red clover tend to degrade fastest and can generally be considered best when used within six months of

**Figure 15-1.** Bagging chamomile blossoms in the drying shed.

harvest under proper storage conditions. Leaf crops tend to be a bit more stable and are generally best used within one year of harvest. Root crops are the most stable and are often still remarkably potent even after a year or more in storage.

Of course, there are many exceptions to these general recommendations. Chamomile blossoms are more stable than most other blossom crops and can store well for up to a year and even more. Lemon balm leaf, even when stored in seemingly ideal conditions, tends to volatilize its aromatics rapidly and is best when used within three to six months of harvest. Root crops vary greatly in their rate of degradation depending on the species. The best practice is to state the date of harvest on herbs you package, try to sell them within a year of harvest, and recommend to your customers that they store them properly and use them in a timely manner to retain the highest percentage of bioactive compounds.

We are continually trying to solve the equation of how much of each crop to grow to maintain cash flow through year-round sales without having excessive inventory expire. Referencing past years' harvest and sales records, making educated guesses, and encouraging our wholesale customers to preorder, guide us in planning our cropping volumes from year to year.

There are a multitude of physical, biological, and chemical activities conspiring against us as we try to preserve the bioactive compounds, flavors, aromatics, and vibrant color that we grow and use these plants for. Microbes are the primary culprits responsible for biodeterioration, and unfortunately they are all but unavoidable given the conditions that plants encounter during their arduous journey from seed to package. Air and soilborne yeast and mold spores have collected on the plants in the field during their growth, and although most of these microbes lie harmlessly dormant on dried plant material in storage because of the relative absence of moisture, there are certain types of fungi that are able to live and multiply with almost no moisture present.

Since we can't effectively remove 100 percent of the moisture from our dried herbs, there can often be

biological activity undetectable by the naked eye gradually diminishing herb quality at best or rendering the product unfit for use or consumption under the worse sets of circumstances. In most cases this biodeterioration is preventable with good hygienic practices in the field and drying environment, as well as properly drying, handling, and storing the dried material.

Microbiological laboratory analyses are used to detect the presence of yeasts, molds, bacteria, and other contaminants on dried botanicals and are mandated by many herb buyers to comply with Good Manufacturing Practices. This microbiological testing counts the number of colony-forming units (CFUs), which are viable microorganisms per gram of product. Standards established by the United States Pharmacopeial Convention and the American Herbal Products Association give recommended tolerances for "plate counts" or concentrations of CFUs. For dried herbs that are tested and found to be within acceptable plate counts, dried at less than 12 percent moisture, and stored properly, biodeterioration can be minimized. This testing also determines if bacteria such as salmonella or E. coli, as well as mycotoxins, are present in the sample, and if so, the lot is deemed unfit for human consumption. Reports from the World Health Organization (WHO) on the amount of herbs being exported from foreign countries that have been rejected because of bacterial contamination and/or mycotoxins are concerning and demonstrate the importance of these analyses for health safety purposes.

Insects are another factor that we have to take into consideration when packing and storing medicinal herbs. Adult insects are relatively easy to spot and remove when processing the plant material, but the possibility exists that unseen eggs have been deposited on the plants before they were harvested and dried. Although unlikely, it is possible that these eggs may survive drying and processing and still be viable in storage. This is a scenario that is hard to prevent without resorting to fumigation, so the most effective method of prevention would be periodically checking herbs in storage to ensure that critters aren't running wild in the bags.

Physical forces such as heat, light, air, moisture, and time can also have a tremendous influence on the quality of herbs in storage. Have you ever left a jar of dried culinary herbs out on the back of the stove for convenience when cooking? If so, perhaps you noticed that those once green, vibrant-looking aromatically potent herbs rapidly turned pale and lost their aromatic and flavorful qualities sitting there exposed to the light. It doesn't matter whether the light is from a natural (the sun) or an artificial source (electricity), light rapidly degrades the quality of herbs exposed to it for even a relatively short period of time.

Heat can also be detrimental to herbs in storage, especially herbs with volatile compounds such as essential oils that are very susceptible to loss through volatilization. Dried herbs are hygroscopic, meaning they readily absorb moisture from the air. Exposure to humidity not only can promote microbiological contamination, but it can also activate enzymes within the plants themselves, leading to decomposition of active metabolites. Air promotes oxidation, which decreases potency, and time exacerbates all of these factors that can negatively affect herb quality.

## How to Store Herbs

Okay, now that we have come to the sad conclusion that nothing lasts forever as far as medicinal herb potency goes, it's time to discuss ways to help maintain the integrity of medicinal and nutritional compounds in botanicals we produce while they are

**Figure 15-2.** It is highly recommended to store small amounts of dried herbs in airtight glass jars, protected from heat and light.

**Figure 15-3.** Fiber drums lined with food-grade plastic bags.

in storage. Ideally we would be able to vacuum-seal our dried herbs in large, airtight stainless steel or glass vessels for storage. Unfortunately this is impractical and cost prohibitive, given the hundreds of pounds or tons that a medicinal herb farm can potentially handle in a given season.

Instead, we have to utilize affordable and effective storage materials such as plastic bags and cardboard boxes or gasketed drums. Food-grade polyethylene bags heat-sealed or closed with heavy-duty twist ties and placed in cardboard boxes for storage are a practical choice for the small organic herb farm. The U.S. Food and Drug Administration (FDA) requires that plastics used in food packaging be of greater purity than plastics used for nonfood packaging. This is commonly referred to as food-grade plastic. Plastics used to package pharmaceuticals are held to an even higher standard than food grade.

Food-grade plastic does not contain dyes or recycled plastic deemed harmful to humans. However, this does not mean that food-grade plastic cannot contain recycled plastic. There is a common misconception that all containers made of white plastic or high-density polyethylene (HDPE) plastic bearing the #2 recycling symbol are food-grade containers. This is not true. If you are considering the purchase of a container from

some place other than a kitchen or restaurant supply store, and the container is not clearly labeled as "food safe" or being made of food-grade plastic, then you should assume that it is not food grade.

Unfortunately, plastic is pretty nasty stuff that we consider a "necessary evil" because of the many chemicals that it is manufactured from, with potential for leaching into material encapsulated within it, as well as into the air we breathe. We like to think of the plastic bags we use to store and transport herbs as temporary storage, and we encourage our customers to transfer the herbs into more stable and less toxic permanent storage containers made of glass or stainless steel until they use them.

We purchase a variety of packaging material from companies such as North Atlantic Specialty Bag Company and Uline. The largest size bags we use are 34 × 48 inches and are made of four-mil food-grade plastic. These bags fit well inside our largest cardboard containers, which are 20 × 20 × 20 inches. This bag/box combination can hold between fifteen and twenty pounds of dried leaf or blossoms or thirty to forty pounds of milled dried roots. This is an ideal size for shipping and handling larger herb quantities for wholesale accounts and also for long-term herb storage. After filling the bag with the proper quantity of herbs, we squeeze as much air as we can out of the bag and close the bag as tightly as possible with a heavy-duty wire twist tie. We then place the bag into the cardboard box and tightly tape the box shut with packing tape.

We also use twenty- and fifty-five-gallon plastic or cardboard fiber drums with gasketed lids for holding large quantities of herbs in storage. These drums can hold two to three of our largest-size bags filled with dried product. Plastic bags, even when tightly sealed with wire ties, are not completely impervious to air and moisture. Storing wire-tied plastic bags inside hermetically sealed steel, plastic, or fiber drums with gaskets is the best method of preventing environmental influences from negatively affecting herb quality.

Used food-grade plastic drums with gaskets can often be found at food-processing facilities, stores, and restaurants; washed; and recycled for use in storing herbs. Oftentimes these facilities would love to see these containers put to good use rather than recycled or deposited in a landfill. In the early years of operation we got dozens of five-gallon plastic buckets with gasketed lids from our local Dunkin' Donuts store. It's quite a contrast to think about what was originally contained in these drums (donut batter) to what we were storing in them (medicinal herbs).

## Where to Store Herbs

The environment where herbs are stored has almost as much influence on maintaining herb quality as the type of packaging they are stored in, especially if the packaging is not airtight. Humidity and oxygen can rapidly compromise potency and stimulate microbial and enzymatic activity. If herbs in storage are exposed to light, whether it be natural or artificial, they can rapidly lose their vibrant color, thereby diminishing overall quality. Excessive heat in storage, even if airtight packaging is used, can volatize essential oils and decrease the herbs' aromatic and medicinal activity.

All of these potential threats can be minimized by selecting a storage location where humidity, light, and heat can be controlled. A clean, insulated cement basement space makes an ideal choice for dry storage, especially if it can be hermetically sealed. Dehumidifiers can keep ambient moisture levels down, lights can be kept off except when working in the space, and the subterranean location helps to keep the room cool during hot summer days. Any clean, dark, dry space that can be hermetically or at least partially sealed off from the elements can serve as dry storage, provided precautions are taken to ensure environmental stability within the room.

People have asked us if they should utilize air conditioners in dry storage rooms that are not located below ground. Our sense is that the energy costs associated with running air conditioners to keep herbs cool in storage would exceed the value of potency retained by their use and therefore we feel this may not be a practical option. We recommend that producers do what they can to hermetically seal packaging and maintain dark, dry conditions in storage, which

Table 15-1. Food-Grade Plastic Bags Used at ZWHF

| Storage Bags | | |
| --- | --- | --- |
| Type | Size | Thickness |
| Large | 34" × 48" | 4 mil |
| Medium | 22" × 36" | 4 mil |
| Packaging Bags for Sales | | |
| Type | Size | Thickness |
| Large | 10" × 20" | 2 mil |
| Medium | 8" × 15" | 2 mil |
| Small | 7" × 12" | 2 mil |
| Paper Window | 4 oz. | |
| Rice Window | 2 oz. | |

are much more important from a quality retention standpoint than maintaining cool temperatures.

Rodent prevention is another important consideration for herbs in storage. There are a number of organically acceptable methods of controlling rodents, including sticky traps, have-a-heart traps, and other nontoxic and effective control materials and methods.

## Maintaining Quality Postsales

After we make a sale, preserving herb quality is in the hands of our customers, many of whom don't realize that if they leave that clear plastic bag of calendula out on the shelf because it looks so beautiful they will be damaging the potent blossoms within a short period of time. We have begun including a card with the following statement in each package of herbs we sell to inform our customers of the importance of maintaining herb quality in storage:

*We thank you for supporting our family farm through your purchase. The herbs in this package have been grown, tended, processed, and packed by hand to provide you, our valued customer, with exceptionally high quality*

**Figure 15-4.** Two-ounce rice bags of Rejuvenation Tea. Photograph courtesy of Kate Clearlight

*botanicals. In order to retain the potent healing and nutritional compounds, we recommend that you remove the herbs from these plastic bags and place them in airtight canisters made of metal or glass. Light, heat, air, and humidity can rapidly diminish the potency of these herbs. We recommend storing them protected from these elements to ensure that they retain their healing attributes.*

*Enjoy! Sincerely, Melanie, Jeff, Lily, and the entire Zack Woods Herb Farm family.*

One of the biggest challenges we have faced regarding herb quality retention is with our wholesale customers who display our herbs in their retail stores. Most of the stores that carry our herbs in their bulk departments display the herbs in clear glass jars so that their customers can see what they are purchasing. Displaying herbs this way is great from a marketing perspective because the herbs can literally help to sell themselves based on their aromatic qualities and aesthetic beauty, but from a practical standpoint these displays can rapidly diminish the value of the products contained within. All we can do is try to educate buyers as to the importance of protecting our products from the elements and hope that they employ creative ways to market our products without compromising their value.

Some stores place the herbs in dark amber glass jars with photographs of the actual herbs affixed to the outside of the jars, which is a great solution. Others place a small sample of the herbs in a small clear container for customers to view while keeping the rest of the dried product in an opaque storage container to protect it from the elements.

**Figure 15-5.** A sixteen-ounce bag of Rejuvenation Tea.

We also wholesale herbs to stores that display our products in our "retail friendly" two- to four-ounce packages in their tea, herb, or grocery departments. For years we used small clear plastic bags with labels for this packaging, but we started noticing that the

vibrant color and aromatic qualities of the herbs in these packages was rapidly being compromised by the store's bright lights. We considered other choices of packaging, including metallic containers, but we really wanted our customers to see the quality of the herbs they were purchasing, and the tins are expensive and seemed like a poor choice from an ecological perspective.

We finally found an affordable, retail-friendly package that could protect the herbs from light while at the same time displaying their contents for customers to see. Our choice of packaging for the retail environment is a "window bag" made of recycled paper and lined with a clear bioplastic film made from non-GMO corn-based polymers; it has a small transparent window on the front of the bag and is resealable with a metal tab, similar to coffee bags (see figure 16-1 on page 232). For a smaller two- or three-ounce retail package we use a similar bag made entirely of a rice-based bioplastic. These bags are not ideal in that they do let some light in through the "window" and they are not completely airtight, but they are the best choice we found for retail markets and far exceed clear plastic bags for protecting herbs from exposure to light. We purchase these bags from the North Atlantic Specialty Bag Company, located in Reading, Pennsylvania.

Our hopes are that in the near future nontoxic, sustainably produced, biodegradable packaging materials will become readily available and affordable to store and market our healing herbs in.

**Figure 16-1.** A co-op display of value-added products.

# CHAPTER 16

# Producing Value-Added Products

Melanie's first experience with value-added herbal products happened when she was sixteen and her stepmother, Rosemary Gladstar, sat her and her twin sister down and said, "Girls, I think it would be a wonderful idea if you helped me start a business." Rosemary had developed many exceptional formulas and healing protocols over her three decades as a community herbalist and had generously shared them with people for years. However, as her life got busier and traveling kept her on the road, she found it harder to meet with people who asked for consultations. Not wanting to turn away people in need, she came up with a solution: create a product line of her favorite formulas and enlist her daughters to be her production team. Not only would this help people access really great herbal medicine, it would put her teenage daughters to work and help them learn invaluable medicine-making skills.

Melanie and her sister jumped at the chance, and with Rosemary at the helm Sage Mountain Herb Products was born. I joined forces with Melanie later on when Melanie's sister moved on and helped develop some new products. Twenty-four years later the company is still alive and kicking, offering wonderful tinctures, creams, and salves. Although we no longer run the company, time spent in the production lab taught us many things, including the power of having valued-added products as part of your business plan.

**Figure 16-2.** Lily selling value-added products at the NE Women's Herbal Conference.

## The Benefits of Having Value-Added Products

"Value-added agriculture," says Melissa Matthewson of Oregon State University,

*entails changing a raw agricultural product into something new through packaging, processing, cooling, drying, extracting or any other type of process that differentiates the product from the original raw commodity. Adding value to agricultural products is a worthwhile endeavor because of the higher returns that come with the investment, the opportunity to open new markets and extend the producer's marketing season as well as the ability to create new recognition for the farm.*[1]

We agree with Matthewson and have found that value-added products can be an extremely helpful way to introduce people to herbs. If you are fortunate to live in an area where there are lots of herbalists, there may be an easy market that you will be able to tap into. If your herbs are of high quality and reasonably priced, herbalists will most likely want them. However, if you live in a place where herbalism isn't as common or mainstream, you may have to approach your marketing more thoughtfully. Guaranteed, if you hand someone unfamiliar with herbs a big bag of nettle leaf, you'll get a funny look coupled with the question, "What do I do with this stuff?"

Hand that same person a delicious tea blend made with nettle and other herbs and more likely than not, he or she will be willing to at least try it. Sometimes we tease that our tea blends are the "gateway herbs" to the farm, but it really is the truth. Once people start drinking our teas, they often start exploring other herbs we carry and want to learn more.

Value-added products can also help by building recognition for your farm, making your farm more versatile, and allowing you to enter new markets. This was true for our farm. At a crucial time when we were just getting off the ground, we were fortunate to get into a very competitive farmers' market. We were able to enter this market in large part because we were *not* veggie farmers. The market already had plenty of farmers selling vegetables and a long waiting list to get in. When we applied, we were seen as somewhat of a novelty because we grew medicinals; we offered fresh, dried, and live plants; and we had two beverage tea blends. This specific farmers' market didn't have anyone offering these types of products, so we were accepted into the market, affording us visibility that we desperately needed at that time as a new farm.

Once we were in the market, handing out tea samples was a way to engage people and entice them to our booth. One day a buyer for the local natural food co-op came through and sampled our tea. The co-op didn't have any locally made tea at the time and none made from plants grown in Vermont. She liked the tea and pitched the idea to her store. Soon the co-op expanded sales of our products and started carrying our bagged herbs and teas in their grocery and wellness sections and it remains one of our top wholesale accounts. Getting into the co-op and many others was a turning point for our farm because it gave us a year-round market to build on. As illustrated by this example, value-added products can be beneficial not only in the sales they bring but also in how they help the public become familiar with your farm and the herbs you grow.

Value-added product can be a great way to introduce uncommon or unusual herbs to the public. By design we grow many familiar and popular herbs, such as nettle, peppermint, echinacea, and valerian. However, we also are constantly trying new things and grow some of the more obscure, less well-known herbs in hopes of finding new plants and medicine to share with our community. And while the echinacea and chamomile will often basically sell themselves, some of our more exotic medicinals require more assistance.

**Figure 16-3.** Live plants being shipped to gardeners and herbalists around the country.

Take, for example, a plant like spilanthes. It is a powerful antimicrobial and antiviral herb that stimulates your mouth with a tingling sensation that increases salivation and improves oral hygiene. It's not commonly known and as a bulk herb doesn't fly off the shelf. However, if we blended it with a little peppermint and calendula and marketed it as an oral antiseptic tea blend or went one step further and made it into a mouthwash, it is highly likely that our spilanthes sales would improve.

There are also intrinsic rewards that come with incorporating value-added products to your farm. Designing new value-added products can be a wonderful creative outlet and also a way to incorporate the talents and interests of other members of your family or farm crew. Farming is a wonderfully dynamic and active career, for sure. However, included in the dynamism are long hours of strenuous and repetitive work during which the mind can wander, and new ideas can take shape. It's one of the things that we love about working with the plants—the opportunity to dream and create new things. If you can harness your creativity to find novel ways to share medicinal plants with your community, your farm will benefit.

We also find that encouraging other members of our family and farm crew to share their talents and ideas helps keep people engaged and happy in their work and brings in new opportunities. For example, we love working with the energetic healing properties of the plants, and our daughter is absolutely crazy about flowers. Out of these two passions we have begun developing a value-added product line of flower essences from some of the plants grown on the farm. Not only does it offer our customers another way to access the healing properties of the plants, it feeds our souls and is a way that our little girl can meaningfully participate in the farm.

Another example that comes to mind happened last season with an employee who enjoys teaching and offering classes. She was an extremely knowledgeable herbalist and gifted speaker. Building on her talents and aptitudes, we created a series of farm tours for neighboring colleges and herb schools, which she organized and led. The tours were a huge success. People had the opportunity to visit the farm, our employee had a new and highly rewarding job, and we could remain in the fields and keep the farm running smoothly. So in addition to thinking of value-added products as "tangibles," you can add value by providing *services* to your community as well. Jeff's agricultural consulting services are another value-added service we have capitalized on and they have become another important facet contributing to our income diversity.

**Figure 16-4.** Window bags are a good way to showcase the quality of dried herbs.

Figure 16-5. Flower essences.

Figure 16-6. Herbal elixirs. Photograph courtesy of Kate Clearlight

# Common Value-Added Products

Herb farmers are fortunate in that herbs can easily and affordably be made into desirable value-added products. Drying your herbs is one of the biggest ways of adding value to your medicinal crops. You can also use the herbs as raw materials for other products. Making soaps; braiding garlic; creating herbal wreaths, jams, flower essences, bath salts, tinctures, teas, salves, potpourris—the sky really is the limit. Being successful with value-added products takes some work but need not be daunting. Some of the major keys to success are designing a strong formula or product, creating a niche market for the product, and becoming fluent in the food business and safety regulations that pertain to your product, if applicable.

Most if not all herbal products used for medicinal purposes fall under the category of "dietary supplement" as defined by the Dietary Supplement Health and Education Act of 1994. The DSHEA defines a dietary supplement as "a product (other than tobacco) intended to supplement the diet that bears or contains one or more of the following dietary ingredients: a vitamin, a mineral, an herb or other botanical, an amino acid, a dietary substance for use by man to supplement the diet by increasing the total dietary intake; or a concentrate, metabolite, constituent, extract, or combination of any ingredient noted in clause (A), (B), (C), (D), or (E)."[2]

This law outlines the regulations for the manufacture and sale of supplements and along with the FDA's Good Manufacturing Practices (GMPs) spells out how processors must manufacture and label their products. Reading through the GMPs and DSHEA information can be overwhelming, but there are organizations such as the American Herbal Products Association (AHPA) that can provide critical information about the process and help inform your work.

# Finding Your Niche

With explosive growth in the herbal marketplace, you may see many herbal products in your local co-ops and farmers' markets. This is a good thing and should not deter you from finding a unique way to express and share your connection to the plants. When developing value-added products, classes, or

**Table 16-1.** Common Herbal Value-Added Products

| Value-Added Product | Definition | Considerations and Regulations |
|---|---|---|
| Dried herbs | Herbs that have had their moisture content removed to ≤10% of original levels | We use a drying shed to dry our herbs. However, drying can be done in many ways. See chapter 14 to explore ways of processing post-harvest. In Vermont dehydrated herbs are defined as a "processed product" by our organic certifying agency. This requires us to apply for certification as both a producer and a processor. It requires more paperwork, documentation, and some fees but is worth it. |
| Teas | A mixture of one or more herbs that is prepared by making an aqueous infusion or decoction. Tea blends can be sold in dried or liquid form. | How a product is categorized dictates how it is regulated. If teas are being used medicinally they would fall under DSHEA regulations. If the teas are strictly beverage, they most likely fall under the umbrella of food regulations. |
| Tinctures | A concentrated liquid extract of an herb or herbs. | Tinctures can be made using alcohol, glycerin, and vinegars. The menstruum and indicated use of the tincture would determine its regulation category. If tinctures are being used medicinally they would fall under DSHEA regulations. |
| Salves and creams | An emollient preparation in which soft and hard oils are used as solvents. The preparation is used topically. | Most commonly labeled as "personal care products." The federal regulatory framework for these types of products and their corresponding categories is ambiguous. See state guidelines for labeling considerations. |
| Flower essences | An aqueous preparation made from the flowers of a plant that captures the plant's vibrational healing energy. Preserved with alcohol. | Used in homeopathic doses. It is unclear how flower essences would be categorized under DSHEA. |
| Food as medicine | Herbal jams, honeys, syrups, wines, oils, etc. are excellent ways to use medicinal crops and are highly desirable in the market. | Food products fall under the jurisdiction of federal and state food regulations. |

services, study the markets and the herbal community and ask yourself, what's missing? What would be unique, inviting, and helpful?

Also take into consideration what you love making or enjoy doing, and see if you can incorporate this into a product or service to offer customers and as a giveback to the planet. "Let the beauty of what you love be what you do. There are a thousand ways to kneel and kiss the earth." This Rumi quote eloquently expresses the need to forge our own way and bring our creativity and passion to our work. When these forces coalesce, wonderful things can and do happen.

Even though it may seem like the market is saturated with herbal products, there are still new things to try and the need for innovation is great, whether it be by offering the first herbal Community-Supported Agriculture (CSA) in your region or providing herbal wedding bouquets (a pretty common request—we kid you not!). The plants will help guide us to what is needed and how we can best find our way. As you are making plans, be bold, try something new. It may take time and will definitely require fortitude, but the possibilities are only as limited as the imagination.

# Herbalism 101—Do I Have to Be an Herbalist?

All this talk of value-added products could have farmers wondering, do I need to be an herbalist? If you are making many of the valued-added products

## Richo Cech from Horizon Herbs: A Sower of Seeds, Dreams, and Good Medicine

*Richo and Mayche Cech began Horizon Herbs in Williams, Oregon, in 1985. Horizon Herbs began in large part because Richo and Mayche couldn't find sources for many of the seeds that they wanted to grow. Undaunted and inspired, they began collecting seeds from around the world, growing them into plants and studying their life cycles. Fortunately for herb farmers and plant lovers alike, the Cechs began sharing both their seeds and the knowledge gleaned from years of working closely with the plants—knowledge about cultivation, conservation, ecology, and medicinal use. Today Horizon Herbs (consisting of four farms ranging in size from five to eleven acres) sells seeds for hundreds of species of medicinal and culinary herbs, cover crops, cacti, trees and shrubs, and vegetables.*

*And while it is inspiring to witness the significant contributions that Richo and his family have made to the herbal resurgence and plant conservation movement in our country, it is equally moving to hear about the genesis of their dream, how they sowed their first real and ethereal seeds, tended them with care, and grew them into a bountiful harvest.*

*It is important for us to look at these starting points, these jumping-off places where there is innovation, risk, and uncertainty, because this is where the creative spirit lives. How often do we listen to a renowned teacher and think, "Wow! That's incredible, but how could I ever do that?" Richo is one of these inspiring teachers, and as he tells the story of selling his first seeds he, by example, encourages people new to this field to find their niche, strengthen their resolve, and begin.*

Richo Cech writes:

I had an early fascination with plants, and while still at home with my mother (before I started kindergarten) I picked deadheads off French marigolds that grew near our doorstop and asked her about the elongated black seeds I found inside. I wanted to know if they would grow into marigolds. My mom said she

thought they would. This was my first investigation into seed saving and replanting, and it went from there. No matter where I went, be it family vacation or exploration into the ravines and river bottoms of Iowa, I collected seeds, dug and delved into the earth, and learned the plants.

In school I took to anthropology, then specialized in archaeology, and ended up doing a three-year stint in East Africa. After returning to the United States in 1977 we (Mayche and I) moved to Oregon in 1978 and settled on an eleven-acre piece with scant sun but lots of water welling up right out of the ground. I had a job planting trees. Planting trees was highly monocultural at the time (Douglas fir after fir after fir), fir as long as you could stand to plant 'em.

But I was looking around me in the woods and seeing some mighty interesting botanicals, such as high-altitude arnica and valerian, devil's club in the wet draws, and under forest canopy, Oregon grape. I wanted to find more about these herbs with the alluring names, so I bought a copy of Maud Grieve's *A Modern Herbal* and was bitten by the herbal bigtime. Seeds were harvested from the wilds and planted in the home gardens. Our gardens became more diverse, not just the three sisters of Native American fame, but their many cousins and other relatives.

We grew elecampane, valerian, calendula, and as many other medicinal herbs from the European tradition that we could find seed for. It was hard to find good seed sources, and somewhere along the path of this long learning a little light went on in my head, and I started to think about a seed company. By the time we moved to Southern Oregon to the town of Williams, where my wife and I had jobs teaching at an alternative school named Horizon School, I was convinced that diverse gardens of medicinal herbs were going to cure all the ills of Earth and society.

I only wanted to garden, and save seeds, and pack seeds, and give seeds away for the goodness of it all. There was a fantastic gathering back then that occurred each fall, known as The Ruch Barter Faire. Everyone would come and bring what they had to trade, squash and beet, dressed in tie-dye, with dusty feet. I sat on a blanket with my wares set out before me, little coin envelopes with medicinal herb seeds inside, handwritten packets, little green bombs of hope that I wanted to send off to explode peace in every garden far and wide. I had a hippie gourd rattle I'd made, carved in African style, and I shook it slowly and prayed for that whole long day, wishing someone would come and buy my seeds, or trade me a beet, or a bauble, or a drink of tea—something! I prayed that whole day long, and nobody wanted any seeds.

But now in retrospect I think I know what was going on. I needed to stoke the cosmic fire, and it was slow to start. The fuel, it would seem, was not perfect. There was some karma that needed to be burned before any profits could be made. My faith needed to be strong to do this work. As the sun set, my faith was tempered like steel in a forge. I sold no seeds that day, but later on, when the idea of planting diversity to heal self and Earth caught fire, and you couldn't even go up to Portland and swing a dead cat without hitting an herbalist, well, sell seeds I did. It was called Horizon Herbs, after the school. I meditated one morning and saw the sun rise above the mountains and birth a sprout from its orange globe, and that became our logo: the rising sun like a seed nestled between the peaks of the mountains, shining with potential, alive.

mentioned in this chapter, it is necessary to know what the herbs are used for and be able to communicate this information to your customers. There are a great many books, programs, webinars, and classes in every state teaching about the medicinal uses of plants, herbal formulation, and product making. As a farmer, if you aren't as well versed in herbalism as you are in soil science or plant propagation, this can be useful knowledge to develop.

However, herb farmers do not necesarilly need to be clinical herbalists. As a rule we don't offer any health consultations. However, like it or not, as an herb farmer you most likely will be asked to explain what a plant is used for. People at farmers' markets may tell you about their ailments and ask for recommendations. And when herbs are in the news you could be asked your opinion, whether you have developed one or not. Even if you only deal with wholesale accounts, having a working understanding of the plants' medicinal activities can help you provide companies with good customer service. So while you don't need to be a "card-carrying herbalist," it is beneficial to cultivate some understanding of herbalism and be able to talk knowledgeably with people about medicinal herbs.

If you aren't already obsessed with all things herbal as we are, you can begin with some simple beginner herb books or online courses that discuss herbal healing modalities as a whole and also delve into the medicinal uses of individual plants. At the very least we recommend you subscribe to one or two reputable herbal journals (not just health magazines) that will keep you up to date on new research, trends in the marketplace, plant conservation needs, and issues facing the herbal community. See the Resources listing for ideas.

**Figure 16-7.** "Let the beauty of what you love be what you do."—Rumi. Photograph courtesy of Kate Clearlight

*Voices from the Field*

## Katheryn Langelier from Herbal Revolution: A Small Farm Specializing in Value-Added Products

*Katheryn Langelier is the founder and owner of Herbal Revolution, a successful product business that developed out of her work as an herb farmer. Having grown medicinals for years, Katheryn created a unique and varied line of products utilizing the crops from her farm, including tinctures, teas, ciders, body products, and a clothing line. Katheryn's herb farm, herbal CSA, and products are an excellent example of how combining farming and value-added products can create a sustainable business model.*

*She attributes the success of her farm and product line to the work she does within her community, which affords her insight into what people need, what their health concerns are, and what plants would be useful for them. Drawing on this information and on inspiration from the plants, she creates products that meet these needs in a safe, effective, and rewarding way. In her work Katheryn has had to negotiate state and federal guidelines governing product safety. Although this has been time consuming and riddled with challenges, there are ways small farms and businesses can find success. Katheryn offers classes that discuss Good Manufacturing Practices (GMPs), sharing resources to help people navigate through these regulations.*

Katheryn Langelier writes:
As a young girl I knew that I wanted to live on a farm. I knew that I wanted to play in the woods and have soil under my fingernails. I didn't know that I was going to end up farming herbs and having a small herbal business until it all happened. For me it was incredibly important to have a relationship with the plants that I use to make products. The plants that I grow are the inspiration for the products that I create. The plants and intuition are what moves me to create the formulas that I blend. I don't feel I could have consciously started a product line if I hadn't started with growing the plants first.

When I first started the business, I was using almost all herbs that I grew and wild-gathered on our land. As the product business has grown I have had to source from other places, and I am currently looking for a larger farm so I will be able to expand our crops.

I am continually inspired and motivated to create because of the relationship I have with the plants that I work with. It is also important for me to spend time thinking about what our communities are in need of. I regularly hear that people have issues sleeping or focusing and are in constant stress and desire relaxation. I believe many people live with digestive, nervous, and immune systems issues. We live in a fairly toxic environment, where people regularly consume foods that have been sprayed with chemicals and grown in lifeless soil.

I think there is a deep loss of connection with the wild. As an herbalist I feel that we can help to introduce the wild back into our communities. We can do this through education, and also through the formulas and blends we create.

Coupled with the creative process of formulation is the need to work through state and federal regulations. At times this has been an incredibly daunting learning process. Those of us that are small don't have the sort of funds to hire consultants to help get us through this lengthy and time-consuming process. There have been many times that I felt like just giving up because of the overwhelming aspects and stress brought on by this portion of the business. Ultimately, I feel that there are a great many things in the regulations that are incredibly important and necessary, and I couldn't imagine not holding myself to these standards of quality. Yet I also feel there should be another set of regulations for smaller businesses. It seems as though the regulations were written for much larger companies that can afford to hire the consultant needed to create documents, pay for all the testing, and so on.

As a sole proprietor I need to wear many different hats, which has been a huge learning curve, since I don't have an educational background in business.

I also found it difficult learning all the things that I should be aware of, such as the GMP regulations and documenting processes. State regulators don't really know much about herbal businesses, and it took me a good many years to find an insurance company that understood what I was doing and could provide me a policy. I didn't find much help from other herbal businesses, so at times it was a frustrating obstacle course.

I, however, love a challenge and try to take advantage of the resources that my community provides for small business owners, as well as seeking help from others with experience in the areas that I lack. When I started there was a dearth of information about how to negotiate these regulations. Now there are more workshops and information on the GMPs, and I feel that it is becoming a bit easier for farmers and small businesses to work their way through them.

Getting to know the practicing herbalists, naturopathic doctors, and other healers in your area can be very useful, too. When more specific and complex questions arise, you can refer people to practitioners who can help. Long ago we started a running list, disaggregated by state, of herbalists, plant nurseries, herbal clinics, conferences, and herb schools to use for referrals. It has been extremely helpful.

While some farmers have little to no interest in herbal product making, other farmers find a natural affinity to it, bringing the crops they are growing right from the field and into the lab or apothecary. There are great companies out there that have found success with this type of model, both at the large and small scale. Take for instance Herb Pharm, a nationally recognized tincture and capsule company that uses twenty to thirty tons of herbs annually; it grows over 60 percent of the herbs they use on their own farms. Another great example is Herbal Revolution in Camden, Maine, that started as a small farm, then grew from there to become an award-winning, high-quality product company offering tinctures, elixirs, and other herbal preparations. There is no "one model fits all" out there. Rather, the herbal community has great potential and diversity in its size and scope for farmers.

**Figure 17-1.** Lily wearing her favorite farm T-shirt in the black cohosh.

# CHAPTER 17
# Marketing

We should start with a disclaimer: Neither of us went to business school, and we have no formal training in marketing. What we have learned about marketing comes directly from our firsthand experience running Sage Mountain Herb Products and ZWHF and also from emulating methods we have seen working with other great companies. Our understanding and approaches come from trial and error (and there have been many), and quite literally from the school of hard knocks. To find a comprehensive examination of marketing strategies and the latest research and trends, you will have to look elsewhere. If you are looking to become the next Fortune 500 in farming, this chapter is probably not for you. What you will find in these pages is the down-and-dirty tale of what we did to get our farm off the ground, how we marketed that first year's harvest of 111 pounds of dried herbs (that was our total farm yield

that year, now we do more than two tons of dried herbs annually), and how we positioned ourselves for expansion.

Even though we weren't formally trained, we did have a plan and a clear vision. It wasn't seamless and at times not elegant and rarely easy, but we are still farming together after fifteen years. Our farm is both successful and profitable; we provide good medicine for people, caretake the land, and support our small family both financially and in a simple lifestyle we love. As with any business, good marketing is absolutely essential for an herb farm. You've got to be able to grow the herbs, and that's certainly the first step, but right on its heels comes the second step—getting the herbs off the shelf and into people's hands.

## Establishing a Vision and a Mission

Establishing a clear vision and mission statement for your farm is a really important step and one that you should devote considerable time working on before you break ground and plow your fields. The vision and mission are the first seeds you plant. It is the intention you put out to the universe and your community; it will become your map and is a guiding light when times get rough. Ultimately your vision and mission will be what you manifest, what you harvest. To continue the metaphor, to get a good harvest you need good seeds. So it is worth the effort to spend time (alone or with your farming partners) visioning and crafting these statements.

So what is the difference between a vision and a mission? Often we use these words interchangeably,

**Figure 17-2.** Divided sky over a newly tilled field.

but they aren't analogous. The vision is your optimal goal or reason for existence. The mission is an overview of how you plan to realize your vision. From the beginning we were very clear about our vision for ZWHF. Our ultimate vision/goal *was and is to be good stewards of the earth, to protect the medicinal herbs that we cherish, and to produce the highest-quality botanicals for our community.* It was extremely important to both of us that stewardship and plant conservation be coupled with the production of medicinal crops. One without the other would not work for either of us. Neither of us would be comfortable with commercially growing crops, no matter how profitable, in a way that hurts the earth or the plant communities. This illustrates why visioning with your farming partners is extremely important because it will uncover not only shared beliefs and strengths but also deal breakers that need to be addressed one way or the other.

The mission is also extremely helpful to articulate because it fleshes out the key approaches you will use to run your farm. It's how you "walk your talk," how you see yourself reaching your goals. For example: Do you want to farm alone or collaborate? Are you focused on key principles or philosophies of farming? Are there areas of medicinals you see yourself specializing in? Do you want to focus on one type of market, need, or niche? You simply can't do everything, and if you jump from one thing to the next haphazardly, you will dilute your effectiveness. It's the mission that helps to crystallize your action steps, by focusing and prioritizing them around your vision.

Later, as you consider taking on new and exciting things, you can filter your decisions through the lens of your vision and mission. If a new endeavor will advance your vision and mission, it's worth consideration. If it doesn't, then you let it go. The mission we developed for ZWHF is twofold: *(1) Use organic, sustainable, and ethical farming practices to produce the highest-quality live, fresh, and dried medicinal herbs; and (2) work in collaboration with other herbalists, farmers, and researchers to develop and share knowledge about the cultivation and preservation of medicinal plants.*

**Figure 17-3.** A clear part of our mission is to farm organically.

Once established, it's important to keep your vision and mission statements prominently visible in your work space and on all promotional materials. Not only does this keep what is important at the forefront of all you do, it helps your community get to know you as more than a number in the Yellow Pages or a name on a Google hit. Your vision and mission will distinguish you from others and help people have a more authentic understanding of what you do and why you do it.

## Which Comes First, the Market or the Herb?

We get this question a lot, and there are no easy answers. On one hand, it can feel a bit daunting to grow a bunch of herbs, not knowing if you can sell them. Conversely, it is very hard to find wholesalers to enter into a purchase agreement if you haven't grown the crop before and can't offer a sample or discuss the crop in detail. Retail and wholesale markets can be tricky because you need to grow herbs to have them to sell, but how do you avoid having surpluses that don't move off the shelf?

It certainly can be a "chicken or egg" phenomenon, and even as we write this book we are in the thick of it; we have quickly sold out of some herbs, are anxious about dwindling supplies of others, and are sitting on a couple of hundred pounds of some bumper crops that we are hoping to move. Frequently the sales follow a pattern, but sometimes what sells out year to year differs. There is no simple

**Figure 17-4.** Bagged herbs ready for market. Photograph courtesy of Bethany Bond

formula, and we have found that sometimes we have to grow some herbs on speculation and put time into marketing or, on the flip side, contract to grow a new crop and figure it out as we go along. We are not adverse to risk (after all, we *are* farmers) so leaping off into the great unknown can be an initially frightening but ultimately rewarding experience depending on how you land.

What helps stabilize this somewhat unnerving juggling act is diversifying the types of crops you grow. Meaning, do not put all of your eggs in one basket! On our farm we often think about crops and their marketability in the following categories: Core Crops, Specialized/Challengers, and Contracted Crops. *Core Crops* are, for the most part, the

tried-and-true crops that are easy to grow and easy to market. They make up a large portion of what we grow and sell, season to season. *Specialized/Challengers* are plants that are highly prized and sought after in the market but present a challenge either to grow or to harvest. Many of these plants are exotics and have highly specific and specialized growing conditions or even if they are common (like chamomile) are extremely labor intensive to harvest. Pound for pound we produce less of these herbs, but their high market prices still make them profitable crops worth the effort they require. Each season we explore ways to increase efficiencies with the Specialized/Challengers in hopes that we can find ways to grow and harvest them in greater quantities to meet the demand and tap into their strong markets. Finally we have the *Contracted Crops* or crops that are grown with a careful eye to marketability. These plants have very specialized markets, and we make sure we can sell them before growing or harvesting them. Japanese knotweed is an excellent example of this. It can be a profitable crop and provides much-sought-after medicine to help treat Lyme disease. However, it is also an invasive species and cannot be harvested without a state permit. We obviously don't grow this plant but have obtained permits to wild-harvest it by the thousands of fresh pounds.

**Table 17-1.** Crop Considerations for Marketability

| Core Crops | Specialized/Challengers | Contracted Crops |
|---|---|---|
| Anise hyssop leaf/flower (*Agastache foeniculum*) | Arnica flower (*Arnica* spp.) | Angelica root (*Angelica archangelica*) |
| Ashwagandha root (*Withania somnifera*) | Chamomile flower (*Matricaria recutita*) | Dandelion leaf/root (*Taraxacum officinale*) |
| Burdock root (*Arctium lappa*) | Elder flowers and berries (*Sambucus nigra*) | Japanese knotweed root (*Polygonum cuspidatum*) |
| Calendula flower (*Calendula officinalis*) | Ginkgo leaf (*Ginkgo biloba*) | Yellowdock root (*Rumex crispus*) |
| Comfrey leaf/root (*Symphytum officinale*) | Goldenseal root (*Hydrastis canadensis*) | |
| Echinacea root/tops (*Echinacea purpurea*) | Hawthorn berries (*Crataegus* spp.) | |
| Garlic (*Allium* spp.) | Passionflower leaf and flower (*Passiflora incarnata*) | |
| Lemon balm leaf/flower (*Melissa officinalis*) | Rhodiola root (*Rhodiola rosea*) | |
| Nettle leaf (*Urtica dioica*) | Schisandra berries (*Schisandra chinensis*) | |
| Milky oat tops (*Avena sativa*) | | |
| Peppermint leaf (*Mentha piperita*) | | |
| Tulsi leaf/flower (*Ocimum tenuiflorum*, syn. *O. sanctum*) | | |
| Valerian root (*Valeriana officinalis*) | | |

We certainly would not do this on speculation. We harvest this type of plant, at this volume, because we have an established relationship with manufacturers. The same can be said of herbs such as angelica and yellow dock root. They are in demand by several extract companies we work with but are not big sellers in the retail arena.

By growing a substantial number of Core Crops you can create a foundation of herbs for wholesale and retail markets, knowing that there is a high probability of selling them. They bring in stable, more consistent sales and help open up other markets and ventures with new plant species. Again we can look to Japanese knotweed (a Contracted Crop) as an example of how work with Core Crops can lead to new markets. As mentioned previously, we didn't work with Japanese knotweed until two years ago. However, we had spent the last fifteen years growing and processing hundreds of pounds of popular roots such as burdock and echinacea.

Therefore, when we were approached to supply a company with several hundred pounds of dried Japanese knotweed, we felt, based on our experience with other core root crops, we could jump into this market. We were also able to show this herb buyer what our roots, in general, looked like in their dried state, the cut-and-sifted grade, and we talked quantity and quality. Because we knew the approximate fresh-to-dried ratios for many species of roots, we had a sense of how much fresh knotweed root we needed to harvest to meet his need. A deal was struck, we obtained the required harvesting permits from the state, and we were off to the races. As it turned out, we easily were able to fill the order and have had the pleasure of working with this great company ever since.

Sometimes a venture doesn't work out as easily, and you will need the stability of Core Crops to weather the storm. One year we tried growing hundreds of pounds of skullcap for one company, only to lose it all to powdery mildew a week before harvest. That year the crop didn't pay off. Thankfully, the other Core Crops balanced out the loss, as we had other species that had reliable harvests and steady sales. We aren't easily deterred and will certainly try growing large crops of skullcap again but at the same time will remain diversified so that we can remain fiscally sound when crop failures occur. There is no magic formula or perfect crop list or rotation. You have to discover what works for your farm based on bioregionality, temperate zones, and personal affinity. That said, paying careful attention to trends in the market and planting a combination of Core, Specialized/Challengers, and Contracted Crops can help position you for success.

## Wholesale versus Retail

Depending on where you live and how you like to spend your time, you may find yourself drawn to the wholesale market or retail or both. We know some folks who prefer to grow a few medicinals in large quantities, selling directly to wholesalers. Other friends love farmers' markets and are located near urban hubs; they are hitting the retail markets hard and are doing well. Other farmers are utilizing current technologies to tap retail and wholesale markets online.

At ZWHF we do a combination of all of these: 51 percent of our sales (2013 sales data) comes from wholesale markets and the other half from retail sales to individuals, either from direct sales off our website or at a few major herbal conferences. We no longer do farmers' markets because our farm is located an hour away from our preferred market, and the time commitment is too much. This 50/50 wholesale/retail split has been a relatively consistent phenomenon for the last five years.

However, earlier in our career most of our sales were retail due in part to lack of volume to enter the wholesale market. We also had more time, early on, to go to multiple markets and conferences. Now that we have expanded the farm and started a family, we tend to keep closer to home and appreciate having a bigger wholesale market. There are costs and benefits to both types of marketing. Which markets you work in will depend on your farm, the herbs you grow, and your individual proclivity and needs.

# The Diverse Marketplace and Advertising Opportunities

Because of advances in technology and the explosion in popularity of social networking platforms, there are so many more ways to market your herbs than ever before. In the beginning we had farmers' markets, conferences, word of mouth, and snail mail. Now the proverbial sky is the limit. Besides advances in websites to include shopping carts and PayPal options, there are virtual farm stands, Facebook markets, PoppySwap, Etsy, Pinterest, and blogging (and a whole host of other platforms) that can seriously broaden your customer base. Using e-mail management tools such as MailChimp or Constant Contact can also be an extremely useful way to keep people informed about the happenings on your farm and what crops are coming into season. We use the "old school" approach of working face-to-face with our customers (which we love) and have also embraced the digital age—having cyber "friends," whom we may never meet but who nonetheless are huge supporters of the farm.

In addition to using these advertising opportunities, we approach product manufacturers and stores in surrounding areas to see if there are herbs they would like us to supply for them. One thing that has been particularly helpful in our area is the localvore movement, coupled with people's desire to have organically grown medicine. Stores and manufacturers, as well as the individual customer, have increased awareness about reducing the industry's carbon footprint and reducing the distance medicine has to travel from farm to market. Not only does it benefit their bottom line by lowering shipping costs, but it is also better for the planet.

While we love the visibility and outreach technology has afforded us, we continue to be very committed to the local herb movement. In the long run we hope that more people will take up herb farming and that there will be a network of growers that can provide their communities with herbal medicine. Eventually,

**Figure 17-5.** Retail vending at herbal gatherings.

as a collective we may be able to reduce the need for transcontinental herb shipments by having an herb farmer in every community.

## Creating Ways to Collaborate and Not Compete

Early last winter we had the good fortune to have our mentors and friends Matthias and Andrea Reisen of Healing Spirits Herb Farm and Education Center come to our home and farm for a visit. Matthias and Andrea were pioneers of medicinal herb farming back in the 1980s, when very few people in the United States were doing it. They began as dairy farmers, but when their dairy farm became hard to manage fiscally, they looked to the plants and began growing medicinals. Using their wealth of knowledge in agriculture and their indomitable spirits and work ethic, they began to explore how to grow and dry herbs on a small organic medicinal herb farm. And that's exactly what they have been successfully

**Figure 17-6.** Our mentors Matthias and Andrea Reisen cofounded Healing Spirits Herb Farm and Education Center, a NOFA-NY Certified Organic Medicinal Herb Farm, in the Finger Lakes Region of New York. Since 1982 there have been three generations living, working, and playing on the farm. They live simply, seeking a life of balance and harmony with all creation. Photograph courtesy of Andrea Reisen

doing ever since—for the last thirty-plus years! So when we began our farm in 1999 we turned to them for help and guidance. We will always be thankful that at that crucial time they lent us a hand, opened their hearts to us, and saw us as collaborators and colleagues, not as competition. Our relationship has deepened over the last twenty years, and to call them family is not an overstatement.

Last winter, while Matt and Andrea were with us, sipping tea and watching the snow fall, we began to reflect on our experiences as medicinal herb farmers and the state of things in the industry. Through our conversation it became clear that we were all noticing a continuing trend of large volumes of medicinal herbs being imported from overseas. We lamented how difficult it is to get bulk herb buyers to buy herbs from small farmers. We talked about the obstacles facing small farmers trying to enter this arena and wondered how these issues could be overcome. Matthias shared his idea of creating a website called HerbsUSA that would be a clearinghouse where domestic farmers could market their herbs directly to the consumer. Jeff and I talked about the desire to start a growers' cooperative in our region, to band small farmers together to share resources and collaborate to meet wholesalers' needs for high-quality herbs. We also thought about approaching big herb distributors and asking them to include domestically grown herb lines in their online shopping carts, which would feature and support small, local farms.

Lots of ideas were kicked around and many seeds planted, but what is important to note, especially in a chapter on marketing, is this essential need for collaboration and not solely competition. There is a market out there for high-quality, organically grown herbs. Herbalists and clients alike are becoming more committed than ever to supporting this. However, if we all compete for shelf space in the one local health food store in our town or all go to the one or two tincture makers in our region and try to sell them the same herbs, we may all suffer. Instead small herb farmers need to band together and look at these other, larger markets and find ways to tap into them. Now is the time to come together to form cooperatives,

*Voices from the Field*

## Jeanine Davis: A Professor and Agricultural Specialist with a Vision for Change

*Jeanine Davis, associate professor and extension specialist for North Carolina State University, is an expert in the field of medicinal herb cultivation and marketing. For the past twenty-six years she has been working with farmers around the country to explore farm diversification. Jeanine has taught classes, held seminars, and provided outreach to farmers, with an emphasis on organic agriculture of medicinal herbs, nontimber forest products, mushrooms, and specialty crops. In addition, she has published numerous articles and a book about growing and marketing woodland botanicals.*

*Having a very comprehensive understanding of the United States herbal marketplace, Jeanine provides insight into what wholesale buyers look for in terms of quantity, quality, and selection of species. Jeanine believes in creating herb growers' cooperatives to help small to midsize farmers meet the national demand for high-quality herbs. At the heart of it she is also a proponent of growers, buyers, and scientists working together to address the main issues facing the herb farming industry today; namely, low wholesale pricing and the lack of specialized equipment and education.*

Jeanine Davis writes:
Cooperatives are the way to make it [small herb farming in the United States] happen. It only makes sense. Although there is clearly the demand, and we should be able to do it, it is still not happening at the level it should be, and the prices the buyers want to pay are ridiculously low. But I'm really hopeful this is

turning around. In the past year we have held three workshops with buyers, who have come out to our workshops and talked to large groups of people about growing herbs for them. This type of collaboration and technical assistance is essential, because to command these higher prices, quality standards need to be met and efficiencies improved.

To put a finer point on it, someone says, "I'll grow herbs for this company," but he or she doesn't know anything about growing herbs. We need to be teaching them [the farmers] how to get the herbs clean enough. They can grow a beautiful herb crop, but the farmers aren't treating it like something you and your children are going to eat. They [the herbs] aren't clean enough. I think it's an educational thing, and I'm hoping the buyers will help us to build up growers.

When I look at my small-scale growers who are producing directly for an herbalist or for very small, local manufacturers it's a different story. They take the time to grow together and to learn. If that grower and buyer will give it three years, they can get to an excellent place, and we can have quality that will beat anything that will come from overseas.

We also need to develop more effective methods and things—equipment we can build. Weed control is an issue, as is proper harvesting, washing, and drying. We need effective and efficient ways to get those crops out of the ground and get them washed and dried in a timely manner. We need to get more of our agricultural engineers and scientists to figure out how we do this more efficiently and do it in a way that is affordable.

encourage and support new farmers to enter the fold, and command this market, not only for the amazing medicines that it will create but also to help preserve the small, sustainable American farm and caretake the land. There are rich opportunities and also difficulties to deal with, but it is worth the effort.

# What Do Wholesale Buyers Want?

One of the most important and most commonly asked questions by people new to herb farming and the wholesale herb market is, "What do wholesale

**Radix Taraxaci**
**Taraxacum officinale**
Dandelion Root

**Plant Material of Interest:** Dried root pieces

**General Appearance:** Cut and sifted irregular root pieces ranging from 0.5–1 mm, ranging from dark to light browns. Outer cortex brownish, inner material light tan to white but not pithy.

**Organoleptic Properties:** Characteristic bitterness to the woody root. Slightly sweet root flavor.

**Marker compounds:** Presence of sesquiterpene iridoids by HPLC/TLC.

**Rejection criteria:** Anything other than cut and sifted root pieces. Colorless plant matter, presence of leaves or other plant parts. Pithy, spongy inner root texture.

**Figure 17-7.** Sample monograph of dandelion root.

Zack Woods Herb Farm
278 Mead Rd, Hyde Park VT 05655
802-888-7278  www.zackwoodsherbs.com

## Certificate of Analysis

Product name:        Angelica
Botanical name:      Angelica archangelica
Product lot #:       20131013ZW
Harvested:           October 2013
Origin:              USA
Plant part and form: Root
Organic cert:        Yes
Description:         A. Organoleptic:
                        1) Color, odor, and flavor: Brown root,
                           cut and sifted; characteristic odor.
                        2) Extraneous matters: None
                     B. Chemical
                        1) Moisture Content:<10%
                        2) Steam Volatile Oil: N/A
                        3) Sulfite: N/A
                        4) Pesticide & Chemical Residues: N/A
                        5) Others: N/A
                     C. General Shipping Guidelines
                        1) In food grade plastic bags
                        2) In cardboard boxes with
                           appropriate labels

There has been no use of chemicals, sewage sludge, genetically
modified materials, or radiation of any type in the cultivation,
harvest, drying, or shipping of this product. The information
provided is in good faith accurate but does not warranty accuracy
of results.

Approved by: Bethany Bond
Date issued: October 30, 2014

**Figure 17-8.** Angelica Certificate of Analysis.

*Voices from the Field*

## Jovial King of Urban Moonshine: A Modern Apothecary with Roots on the Farm

*Urban Moonshine is a forward-thinking, innovative company located in the heart of Burlington, Vermont. Founded in 2009 by Jovial King, Urban Moonshine makes high-quality organic bitters and herbal tonics that have national distribution and have received accolades from herbalists, such healing gurus as Dr. Andrew Weil, and cocktail connoisseurs alike. Not only do their products taste delicious, they are incredibly healing and have been generating increased awareness in the mainstream about the daily use of bitters and herbal tonics to improve wellness.*

*One of their slogans, "Farm to Bottle," speaks of their commitment to supporting farmers. True to their word, Urban Moonshine buys from small, organic growers (like us and others), even when going to mass wholesalers would be cheaper and easier. They do this in part because locally grown artisanal herbs are higher in quality than those processed commercially by machines.*

*As Urban Moonshine has grown and expanded, they have remained committed to quality and have, like other companies, navigated the morass of GMPs and regulations. As farmers who work with them, we have become increasingly aware of the scrutiny they and, as a result, our bulk herbs are under as regulatory agencies require testing, documentation audits, and detailed paper trails. Having an understanding of and being responsive to the requirements facing producers is important for herb farmers who plan to enter the wholesale arena. Farmers and producers working together to meet authentication, cleanliness, and organic standards is essential.*

Jovial King writes:
We are very committed to certified organic. I find it fascinating that people care so much about organic in the grocery aisle, and when they hit the supplement section it goes right out the window. The supplement section is the most highly processed unnatural section in natural food stores. We need to put as much thought into where our medicine comes from and how it was produced as we do our food. It's even more important in certain ways because a lot of herbal products are made from a large concentration of plant materials. We are not just talking about one head of lettuce; we are talking about pounds of herbs going into one single bottle—where did those herbs come from, and how were they grown?

In the beginning, before the GMPs came into play, I would buy herbs from any cool organic herb farmer that had high-quality botanicals. But now that we are an FDA-regulated, GMP-compliant certified-organic company, we have very strict guidelines for quality. So our main two deal breakers these days are documentation of organic certification and the amount of herbs that someone is able to provide us per year. Unfortunately, because of the amount of documentation and testing we have to do for each herb and each vendor, we just can't buy a few pounds of this or that herb from the guy down the street. We get calls every few weeks from herb farmers that are just starting out and are like, "Hey, I'm growing some burdock. Do you guys want to buy any? It's really great stuff, and it's totally organic." We have to ask, "Well, is it organic, or is it certified organic?" There is a big difference. We need the documentation; we have to have the paper trail for our inspectors.

For GMP compliance we have to keep a file on each vendor we work with, documenting why we think they are a quality vendor. The FDA is insistent about this; it comes up every time they inspect us. It's just not enough for us to say, "We really like them, and they grow great herbs!" We now have to, at a minimum, test three items from each of our vendors to make sure that what they say is honey, dandelion root, roses is *really* honey, dandelion root, roses, and so on. We basically have to build an evidence file of why we can trust them to supply us with quality goods. The testing is expensive, and we already have a lot of vendors, so when we get a call from a farmer that wants to sell us ten pounds of burdock, I really wish I could say yes, but to buy that $100 worth of burdock, I have to qualify him as a vendor, start a file on his business, and spend money on testing to make sure he is really selling what he says he is selling.

It's a whole new world in the American supplement industry; sad but true. The GMP regulations are changing the face of herbal medicine in this country. We have to work really hard and go above and beyond to stay GMP compliant as a small company—it's no joke. There is a lot of paperwork and time spent on quality control, the files we keep are immense, and when the FDA comes in for the inspection they take days (four days on our last inspection) going over every single piece of paper, and if they find anything missing or forgotten you get written up for it. The GMP regulations themselves are not bad. It's a great idea to require lot numbers and batch records, but it is the one-size-fits-all approach that is so devastating to the herbal community.

Should the herbalist making medicine out of her garden for her community face the same level of regulation as a national multimillion-dollar brand? No way. They make it much harder to enter the marketplace legitimately, almost impossible for a small-time herbal company. However, I will say that the regulations will help to clean the industry up and create more corporate responsibility when it comes to safety and transparency. It is still the Wild West when you look at a lot of the junk that people are selling under the guise of "health supplements."

buyers want?" With the new requirements outlined by the FDA in their GMPs, wholesalers are under increased pressure (especially when organically certified) to verify that their herbs are correctly identified and harvested and processed properly. Herbs should not be contaminated by other botanical species, pesticides, fungicides, heavy metals, and other foreign materials. In addition, each wholesaler will have specifications regarding packaging and shipping, lot tracking, plant authentication, and sometimes chemical testing and verification. Because each company is different, it is essential for herb growers to have a strong working relationship and clear communication with their buyers. In our experience some of the key pieces to put in place as you begin reaching out to wholesale buyers are these:

- Have your organic grower or wildcrafter certification and documentation on file and be able to supply that to the company.
- Create a system to track harvest time, date, location, plant part(s) harvested, and lot numbers for each species.
- Provide a Certificate of Analysis and/or a Plant Identification Sheet for each species you supply.

## You Found What?

One large tincture manufacturer recently shared with us this list of items that they have found in herb shipments received over the years. Some of them led to the rejection of the crops. It is important to check crops carefully because you *do not* want these things in your packages.

Following is a list of the interesting and somewhat alarming foreign objects (no kidding) found in a large herb company's bulk herb purchases from the last few decades: moths and larvae, insects too numerous to list, snails and slugs, dirt, stones, worms, feathers, human hair, animal hair, grass and grass seed, weeds too numerous to list, rogue plants, twigs, pods, snakeskin, feces (snail, insect, and bird), mold spots, rotten spots, cocoons, pine needles, eggshell, lichen, moss, pumpkin seeds, galls, Styrofoam, nylon string, foam rubber, paper, plastic, nails, cloth fibers, paint flakes, bus ticket to Bombay, rubber bands, glass, metal slivers, gel from a burst freezer pack, staples, piece of a cassette, wire, paring knife, fishhook, bark, mulch, keys, reading glasses, and bolts.

- Be willing to provide whole plant samples for identification and microbial testing if required by the company.
- Before growing or harvesting a crop for a company, request a plant monograph and/or harvest specifications so that you are clear about what the buyer is requesting.
- Be aware that your crops can be rejected, at your expense, if GMP requirements are not met.

While the GMPs theoretically are there to improve the system, there are still places where they could be improved, and they can cause a lot of stress on manufacturers large and small. As suppliers we try to be sensitive to the strain that these types of regulations put on buyers and work together with them to make compliance easier by providing clear documentation, as well as high-quality herbs.

In the end the most important aspect of marketing, whether you are selling half a pound of peppermint to a neighbor or two hundred pounds of yellow dock to a manufacturer, is to make sure to showcase what makes your herb farm unique. A key component of this marketing will most likely be high quality. Unlike many of the herbs on the mass market that are cut by machine, dried on the ground, and bailed for shipment, your herbs, if grown and processed well, will contain the vitality of your land, will be strong in color and aroma, and will contain potent medicine. There is no comparison; it's like comparing artisanal cheese and Cheez Whiz. And while marketing, like farming, is highly personalized and farmers will approach it in different ways, we have found the best marketing strategy is to let the herbs speak for themselves and to continue to help educate people on the benefits that small farms have, not only for people using herbs but also for the plants and the land.

Photograph courtesy of Bethany Bond

# PART TWO

# Herbs to Consider Growing for Market

There are literally thousands of species of medicinal plants in use and in commerce around the world. Selecting which species to grow on our own farms and in our own gardens is not always easy. For growers who live in regions with long growing seasons, good soils, and ample water for irrigation, you have a wider range of choices than we have here in northern Vermont. Here we are fairly limited in what we can grow based on our average 150-days-per-year growing season in USDA hardiness zone 4b. Luckily our soils are average to good, and we usually have ample rainfall and irrigation water.

We are also fortunate in that we live in a very "herb-centric" state, meaning that Vermont seems to have a higher than average percentage of herbalists and herbal product businesses per capita than the average U.S. state. We currently grow, process, and market over fifty species of plants for bulk herb sales and another twenty or so species that are a bit more challenging to grow in a production setting or are less marketable. These "extras" we grow for live plant sales or specimen plants in our flower beds. We chose not to profile these "extras" because we have not had enough experience with growing, processing, and marketing them in large quantities to share useful data.

In part two we have chosen to profile the fifty species of medicinal herbs that we have had consistently good success with growing as well as marketing for bulk herb sales. Most of these fifty species have been in our rotation for the fifteen years we have been farming, so there is a good foundation of experience from which we have sourced the information we are presenting. When we talk about plants with our friends, family, or customers (which we love to do), Melanie has a habit of stating, "That's my favorite plant," regardless of which species we are discussing. It's cute to tease her about that, but she is really on to something because it's hard to choose favorites based on the incredible and unique qualities that each plant has to offer. So here are fifty Herbs to Consider Growing for Market, otherwise known as "Melanie's favorite plants."

The data shared in these profiles with regard to plant spacing, yields, moisture ratios, and pricing are based on the data we have kept on our farm for the last fifteen years and is partially augmented with data that is currently available in the industry. Please bear in mind that yield data are based on our row-spacing configurations, which are presented in the text of each profile. Therefore if we say one-eighth of an acre of something yields a certain number of pounds, this doesn't necessarily mean that the whole entire eighth-of-an-acre field was covered in plants. These are estimates for our region and should *not* necessarily be taken as gospel. There are many variables that can and will make these data points fluctuate, and farmers in Iowa or Alaska will most likely have different experiences with these crops from ours. Unlike veggie and other types of farming, where there are regional databases containing crop budgets and yield data for everything from artichokes to zinnias, there is a significant dearth of information of this kind for medicinal herbs. In fact, the USDA has an extensive database of crop data for thousands of species of commercial crops—even some very obscure ones—but interestingly enough, there is virtually no available data for the production of medicinal plant species. I like to joke that there is a conspiracy here being heavily influenced by the pharmaceutical industry, but it may be no joke.

At first we were hesitant to include our data, as some of it may quickly become dated, and we did not want to be misleading. However, in the end we felt it was extremely important to begin getting these numbers out there and to implore other farmers to share what they glean from their own experiences. Perhaps by working together with our local agricultural networks we can begin to develop a broader database for these crops as well as many others so that growers in the future won't be faced with the same challenges we had regarding lack of reference materials.

We wish you success with growing, harvesting, and marketing your own medicinals, and we hope this information helps you to contribute to the demand for high-quality botanicals produced locally on small organic farms.

# Alfalfa                                    *Medicago sativa*

## Life Cycle

Alfalfa is an herbaceous perennial legume that is hardy to USDA zones 3 to 11.

## Plant Description

Alfalfa is a member of the Fabaceae family and is native to Iran, Central Asia, and Siberia. This perennial has a long taproot supporting an upright stem with clover-shaped leaves that grows one to three feet tall. The leaves vary from bright green to chartreuse yellow, with light-purple blossoms forming in mid- to late summer. It is highly attractive to bees, butterflies, and many other pollinators and is commonly used for forage, hay, and silage.

## Growing Conditions

Alfalfa grows well in many climatic conditions and soil types but is partial to sandy loam. The plant is somewhat drought tolerant and can grow in many different growing conditions. However, alfalfa does not grow well in compacted soil and prefers soil that is well drained, with high fertility and high pH (>6). Alfalfa does best in full sun with consistent soil moisture.

**Figure 18-1. Alfalfa in bloom.** Photograph courtesy of Michelle Tribe

## Propagation

Alfalfa's seeding rate is approximately twenty pounds per acre; it should be drilled or planted with a broadcast seeder. It is recommended to prepare fields with lime if needed based on soil tests, and alfalfa also benefits from increased fertility, as it is a fairly heavy feeder.

## Planting Considerations

Alfalfa is a highly beneficial cover crop. It reduces erosion and has deep taproots that improve soil structure by breaking up compaction and improving water filtration and drainage. Alfalfa is a legume that fixes nitrogen in the soil and when tilled under provides soil with biomass and improves tilth. Alfalfa is also an important forage plant for many wild animal species.[1]

## Medicinal Uses

Leaves and flowers are highly nutritive and contain chlorophyll, protein, and vitamins A and C. Alfalfa is used to help lower cholesterol and help regulate and balance estrogen levels. It is often taken as a tea or in formulation with other tonic herbs.

## Harvest Specifications

The aerial parts (leaf and flower) of alfalfa are harvested at its most nutritious stage, which is when it reaches full maturation and begins to bloom. The aerial parts of the plant can be harvested with a sickle bar mower, scythes, or field knives. It is fairly easy to harvest alfalfa, but care should be given to process it promptly, as it can begin composting quickly if left too long in direct sunlight. It is common to get multiple harvests per season.

## Postharvest and Drying Considerations

To dry alfalfa, spread on drying racks in a single layer with good airflow. It dries in approximately two days under optimum conditions at temperatures of 100 to 110°F. Once it's dried, garble alfalfa on quarter- to half-inch stainless steel mesh and remove stems. Sell leaf and flowers together.

## Pests and Diseases

Common issues with alfalfa include root nematodes, aphids, and leafhoppers. It is also susceptible to some stem and root rot if not planted in well-drained areas. That said, we have had very few issues with pests and diseases with alfalfa. Proper field management and crop rotation mitigate most issues.

## Yields

Three hundred pounds of dried alfalfa leaves and flower per one-eighth-acre bed (multiple harvests per season). Moisture ratio for alfalfa tops is approximately 5:1 fresh:dried.

## Pricing

Retail price for one pound organic:
- Dried alfalfa leaf: $9 to $24
- Fresh alfalfa leaf: $9 to $14

# Angelica        *Angelica archangelica*

## Life Cycle

Angelica is a majestic biennial hardy to USDA zones 3 to 9.

## Plant Description

Angelica belongs to the Apiaceae (parsley) family and is believed to have originated from northern

Europe, Asia, and Greenland. This biennial is often referred to as the queen of the garden and is lovely to behold. During the first year, growth occurs primarily in the roots and leaves. In the second year the plant produces hollow fluted stems six to eight feet in height that support globe-shaped umbels of cream-colored flowers. In the second year angelica goes to seed, then dies back in the fall. However, it is not uncommon for this biennial to perennialize for three to five years, and it self-seeds quite readily.

The pinnate leaves of angelica are bright green and finely toothed. They grow spirally off the main axis in groups of three leaves that branch off three to five times from a single leaf sheath. The plants may grow up to eight feet in height during the second year. The hollow stem is ribbed and ranges in color from purple at the base to light yellow at the crown. The semiround umbels contain greenish-yellow flowers that are delicate and small. Highly aromatic, angelica flowers are a favorite among pollinators. In July when they bloom, our fields of flowers are buzzing with activity. All parts of the plant contain resin canals and are rich in essential oils, which is one reason angelica is a prized medicinal and culinary herb.

## Growing Conditions

Angelica thrives in full sun and grows well in partial shade. Often found growing in wetlands and on stream banks, angelica tolerates a substantial amount of moisture and grows best in rich soils. For cultivation angelica grows best in nutrient-rich humusy soil. It likes moist soil and we have found that angelica grows well in full sun in Vermont. In warmer, sunnier climates angelica will grow well in partial shade.

## Propagation

Angelica can be propagated by direct seed or transplant. The seeds are light-dependent germinators, and stratification is recommended to assist in breaking dormancy. Direct seeding of angelica can be done in the fall or early spring into well-prepared, moist soil. To grow from transplants use fresh, stratified seeds and sow thinly on top of the soil or barely cover. Keep seeds moist until germination. Angelica

**Figure 18-2.** Angelica.

takes approximately three weeks to germinate. It is important to use properly stored fresh seed as older seed rapidly loses its viability.

## Planting Considerations

Because angelica is a very tall species that will remain in your beds for at least two years, consideration should be given to bed layout. Angelica can be planted at the edge of a bed to act as a windbreak and provide partial shade for plants like skullcap that do better out of full sun. Recommended bed spacing for angelica is sixteen inches in the row and twenty-eight inches between rows. Angelica grows best in rich, moist soil, but excessive nitrogen can cause the plants to put more energy into aerial growth, limiting root production.

## Medicinal Uses

Angelica is a favorite bitter herb for warming the body, for stimulating digestion, and as a bronchial aid and expectorant. For women the decongestive, warming properties of angelica can help alleviate bloating, painful menstruation, and cramping. The roots are commonly used in teas, elixirs, extracts, and bitters. Angelica's unique flavor and aromatic properties have been used medicinally and as a culinary for centuries. The stems and seeds are often candied and used to create flavorings. All parts of the plant are used in the production of essential oil; however, the roots have the highest concentration. Additionally, the stem, leaf, and seed have been used by native peoples of northern climates as a curdling agent to make cheese and as a food source rich in vitamin C. Flower essences are also made from angelica blossoms and are used to strengthen people's spirit and help them feel more connected and centered.

## Harvest Specifications

Roots are harvested in early spring of the second year, preferably while the plants are still dormant before significant aerial growth occurs. Waiting until the second year allows the roots to develop more fully and reach maximum size. If roots continue to grow past the spring, they can become hollow at the center and pithier. Not only does this diminish yields, but the roots become less aromatic and not as medicinally potent. The roots are fleshy and large, branching out from a thick centralized stalk, and can grow one to two feet deep. As a result, it is helpful to mechanically loosen the surrounding soil before digging the roots. This can be accomplished by running a shank or cultivator up the row on either side of the angelica. Once the soil is loosened, roots can be dug out by hand with spading forks or mechanically, using a bed lifter or potato digger.

## Postharvest and Drying Considerations

Soil often gets compacted in the central crown of angelica where the roots begin to branch off. Therefore it is helpful to divide the roots before washing. Angelica roots are relatively soft and can be chopped easily with a field knife. After splitting the roots wash them thoroughly before drying. Mill roots after they are fully dried. Angelica roots dry under optimum conditions at temperatures of 100 to 110°F in three to four days.

## Pests and Diseases

We have found that angelica grows easily, with no pest or disease pressure; in fact, angelica is one of our premier insectary plants, attracting an incredible array of parasitizing wasps and other predatory insects. In Europe large-scale angelica growers have reported difficulties with the fungal disease *Rhizoctonia crocorum*. This purple spotting on the roots can often lead to root rot. Other growers have experienced grub and rodent damage. Crop rotations and good cultural practices are recommended to avoid pest and disease issues.

## Yields

Three to four hundred pounds of dried angelica roots per one-eighth-acre bed. Moisture ratio for angelica roots is approximately 4:1 fresh:dried.

## Pricing

Retail price for one pound organic:
- Dried angelica root: $14 to $37
- Fresh angelica root: $12 to $21

## Commonly Imported From

Croatia and Bulgaria

# Anise Hyssop     *Agastache foeniculum*

## Life Cycle

Anise hyssop is an herbaceous perennial hardy to USDA zones 3 to 9.

## Plant Description

Anise hyssop is a popular ornamental plant, a favorite nectar source for pollinators, and a delicious medicinal herb. It is a member of the mint family, Lamiaceae, and has fragrant leaves and spiky purple flowers. The aerial parts of the plant have a sweet licorice or anise flavor and are delicious, as well as pleasing to the eye. Thriving in well-drained soil and full sun, anise hyssop originated from the fields and prairies of North America and can be found throughout America and Canada. Anise hyssop grows in erect, upright bushes that can be three to four feet in height and one to two feet wide. When the plant comes into bloom in late summer it attracts pollinators, including bees, hummingbirds, and butterflies.

## Growing Conditions

Home to the prairies, fields, and woodland edges, anise hyssop grows best in full sun with medium to moderate moisture in soils with a pH of 6 to 6.5. This plant is easy to grow and does not require much specialized care. However, it is sensitive to overwatering and does not grow well in poor or overly rich soil.

## Propagation

Anise hyssop can be direct-seeded in the fall or spring or grown from vegetative cuttings or root divisions. On our farm we seed into plug trays in the greenhouse and transplant after the seedlings become well established and well rooted. Once transplanted, if left to its own devices anise hyssop can readily self-seed and produce numerous volunteers. Seeds germinate easily in one to two weeks.

## Planting Considerations

As a perennial, once established anise hyssop can be harvested year after year for at least three seasons. Therefore, it should be planted in an area where it can take up residence for a while, and because of its striking beauty it is nice to showcase it in your more visible fields. After approximately three years of growth and continual harvests, the plants' vigor wanes and they should be replanted in a different location. After the first season anise hyssop plants grow together and form a solid hedgerow of leaves and flowers, making it a useful border plant. Also the color combination is striking, and the pollinators have a bounty of blossoms. Recommended bed spacing for anise hyssop is twelve inches in the row and twenty-eight inches between rows.

## Medicinal Uses

Cooling and refreshing, anise hyssop makes a delicious tea, which has been traditionally used for many things: as breath freshener, to buoy the spirits, to gently relax the nervous system, and to settle gastric upset. Known to have antimicrobial properties, anise hyssop has been used topically in washes to treat abrasions and infections. It also has been used by

**Figure 18-3.** Anise hyssop.

some herbalists, in formulas, to help aid in the treatment of respiratory infections. Another wonderful way to use anise hyssop is to include it in incense or smudge mixtures; it adds a sweet and pleasing aroma.

## Harvest Specifications

Anise hyssop should be harvested when the plant is in bloom. The medicinal properties, as well as the flavor, are at their peak when the plant starts to flower. Using a field knife or sickle bar mower, cut the entire aerial part of the plant, leaving five to six inches at the base for regenerative growth. It is possible to get two harvests a season from well-established plants. Harvest the leaves and flowers on a bright and sunny day, when the plants' volatile oils are at their highest levels. It is also recommended to harvest all aerial parts when the plant is dry and free from dirt or debris, preferably not after rain events that could splash soil onto the lower leaves. Contamination can also be reduced by mulching rows and taking care to harvest leaves above the dirt line.

## Postharvest and Drying Considerations

Anise hyssop does not have a high water content, so it can be dried more quickly than other mints, such as peppermint or lemon balm. That said, it is recommended to start drying at cooler temperatures of 80 to 90°F (if possible), then finish drying with temperatures not exceeding 100°F. When racking anise hyssop, place the herbs in a single layer to allow for good airflow and uniform drying. Under optimum conditions, anise hyssop should be dry within forty-eight hours. To process, garble the herbs over a garbling rack made out of quarter- to half-inch stainless steel mesh and remove stems. Blossoms will crumble off with the leaves and should be mixed together. The deep green leaves and bright purple flowers make for a vibrant blend.

## Pests and Diseases

Some commercial growers report beetle-pest pressures as well as difficulty with rusts and powdery mildews. Crop rotations and good cultural practices are recommended to avoid disease and pest issues. Deer do not eat anise hyssop and some growers have even reported that it can have a repellent activity when planted near species that are more desirable to deer.

## Yields

150 to 160 pounds dried leaf and flower per one-eighth-acre bed. Moisture ratio for leaves is approximately 7:1 fresh:dried.

## Pricing

Retail price for one pound organic:
- Dried anise hyssop leaf/flower: $20 to $25
- Fresh anise hyssop leaf/flower: $12 to $14

# Arnica                                    *Arnica* spp.

## Life Cycle

Arnica is an herbaceous, low-growing perennial that is hardy to USDA zones 2 to 9.

## Plant Description

Arnica is in the Asteraceae family and has lance-shaped leaves that are light green and somewhat rough in texture. The plant grows to approximately two feet in height and puts out bright yellow flowers starting in mid-July and continuing through the summer. Once established, arnica produces an extensive number of seeds that can be harvested for planting. It also spreads prolifically by rhizome growth. If left to its own devices, arnica will quickly fill in open spaces. There are several species of arnica. We grow *Arnica chamissonis* because it does better at our elevation; growers who live at high elevations may have greater success growing *Arnica montana*, which prefers a more acidic soil pH. There is a strong market for both species, and they are used interchangeably by many herbalists.

## Growing Conditions

Native to the western regions of the United States and Canada, arnica can be found growing in the wild in acidic and rocky soils, usually in the higher elevations of mountainous terrain. Arnica is hardy to as low as zone 2 and when cultivated grows well in well-drained, loamy soil and in full sun. It should be noted that while arnica is an avid spreader and will take over an area, it is not aggressive enough to smother out perennial weeds. Therefore, careful weed elimination during bed preparation, prior to planting arnica, is highly recommended. It can be difficult and time consuming to remove weeds that are interspersed with the lovely, creeping arnica.

## Propagation

Arnica seeds are only viable for a short time, and it is recommended to use fresh seeds when planting. While there are mixed opinions about whether arnica needs to be stratified, we find it helpful to stratify the *A. chamissonis* seeds for up to two weeks in early spring in an unheated greenhouse. Seeds are light-dependent germinators and are sown on the surface or covered lightly with potting medium. Seeds germinate in three to four weeks and are transplanted out once the plants are well rooted. Arnica can also be propagated easily from root divisions or vegetative cuttings, which is our preferred method of making starts.

## Planting Considerations

Arnica is a short, creeping perennial that requires full sun. Avoid planting it next to tall species, such as nettle, echinacea, or angelica, that can easily shade it. It takes at least a year to establish a solid patch of arnica, and it is often difficult to weed rogue plants out after the arnica begins to spread. Plant your arnica crop next to herbs that don't easily self-seed

**Figure 18-4.** Arnica flower.

and are more self-contained. We like to plant arnica next to our rhodiola and yarrow. These plants tend to grow companionably next to one another without "migrating" into one another's beds. Recommended plant spacing for arnica is six inches between plants, in triple rows with fourteen inches between rows. During the first year arnica will spread by rhizomes to fill in all open spaces between transplants, making a dense green mat of living plants.

## Medicinal Uses

Arnica flowers have been used for centuries to make topical applications such as creams, oils, salves, and liniments to reduce inflammation and pain caused by trauma and overexertion. Arnica is commonly used to treat damaged tissue, bruising, sprains, breaks, and strains. Flowers can be used in their fresh or dried stages, and some herbalists also use the leaves in combination with the blossoms. On the farm we make solar infusions of arnica oil by using fresh blossoms and olive oil. We find arnica oil extremely beneficial for alleviating the muscular tension and fatigue that comes from long hours of repetitive and strenuous work. (As ironic as it sounds, we often apply arnica oil to our backs before going into the arnica fields to pick its low-growing blossoms.) Homeopathic preparations of arnica are also used internally to reduce inflammation and aid in healing soft-tissue injuries. *Caution:* Arnica should not be used on open wounds or broken skin. Arnica is not used internally except in homeopathic or extremely low doses as specified by a knowledgeable practicioner.

## Harvest Specifications

With a healthy plot of arnica it is possible to get a light blossom harvest in the first year. However, the substantial yields come in subsequent seasons. In Vermont arnica begins to flower in late June, then continues to produce blossoms throughout the summer. In places in the western United States and Canada arnica has been known to flower into early fall. Harvesting arnica blossoms is a backbreaking but rewarding endeavor. We hand-harvest the flowers by popping the blossoms from the stems when they are just beginning to fully open. We have not found a way to mechanize or speed up harvest because the plant has a staggered blossoming pattern: Some flowers are open and ready to harvest, others are in closed-bud stage, and others may be starting to go to seed. While this is a successful reproductive method for the plant, it makes the harvest a bit trickier. Using a chamomile rake often pops off immature buds and is not effective. We find hand-harvesting multiple times during flowering is the way to go. During the peak of harvest we are picking every three to four days.

Once the blossoms have passed we let the plant grow on for another week, then cut the top two to three inches of the plant for leaf harvest. We harvest when the leaves are dry and clean, which can be tricky because it grows so close to the ground. We have limited demand for leaf, so we only do one cutting, allowing the plants to put on strong growth and maximize blossom potential.

## Postharvest and Drying Considerations

Arnica blossoms should be spread out in a thin layer on racks with good airflow. The temperature should be no higher than 100°F, and blossoms, when possible, should be dried in relative darkness. Do not be alarmed if your dried arnica looks as if it puffed out and went to seed. This fuzzy, seedy look is characteristic of arnica and is not a problem. When we first started drying arnica we thought we had done something terribly wrong or harvested the blossoms too late. After several attempts and still with no success at keeping the arnica from going to seed, we got desperate. Determined to figure it out, we picked completely closed buds and dried them and *poof*! More seed. That's just arnica; it's how she flowers. The key is to dry the blossoms, fluff and all, quickly and completely. Also be careful when removing the blossoms from the rack that the fluff doesn't fly away. Dried arnica is light and can easily take flight, so remember to shut off the fan. Arnica dries in approximately three days.

Leaves are dried in a single layer on racks in temperatures of 95 to 100°F. They have relatively low moisture content and dry easily in two to three days. Once they are dry, run the leaves over a garbling rack made out of quarter- to half-inch stainless steel mesh and remove stems.

## Pests and Diseases

With proper crop rotation and soil maintenance we have seen little to no pest or disease pressure. Hydroponic growers in Europe have reported issues with *Sphaerotheca fuliginea* and *Entyloma arnicalis* that can stunt leaf growth.

## Yields

Sixty to seventy pounds dried arnica flower per one-eighth-acre bed (when harvesting in the second or third season once blossom harvest is more substantial). Moisture ratio for arnica flowers is 6:1 fresh:dry.

## Pricing

Retail price for one pound organic:
- Dried arnica blossoms: $60 to $80
- Fresh arnica blossoms: $26 to $30

## Commonly Imported From

Romania and Canada

# Ashwagandha                                   *Withania somnifera*

## Life Cycle

Ashwagandha grows prolifically as a perennial in tropical locations. However, in Vermont and other temperate zones, ashwagandha is grown as an annual and will not overwinter outside without elaborate protective measures.

## Plant Description

Ashwagandha is a beloved ayurvedic herb that comes to us from India and is commonly found growing in the wild in Pakistan, Sri Lanka, and areas of Africa. Ashwagandha belongs to the Solanaceae family. Growing to about three feet in height, ashwagandha has simple, elliptical, blade-shaped leaves that are dark green in color. The entire plant, including the stems and leaves, is covered with soft light hair; flowers are greenish-yellow and are small and nondescript. In the fall ashwagandha sets bright red berries approximately one-quarter inch in diameter.

## Growing Conditions

In the tropics ashwagandha grows as a perennial in a range of environments that differ in rainfall, soil type, altitude, and temperature. In Vermont, where it grows as an annual, we cultivate ashwagandha in well-drained soil in full sun. Ashwagandha likes a dryish though not droughty soil and can be cultivated similarly to the way tomatoes and eggplants are.

## Propagation

Ashwagandha seeds are light dependent and are direct-seeded for cultivation (five pounds of seed per acre) in tropical places that have long, warm growing seasons. However, in cooler temperate zones it is highly recommended to start seeds indoors in early spring and transplant into prepared fields. Ashwagandha seedlings can be susceptible to damping off, so it is helpful to cover seeds with a fine dusting of

**Figure 18-5. Ashwagandha in berry.** Photograph courtesy of Larken Bunce

**Figure 18-6. Ashwagandha.** Photograph courtesy of Bethany Bond

vermiculite, sand, or potting medium after sowing. It is important to keep seeds moist, but do not over-water. Good airflow over the seedlings can also help prevent damping off. We find that ashwagandha germinates better in warmer soils, and we do our best to mimic its tropical homeland. To that end we place seed trays on heat mats, and the seeds usually begin to germinate in two weeks. Ashwagandha is a sporadic germinator, and seedlings emerge in waves. We find it helpful to prick out seedlings into plug trays as they mature but to keep the seed trays for at least a month to allow more seeds to germinate. Before transplanting we carefully harden-off ashwagandha and time planting for when there is a snap of warmer, dryer weather and the threat of frost has passed.

## Planting Considerations

Like many of the Solanaceae family, ashwagandha can be susceptible to pests and diseases. Care should be given when laying out ashwagandha beds to avoid planting next to other members of the Solanaceae family or in soils that have recently held members of this family of plants. This helps to avoid spreading disease and pests. Because ashwagandha is an annual, it should be moved season to season within a field rotation. Recommended plant spacing for ashwagandha is twelve inches between plants and twenty-eight inches between rows. We plant two rows per bed and have begun experimenting with planting ashwagandha in raised beds on black plastic mulch to increase soil warmth and reduce weed pressure.

## Medicinal Uses

The roots of ashwagandha are used to help build energy, reduce nervous tension, and considered a strong reproductive tonic for both men and women. A unique property of ashwagandha is that while it is energizing to the system, it does not over stimulate the body and is often used to help treat insomnia. This prized adaptogen has been used traditionally in ayurvedic practices in tea form, often brewed with other warming herbs like cinnamon, cardamom, and ginger. Ashwagandha has a strong flavor and can be taken in extracts and capsules.

## Harvest Specifications

Roots are harvested after the first killing frost when the tops have died back. The aerial parts of the plant should be mowed down before digging to ease the process. Ashwagandha roots are branching and multistemmed, growing on average one foot in depth. They can be dug by hand with a spading fork or mechanically with a bed lifter or potato digger.

## Postharvest and Drying Considerations

Soil often gets compacted in the central crowns of ashwagandha where the roots begin to branch. Therefore, it is helpful to divide the roots before washing. Ashwagandha roots are relatively hard and somewhat woody and may require a small hatchet or knife and mallet to chop. After splitting the roots, wash thoroughly before drying. Mill roots *before* drying, as they can be difficult to chip after they are fully dried due to their extremely dense nature. Dry chipped roots on screens at temperatures of 100 to 110°F. Ashwagandha roots dry under optimum conditions in three to four days.

## Pests and Diseases

Ashwagandha can be susceptible to bacteria, fungi, viruses, phytoplasms, nematodes, and pests. One of its primary pests in tropical areas is the spotted beetle, *Henosepilachna vigintioctopunctata*. In temperate zones we have seen pressure from *Leptinotarsa decemlineata*, the Colorado potato beetle (CPB). That said, pests and disease have not been of huge concern on our farm. Proper crop rotation and soil maintenance have helped create a healthy growing environment, and ashwagandha crops have grown well with little difficulty from disease or pest pressure. In addition to implementing good growing practices, proper application of *Bacillus thuringiensis* (also known as Bt, a commonly used biological pesticide approved for use on organic farms) can help eliminate beetle populations. Bt should be used sparingly, as it can eliminate beneficial insects as well. Raw neem oil spray is our preferred method of managing CPB pressure. See chapter 11 for more detailed application instructions.

## Yields

One hundred twenty five pounds dried root per one-eighth-acre bed. Moisture ratio for ashwagandha root is 3:1 fresh:dried.

## Pricing

Retail price for one pound organic:
- Dried ashwagandha root: $12 to $24
- Fresh ashwagandha root: $11

## Commonly Imported From

India

# Astragalus

## *Astragalus membranaceus*

## Life Cycle

Astragalus is a long-lived herbaceous perennial that is hardy to USDA zones 4 to 11.

## Plant Description

Astragalus is a member of the Fabaceae family. Also known as milk vetch, astragalus is native to China but grows well in many cool temperate zones. The Latin name for astragalus has recently changed from *Astragalus membranaceus* to *Astragalus propinquus*. We prefer to use the older classification (*Astragalus membranaceus*) in this book as people are more familiar with it. Astragalus is a legume, and like other legumes it has typical "pea" flowers that are cream colored and form on the leaf axis. In our region astragalus flowers in late July and early August. Its alternate leaves consist of small leaflets that are bright green and fernlike. At maturity, plants are approximately three feet tall. Astragalus produces a substantial multi-branched taproot that is fleshy and white and can be harvested after four to five seasons.

## Growing Conditions

Astragalus should be grown in full sun in well-drained soil with good fertility. Astragalus is a deep-rooted plant; therefore, deep tillage to break up any hardpan is highly recommended. As a legume astragalus has nitrogen nodules on its roots, so it can produce some of its own source of nitrogen. However, astragalus should be planted in beds with high organic matter and good P and K levels to increase yields and plant vigor. Some growers report success growing astragalus by sowing seed directly into fields in early spring. Because astragalus tends to be slow growing, here in Vermont we like to give it a head start, by propagating seedlings in the greenhouse and then transplanting out. To get good, mature, potent roots astragalus must grow for at least four seasons. Care should be given to prepare beds well and establish a solid crop in the first season.

## Propagation

Astragalus seed is a cool-soil germinator that requires scarification. To prepare the seeds, scarify them by rubbing with medium grit sandpaper to nick the seed coat. The seed coat needs to be compromised, but be careful not to damage the inner seed germ. Once they are scarified, soak the seeds in a diluted seaweed solution overnight to swell them, then sow into flats. Astragalus is not a light-dependent germinator. Seeds are planted one-quarter to one-half inch deep and should begin to germinate within ten days. Some growers report success with stratifying (cold-conditioning) the seeds before scarifying them, but we have found scarifying alone without stratification produces very high germination rates.

**Figure 18-7.** Astragalus. Photograph courtesy of Larken Bunce

## Planting Considerations

Wherever you plant astragalus it will be for many seasons (at least four before harvest), so bed preparation is essential. It is important to eliminate perennial weeds with cultivation and cover crops and to build fertility before planting. For astragalus bed preparation we use a chisel plow to loosen the subsoil, allowing the astragalus taproots to grow more easily and produce higher yields. If weeds are particularly pernicious we use cover crops to smother them out. Recommended plant spacing is twelve inches in the row and twenty-eight inches between the rows.

## Medicinal Uses

Astragalus is a superior adaptogenic tonic; it is one of the most commonly used herbs in Asia and is also extremely popular in the United States. Rich in immune-stimulating polysaccharides, it is used to treat many immune deficiencies and is a very effective supportive therapy for people undergoing cancer treatment. Studies show that astragalus helps the body deal with the toxicity and side effects caused by chemotherapy and radiation, while not diminishing their effectiveness. Astragalus is extremely nourishing, helps the body deal with stress, and stimulates white blood cell production in bone marrow. Studies have shown that regular use of astragalus helps reduce tumor growth, and its antioxidant properties strengthen the blood and improve cardiac function.

A favorite way to use astragalus is cooking the roots to make a nourishing broth. The root is boiled with other vegetables and medicinal mushrooms, then removed before serving, as it is fibrous and hard to chew. Astragalus roots are commonly powdered to make capsules and lozenges. Roots can also be tinctured and used in herbal extracts and longevity elixirs. Astragalus is a pleasant-tasting herb and is not only highly medicinal but versatile in how it can be used in formulation.

## Harvest Specifications

Astragalus roots should be harvested when they are four to five years old or more. The roots at maturity grow up to three feet long and are fleshy and white. Harvest the roots when the tops die back and plants begin to enter dormancy; this is when polysaccharides are at their highest levels. Mow down the tops before digging, and loosen the soil by running a chisel shank or cultivator up the row on either side of the astragalus. Once the soil is loosened, roots can be dug out by hand with spading forks or mechanically, using a potato digger.

## Postharvest and Drying Considerations

Soil often gets compacted in the central crown of astragalus where the roots begin to branch off. Therefore it is helpful to quarter the roots before washing. Astragalus roots can be chopped with a field knife. After splitting the roots, wash thoroughly before drying. Mill roots after they are fully dried. Astragalus roots dry under optimum conditions, at temperatures of 100 to 110°F, in approximately three to four days.

## Pests and Diseases

The most common diseases impacting astragalus are powdery mildew and root or crown rot. Proper drainage and weed control can greatly reduce this. Leafhoppers and aphids can also damage foliar growth. Consider companion planting near herbs that repel harmful insects. Creating borders of highly aromatic plants such as spearmint, cilantro, and basils can be beneficial. Proper soil management and crop rotations also help reduce pest and disease pressure.

## Yields

One hundred pounds dried root per one-eighth-acre bed. Moisture ratio for astragalus root is 4:1 fresh:dried.

## Pricing

Retail price for one pound organic:
- Dried astragalus root: $14 to $24
- Fresh astragalus root: $13

## Commonly Imported From

China

# Black Cohosh  *Cimicifuga racemosa* syn. *Actaea racemosa*

## Life Cycle

Black cohosh is an herbaceous perennial that is hardy to USDA zones 3 to 9.

## Plant Description

Black cohosh is a member of the Ranunculaceae family and is a striking woodland perennial that is native to North America. It has a large range, spanning from northern New England and Ontario to the Appalachian Mountains and as far west as Oklahoma. In Vermont black cohosh is a native plant that grows in the dappled shade of hardwood forests but can also thrive in more open environments and is often field-grown in full sun. It has compound pinnate leaves that are toothed and divided into two to five leaflets growing in groups of three. The foliar part of the plant is approximately five feet tall when mature.

In midsummer black cohosh produces elegant plumes of white racemes (flowers) that are up to seven feet in height. These beautiful flowers mature into seed stalks that produce a distinctive rattle when shaken, giving rise to black cohosh's other name: rattlesnake plant. Black cohosh flowers give off a peculiar odor that attracts its pollinators. Seeds mature in late summer to early fall and can be planted immediately after harvest, allowing for the temperature fluctuations needed to stratify them. The roots of black cohosh are dark brown on the outside, cream-colored to white on the inside, and have pronounced knobs or swellings at the crown. These root crowns can easily be divided for replanting.

## Growing Conditions

Native to the woodlands, black cohosh likes soils that are rich in organic matter, well drained, and not too acidic. Normally found in fungally dominant soils in the understory of hardwood forests, black cohosh does well in partial shade but can be successfully grown in full sun in cloudier states such as Vermont. Some questions remain about whether there are medicinal differences between the woods-grown and the field-grown cohosh. We grow black cohosh both ways and have been happy with the quality of roots coming from the field. Black cohosh benefits from well-prepared, raised beds and mulching. This helps to improve drainage and allows for healthier root growth.

**Figure 18-8. Black cohosh in bloom.** Photograph courtesy of Larken Bunce

## Propagation

Black cohosh can be grown successfully from seed though germination rates can be low and sporadic. Fortunately, mature plants produce copious amounts of seed so growers should plant heavily to compensate for slow germination. The seeds need to go through a sustained stratification of warm temperatures followed by cool temperatures to mimic what occurs in nature. If you are direct-seeding black cohosh it is recommended to seed in late summer to early fall; this will allow for a natural stratification. Black cohosh can also be grown from rhizome divisions or rootlets. Plant roots in the fall or spring with the rhizomes' tops at least two inches below surface. In woodland beds mulch with two to three inches of leaves.

## Planting Considerations

Black cohosh should be companion planted with other woodland medicinal treasures such as goldenseal, ginseng, wild ginger, Solomon's seal, and blue cohosh. Like goldenseal and ginseng, black cohosh is being overharvested in the wild and is on United Plant Savers' "At-Risk" List. Cultivating black cohosh not only provides important herbal medicine, it is also beneficial from a conservation standpoint because it takes the pressure off wild plant populations. Recommended spacing for row crops of black cohosh is eighteen-inch plant spacing within the row and twenty-eight inches between rows.

## Medicinal Uses

The roots of black cohosh are commonly used to treat painful menses or to bring on delayed menstruation, and it is a favorite with menopausal women because it helps regulate hormones with its estrogen-like action in the body. Black cohosh also stimulates uterine contractions and is used in combination with blue cohosh to aid in childbirth. In addition to its hormonal properties and interactions with the reproductive system, black cohosh relaxes smooth muscle and is a strong antispasmodic. It has been traditionally used to treat spastic coughing and tinnitus. Black cohosh is prepared and administered most commonly as a tincture, as a decoction, or in capsule form.

## Harvest Specifications

The roots of black cohosh are harvested after the third year during the fall or early spring, when the chemical constituents are at their highest concentrations. The roots are dark brown and black in color, and growth is more lateral than deep. The majority of the root mass is centered at the knobby crown, the nexus where the stems meet the root. The roots at the crown are larger and denser, with smaller rootlets of spaghetti-like thickness branching off from the main rhizome. When digging roots wait until the tops die back in the fall, then mow the aerial parts down. It is not necessary to use a shank between rows to loosen the soil as the roots grow shallowly. Black cohosh roots often grow in a dense mat and can be dug easily by hand with a spading fork or mechanically using a bed lifter or modified potato digger. These root mats often get caked with soil, so it is a good idea to thoroughly bang soil off roots in the fields.

## Postharvest and Drying Considerations

Chop the central crown of black cohosh into several pieces before washing. The roots are relatively hard but can be divided with a handheld ax or field knife and mallet. After splitting the roots, wash thoroughly, then chip with a root chipper before drying. Spread chipped roots in a thin layer on racks, and dry in partial shade at temperatures of 100 to 110°F. Under optimum conditions black cohosh roots should dry in four to six days.

## Pests and Diseases

Black cohosh can be susceptible to leaf spots and root rot. Both can be easily avoided with good cultivation techniques. To avoid leaf spots provide adequate plant spacing and reduce significant weed pressure to allow good airflow. Root rot is most common in beds that are weedy and not well drained. Good bed preparation, cover cropping, and crop rotations can build a strong growing environment where rot is less inclined to become established. Some farmers report having pest pressure in the form of deer, rodents, slugs, and some species of beetles. This has not been an issue on our farm, but some growers may need to consider fencing or biological controls.

## Yields

One hundred pounds of dried root per one-eighth-acre bed. Moisture ratio for black cohosh root is 3:1 fresh:dried.

## Pricing

Retail price for one pound organic:
- Dried black cohosh root: $17 to $26
- Fresh black cohosh root: $10

# Blue Vervain                    *Verbena hastata*

## Life Cycle

Blue vervain is an herbaceous perennial that is hardy to USDA zones 3 to 9.

## Plant Description

Blue vervain is an upright plant with square, greenish-red stems and opposite, lance-shaped leaves that are coarsely toothed. Despite the square stem and opposite leaves, blue vervain is not a mint but rather a striking member of the Verbenaceae family that is native to North America and found throughout Canada and the United States. Blue vervain grows four to five feet tall, and has clusters of two- to five-inch spikes that are composed of individual purple

**Figure 18-9. Blue vervain.** Photograph courtesy of J. N. Stuart

**Figure 18-10.** Photograph courtesy of Larken Bunce

flowers. These small flowers are less than a quarter inch wide and bloom intermittently, a few flowers at a time, starting at the base of the spikes, then spiraling upward. Vervain is a favorite among pollinators and attracts both bees and butterflies.

## Growing Conditions

Blue vervain likes full sun but can tolerate partial shade and in the wild grows well in open fields, meadows, prairies, and stream- and riverbeds. Blue vervain does well in well-drained soil containing good fertility and organic matter. It is not drought tolerant and needs an adequate amount of water.

## Propagation

The seeds of blue vervain germinate well if stratified and are sown in early spring. The seeds are not light-dependent germinators and should be lightly covered with potting medium after planting. Once stratified, the seeds should germinate in two to three weeks. In Vermont we find it most helpful to sow seeds into flats, then transplant well-established plugs.

## Planting Considerations

Blue vervain can grow very tall and is striking when it goes to bloom. It makes a lovely hedgerow and is good to plant on the edges of beds. Blue vervain is not susceptible to pests and diseases so it can be planted next to most perennials. The only consideration would be to make sure it is not shading out lower-growing plants that require full sun. We commonly plant it next to echinacea and black cohosh. Blue vervain can also be propagated from root divisions or vegetative cuttings. Plant blue vervain with twelve-inch spacing in the row and twenty-eight-inch spacing between rows.

## Medicinal Uses

The leaves and flowers of blue vervain are traditionally used as a nervine (a nervous system tonic) and can be especially effective at easing nervous tension that manifests as stiffness or rigidity in the neck, shoulders, and back. Blue vervain can also help provide relief to women who experience irritability around their menstrual cycle, and it is useful in alleviating nervous depletion and exhaustion. As a nerve tonic blue vervain is valued for its slightly sedative effect and has been used for anxiety and depression by helping the body relax and release wired energy and stress. Some people report feeling more expansiveness and feeling calmer after taking blue vervain. In addition to its nervine properties blue vervain is used to reduce fevers by promoting sweating, and it also contains bitter properties that help to ease sluggish digestion. The flowers and leaves are used internally as a tea, tincture, or flower essence. They can also be used topically in liniments, oils, or salves.

## Harvest Specifications

The medicinal properties of blue vervain are at their peak when the plant begins to flower in mid- to late summer. Harvest both leaf and flower by using a field knife or sickle bar cutter to cut the entire aerial part of the plant, leaving approximately twelve inches at the base for regenerative growth. In some areas it is possible to get a second cutting from well-established plants. Harvest the leaves and flowers on a bright and sunny day, when the plants are dry and free from dirt or debris, preferably not after rain events that could splash soil onto the lower leaves.

## Postharvest and Drying Considerations

Blue vervain is easy to overdry and can lose its medicinal properties quickly, so it should be processed quickly at lower temperatures. Dry on racks in temperatures of 90 to 95°F. When racking blue vervain, place the herbs in a single layer to allow for good airflow and uniform drying. Under optimum conditions blue vervain should be dry within forty-eight hours. To process, garble the plants over a garbling rack made out of quarter- to half-inch stainless steel mesh, and remove stems. Blossoms will crumble off with the leaves, and they should be mixed together. The deep green leaves and bright purple flowers make for a vibrant blend.

## Pests and Diseases

Blue vervain is minimally susceptible to insect damage and foliar disease, and the health of this crop can be maintained by good cultivation practices and crop rotations.

## Yields

One hundred twenty pounds of dried leaf and flower per one-eighth-acre bed. Moisture ratio for blue vervain leaf and flower is 4:1 fresh:dried.

## Pricing

Retail price for one pound organic:
- Dried blue vervain leaf/flower: $11 to $20
- Fresh blue vervain leaf/flower: $8

# Boneset                    *Eupatorium perfoliatum*

## Life Cycle

Boneset is an herbaceous perennial that is hardy to USDA zones 3 to 9.

## Plant Description

Boneset is a lovely member of the Asteraceae family that is native to the southern and eastern United States. It has an interesting leaf structure that is known as perfoliate, meaning the base of the leaves wrap around the stem. Boneset leaves are a light greenish-yellow color and approximately eight inches long. Growing two to four feet tall, boneset produces beautiful white flowers in late summer through early fall. Flowers grow in flat clusters that are four to seven inches in diameter and composed of individual disk-shaped florets. Boneset flowers have high nectar content and attract many pollinators.

## Growing Conditions

Boneset likes to grow in full sun and moist soils. In the wild it is found growing in swampy areas and swales and can tolerate high concentrations of

**Figure 18-11. Boneset.** Photograph courtesy of Superior National Forest

moisture. Boneset does not require extra nutrients or pampering and grows well in moderate fertility.

## Propagation

The seeds of boneset are light dependent. It is best to sow seeds on the surface and keep them moist. Although it is not necessary to stratify, some growers have found that this increases germination rates. Boneset readily self-seeds and can be direct-seeded into beds or indirect-seeded for transplants.

## Planting Considerations

Because boneset tolerates low fertility and high-moisture contents, it is a perfect choice for planting in hard-to-farm places on your land. Boneset will readily self-seed, and though it is a vigorous grower, it is not invasive. If there are marshy places on your land, consider establishing a stand of boneset. It can grow relatively easily with little fuss and can grow well and hold its own with other weedy species. When cultivating, plant boneset at twelve- to eighteen-inch spacing within rows and twenty-eight inches between rows.

## Medicinal Uses

Boneset leaves and flowers are highly effective at relieving the aches and pains associated with the flu. They can also be used to bring down fevers by helping the body sweat. In addition, boneset helps the bronchial system by loosening phlegm and aiding with expectoration. Boneset has anti-inflammatory properties internally and can be used as a poultice to bring down swelling, treating strains and soft-tissue trauma.

## Harvest Specifications

Aerial parts of boneset are harvested when the plants are just beginning to flower and not all buds have opened. The tops should be harvested when the plant is dry and clean. Using a field knife or sickle bar mower to cut the tops, harvest the entire aerial part of the plant, leaving the bottom six to twelve inches to regrow. Once plants are established it is sometimes possible to get two cuttings in a season.

## Postharvest and Drying Considerations

Boneset is an herb that can lose its vitality and chemical constituents quickly, so it should be processed in an efficient manner and stored carefully. Dry on racks in temperatures of 95 to 100°F. When racking boneset place the herbs in a layer no more than one to two inches thick to allow for good airflow and uniform drying. Under optimum conditions boneset should be dry within forty-eight hours. To process, garble the plants over a garbling rack made out of quarter- to half-inch stainless steel mesh, and remove stems. Blossoms will crumble off with the leaves, and they are commonly sold mixed together.

## Pests and Diseases

No major pests or diseases have been reported on boneset.

## Yields

One hundred pounds of dried leaf and flower per one-eighth-acre bed. Moisture ratio for boneset leaf and flower is 4:1 fresh:dried.

## Pricing

Retail price for one pound organic:
- Dried boneset leaf/flower: $12 to $20
- Fresh boneset leaf/flower: $10 to $11

# Burdock                                 *Arctium lappa*

## Life Cycle

Burdock is an herbaceous biennial hardy to USDA zones 3 to 7.

## Plant Description

A much-maligned member of the Asteraceae family due to its prickly burrs and aggressive growth habit, burdock is a grand plant and medicinal treasure trove. Even though it is extremely common throughout the United States, it is an introduced species that is native to Eurasia. Burdock grows three to four feet tall in its first year, producing large leaves that are deep green, heart shaped, and somewhat coarse and fuzzy. In the second year burdock sends up its flowering stalks, which can easily reach seven to eight feet tall. Burdock's flowers are pinkish-purple and are found on the top of its prickly burrs. The taproots of burdock are dense and fleshy in the first year, growing up to two feet in depth. In the second year roots become fibrous and woody, with a hollowed-out center, and their medicinal value declines.

**Figure 18-12.** Burdock in the field.

## Growing Conditions

Burdock is a true pioneer species and is a very sturdy biennial that will grow well in many conditions. It is not finicky in its sun requirement, growing in full sun, partial shade, and full shade. It can grow with moderate moisture but also does well in damp soils. Commonly found on land that has been disturbed, burdock can grow well in poor, rocky soil. However, for propagation burdock thrives in soils rich in organic matter with good, deep tillage.

## Propagation

It is recommended to stratify seeds to increase germination. Burdock can be direct-seeded in early spring, with germination occurring in one to two weeks. Plant spacing for direct seeding is approximately two inches apart within the row; seedlings are then thinned to two to four inches between plants. In each twenty-eight-inch-wide bed we plant a triple row with fourteen inches between the rows. This can be done by banding multiple hand seeders together or making multiple passes with a single seeder.

## Planting Considerations

Some people will be shocked if you tell them you are growing burdock. They will immediately think of the troublesome burrs that stick to clothes and pets and may know very little about burdock's rich medicinal value. Fear not: It is easy to grow burdock without creating pesky populations of burrs. When burdock root is grown for market, the roots are harvested before the plant produces its aerial reproductive parts (burrs). The same is true for leaf production. The leaves are larger and more vibrant in the first year of growth and can also be harvested before burrs are produced. Some farmers, however, grow burdock for seed production. In this case it is recommended to plant burdock in an out of the way location or at least have an understanding that burrs are tricky things to contend with. Harvest seeds as soon as they are ready, then mow down plants immediately to prevent spreading. Another consideration when growing burdock roots is bed preparation. Burdock produces long taproots and does best in raised beds or in fields that have been prepped with a chisel plow. We grow burdock similarly to the way vegetable growers grow carrots.

## Medicinal Uses

There is a lot to love about burdock. Among herbalists it is a favorite restorative tonic and highly nutritive food source. Rich in vitamins and minerals such as iron and magnesium, burdock root is a super food that can be eaten in soups, steamed like parsnips, or added to salads and stir-fries. Burdock root is used to tonify the liver and kidneys and is used also to help purify and strengthen the blood. Because it helps improve the detoxifying functions of these systems it is highly effective in the treatment of various skin maladies, including acne, psoriasis, and eczema. Burdock roots also have bitter properties that stimulate bile production and improve digestion, making it a favorite ingredient for bitters and digestives. The leaves of burdock, while less popular than the root, are also medicinal. They are antimicrobial, can draw out toxins, and are used topically for poultices to treat boils, rashes, bites, and infections. The seeds are used to stimulate immune response, and some herbalists use them to treat colds and flus.

## Harvest Specifications

On our farm we focus primarily on root production, collecting leaves only for personal use. We have not found a sustainable market for the leaves and seeds. We harvest the roots in the fall of the first year. Growing burdock roots for more than one season results in rapidly declining quality in the roots as they tend to turn pithy in the core. The taproot grows up to two feet in depth and can be dug by hand with a spading fork or mechanically using a modified potato digger run as deep as possible. Prior to digging roots it is helpful to mow down the aerial tops and loosen the soil by running a shank or cultivator up the row on either side of the burdock. Leaves are harvested in the first year by pulling them off the hollow stalks. Place leaves in a single layer on racks, and dry at temperatures of 90 to 100°F. Seed harvest is a bit trickier because the burrs take some effort to break up and they produce chaff that can be harmful to breathe and that sticks to everything. It is a major irritant to eyes, lungs, and skin. It is a wise precaution to wear a respirator and protective suit when harvesting seeds. Choose a mature plant, and thresh over the side of a pickup truck into the truck bed. It is best to do this in the field where the plant is growing so as not to spread to other fields. Run seed and chaff through a clipper seed cleaner to separate the seed or winnow by hand in the wind or using a fan. The larger seeds have better germplasm and are better for planting, so they can be sorted by size for higher germination rates.

## Postharvest and Drying Considerations

Because burdock roots are long taproots and only branch off minimally, they do not collect dirt the way other root crops do and are very easy to clean and process. Simply chop large roots into pieces (either diagonally or in cross sections) and wash. Burdock roots are relatively soft and can be chopped easily with a field knife. After chopping the roots wash them thoroughly, then dry at 100 to 110°F. Burdock roots dry under optimum conditions in four to six days. Mill or chip roots after they are fully dried.

## Pests and Diseases

No major pests or diseases are reported for burdock root, leaf, or seed.

## Yields

- One hundred fifty pounds of dried root per one-eighth-acre bed. Moisture ratio for burdock root is 5:1 fresh:dried.
- Six hundred pounds of dried leaf per one-eighth-acre bed. Moisture ratio for burdock leaf is 6:1 fresh:dried.
- Twenty-five pounds of seed dried per one-eighth-acre bed.

## Pricing

Retail price for one pound organic:
- Dried burdock root: $12 to $20
- Fresh burdock root: $8 to $13
- Dried burdock seed: $37
- Fresh burdock seed: $21

# Calendula                    *Calendula officinalis*

## Life Cycle

Calendula is an annual at temperate climates.

## Plant Description

Calendula is often referred to as pot marigold, is a member of the Asteraceae family, and is native to southern Europe. There are over one hundred varieties of calendula, and the flowers can range from small yellow blossoms to big sunbursts of flowers up to three inches wide and bright orange in color. The aerial part of the plant is upright and bushy, growing two to three feet tall with bright green leaves. The flowers are sticky and fragrant, containing medicinal resins. For commercial growing we use the "Erfurter Orangefarbige" variety because these plants produce larger, dense flowers that have a high resin content and are high yielding.

## Growing Conditions

Calendula does well in full sun and in well-drained soil that is rich in organic matter. It does not require specialized nutrients but likes good fertility and soil organic matter. This strikingly beautiful plant is one that is great to showcase! People and pollinators will

**Figure 18-13.** Calendula blossoms. Photograph courtesy of Bethany Bond

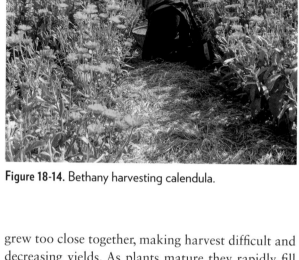

**Figure 18-14.** Bethany harvesting calendula.

flock to its lovely color and sunny disposition. Plant calendula where you can see it and also where there is easy access to drying facilities, as harvesting will be done every few days from early summer to early fall.

## Propagation

Calendula seeds should be sown approximately one-quarter- to one-half-inch deep or covered lightly with potting medium. On our farm we grow plugs in the greenhouse and transplant them into well-prepared beds. This gives us a jump on the season and helps the calendula get established and outcompete weeds. Calendula can be direct-seeded into the field, but extra care should be given to prepare the bed and eliminate weeds. In the short, cool, wet springs of Vermont we have had more success transplanting plugs than growing by direct seeding. We grow plugs for four to six weeks, then harden them off thoroughly before transplanting.

## Planting Considerations

Calendula plant spacing is one foot in single rows, and at least thirty-six inches between rows. We experimented with planting two rows per bed at twenty-eight-inch spacing and found that the plants

grew too close together, making harvest difficult and decreasing yields. As plants mature they rapidly fill in the space within the row, making a dense hedge of calendula. We find it helpful to give a little extra space between rows to avoid breaking off stems of neighboring plants as we harvest. Creating raised beds can also be beneficial to make blossom harvesting easier. While calendula is certainly a joy to harvest, it can be a bit backbreaking. So we are exploring ways to increase the height of the beds to make an easier harvest.

As a member of the Asteraceae family, calendula can be susceptible to aster yellows, a viral disease carried by leafhoppers that causes deformities to the flowers. While aster yellows rarely knocks out an entire field, it can impact crop yields, and care should be taken to prevent and eliminate it. Other crops, such as echinacea, are also impacted by this disease. As a result, in crop rotations we do not interplant echinacea or calendula in the same field. We also do not plant calendula in a field directly following echinacea and vice versa. This can help reduce spreading aster yellows from one crop to the next.

Another thing to consider when growing calendula is that it is a hybrid and will reseed itself. However, the resulting progeny may not be "true"—meaning that not all members of the second generation are the same variety as the parent generation. Letting a crop of "Erfurter Orangefarbige" self-seed will not result in a pure crop in subsequent seasons. It is better to start with new seed or transplants each season. Another thing to take into consideration is weed control when growing calendula. Hand weeding and mechanical cultivation are good to do at the early stages of growth but are not needed or recommended later in the season. Mulching beds is a good way to prevent weed pressure. The stems of calendula are fragile, so it is good to limit tractor cultivation over them when plants are large to avoid breaking the plant when it is in flower. Calendula plants, once established, will grow together and shade out most weeds.

## Medicinal Uses

Calendula blossoms are used both internally and externally and have strong fresh and dried markets. Externally, the flowers are used to heal burns, cuts, and skin abrasions. Not only does calendula promote cellular healing, it is also antiseptic and antimicrobial and helps fight infection. That is why it is a favorite for both first-aid kits and cosmetic purposes and is used in all sorts of topical applications such as liniments, salves, oils, creams, and serums. Calendula can also be used antibiotically to clean wounds and is often paired with such herbs as spilanthes, myrrh, and peppermint to use as a mouthwash. Internally, calendula is a strong lymphatic, excellent for tonifying the lymphatic system and flushing toxins. Calendula is a cooling herb that has anti-inflammatory and bitter properties and can be taken internally in tea or tincture form.

## Harvest Specifications

Calendula is harvested when the blossoms are fully open but before they begin to go to seed. Harvesting is best done by hand. Some people use blossom combs to harvest. This can be problematic in that combs are not selective and capture unopened buds and seed heads, which will need to be picked out before drying (a time-consuming process.) Calendula

is a very generous bloomer and will produce new flushes of blossoms every few days. Keeping up with the harvest is important, not only to get the blossoms at their peak but also to maintain the blooming cycle. If older blossoms are not removed and they begin to set seed, the plant will stop growing and producing flowers. During harvest be sure to deadhead blossoms that are beginning to go to seed. Calendula harvest should be done during the heat of the day, when the blossoms are fully open and the resin concentrations are high; fingers should get nice and sticky when picking. Harvest blossoms every two to three days from mid-July to the first killing frost in the fall. For large plots of calendula it can be helpful to divide fields and stagger harvest days to ease back pain from long hours spent picking.

## Postharvest and Drying Considerations

Drying calendula can be tricky. It is heartbreaking to open a beautiful bag of "dried" calendula only to find it has begun to mold because of some moisture left in the blossoms. It really only takes one or two "wet" blossoms to spoil the whole batch. The key to proper drying is careful monitoring. Different parts of the blossoms dry at different rates. The petals quickly lose their moisture and will be dry in a couple of days. The centers of the flowers are much denser and require more time. For optimum color and constituent retention, dry in complete darkness at temperatures of 95 to 100°F. Good airflow is also key; make sure fans are blowing directly over the racks. Placing calendula on racks in a single layer also maximizes drying.

Depending on conditions it can take a week or more to get blossoms completely dry. To determine when they are done, it is important to sample many flowers. First rub the petals. They should be crisp and dry and come off easily. Then split open the centers, which should be dry but also somewhat pliable. The centers contain a lot of resin, so they won't "snap" per se, but the stem directly below the center should be easy to break off. Finally, there is a sound that calendula makes when it is dry. As you run your hand through the blossoms, they should make a rustling sound, almost like the sound made when you walk through dried leaves. If the batch of calendula doesn't make that sound, and is more pliable and soft, chances are it's not done. This listening technique is especially helpful to use when checking calendula that has been put in storage. Even the most experienced herb drier can misstep with calendula, so we make it a standard practice to check all dried calendula in storage every couple of days for the first couple of weeks. Usually, if you catch it quickly the calendula can be reracked to finish drying properly and saved.

## Pests and Diseases

The only pest or disease that we have experienced with calendula is aster yellows, a disease carried by leafhoppers. There is no treatment for aster yellows. The best course of action is to carefully remove diseased plants and use proper crop rotations to limit the spread of leafhoppers. Some farmers are exploring applications of neem, but no conclusive results have been reached. Occasionally, powdery mildew will form on the calendula leaves. We find that good field management and good crop rotations help minimize this.

## Yields

230 pounds of dried calendula flowers per one-eighth-acre bed. Moisture ratio for calendula flowers is 5-6:1 fresh:dried.

## Pricing

Retail price for one pound organic:
• Dried calendula flower: $21 to $40
• Fresh calendula flower: $12 to $21

## Commonly Imported From

Egypt

# California Poppy — *Eschscholzia californica*

## Life Cycle

California poppy is classified as an herbaceous perennial, but in colder climates it behaves like an annual. It is found in all temperate zones.

## Plant Description

California poppy is a lovely member of the Papaveraceae family and is native to California and the western United States. The plant is low spreading, reaching a height of one to two feet. The leaves are a blue-green color and are fernlike in shape, composed of finely divided lobes in groups of three. Foliage grows in a dense mat before producing long stalks with single flowers. Flowers have four satiny petals that are bright yellow and eventually produce long, slender seedpods during the reproductive stage.

## Growing Conditions

California poppies may be delicate looking, but they are actually very rugged plants. In their natural habitats they grow well on rocky soil in full sun and are drought tolerant. When California poppy is grown for production, it does well in soil that has moderate fertility, is nitrogen rich, and is well drained.

## Propagation

Despite their ability to thrive in warm environments, California poppies must be sown in cold soil. They do not transplant well and should be direct seeded in late fall or early spring. It takes two to three weeks for germination. Once established, California poppies will readily self-seed. When growing for production many farmers treat California poppy as an annual and harvest the entire plant after one season.

## Planting Considerations

California poppy is a delightful plant and good for pollinators. Consider planting with other annuals, such as chamomile and calendula. These plants also do well in cool soils and prefer good drainage. Planting your annuals together can be beneficial, in that it makes crop rotations easy, allowing you to move sets of annuals to new fields each season. Recommended plant spacing is eight to ten inches within the row and fourteen inches between rows. We plant a three-foot-wide bed by broadcast seeding it.

## Medicinal Uses

California poppy is the nonaddictive cousin of the opium poppy and is a very useful nervine. The leaves, flowers, and seeds are analgesic and antispasmodic, both qualities that make it highly sought after. California poppy also has a calming and sedative effect and is used to reduce anxiety, promote sleep, and ease nervous tension and hyperactivity. Extremely safe, it is often used for children and can be taken in teas or tinctures.

## Harvest Specifications

The leaves and flowers are harvested when in bloom. Cut the tops down with a field knife and dry on racks in a single layer. If seedpods are also desired, leave some plants to mature. Harvest seedpods when longitudinal veins on the pod become pronounced and the pods begin to turn light brown. Dry pods on racks.

## Postharvest and Drying Considerations

As with many aerial harvests California poppy dries best in temperatures of 95 to 100°F. Tops should be dried whole, in partial shade, and with good airflow.

**Figure 18-15.** California poppy in bloom. Photograph courtesy of Larken Bunce

**Figure 18-16.** California poppies. Photograph courtesy of Larken Bunce

Under good conditions California poppy can dry in two to three days. Sometimes seedpods take longer. Be sure to break heads open to check that they are completely dried. Leaves and flowers can be garbled after drying by rubbing them over quarter- to half-inch stainless steel mesh and picking out the stems. Once garbled, mix leaves and blossoms together.

## Pests and Diseases

Leafhoppers, aphids, and powdery mildew can sometimes be a problem on California poppy. Crop rotations and good cultivation practices are recommended to avoid pest and disease issues.

## Yields

Yield data is challenging to come by, and we are still working on our production numbers for this crop. Some farmers report grossing $7,000 per half acre for dried leaves and flowers.

## Pricing

Retail price for one pound organic:
- Dried California poppy leaf/flower: $20
- Fresh California poppy leaf/flower: $10

## Commonly Imported From

France

# Catnip                                              *Nepeta cataria*

## Life Cycle

Catnip is a short-lived herbaceous perennial that is hardy to USDA zones 3 to 9.

## Plant Description

Native to Europe, this perennial was introduced to North America and is a favorite among felines and herbalists alike. A member of the Lamiaceae family, catnip is a mint with classic square stems and alternating leaves. Catnip has toothed heart-shaped leaves that are fuzzy and greyish-green in color. It is highly aromatic and grows three to four feet tall, making a great border and companion plant. In mid- to late summer catnip produces dense spikes composed of small individual flowers that are white with a lovely light purple undertone. On average catnip lives around five seasons but is most prolific during its second and third years of growth.

## Growing Conditions

In the wild, catnip grows in dry soils and is commonly found on the edges of fields, on stream banks, and on disturbed land. It grows well in full to partial sun and likes well-drained soil with good fertility and organic matter.

## Propagation

The seeds of catnip are light-dependent germinators and should be planted on the surface or covered lightly with potting mix. Seeds should be stratified for best germination results. Catnip can be direct-seeded in early spring or grown from transplants or vegetative cuttings. On our farm we prefer growing catnip from transplants; this makes for fewer weeds and a cleaner harvest. Once established, catnip will spread vegetatively and will also self-seed.

## Planting Considerations

Although this perennial can live for five years or more, catnip yields tend to be lower in older plants. Yields are consistently higher in plants in their second and third years of growth but after that begin to wane. Being highly aromatic and attractive to pollinators, catnip can be a good companion plant to repel pesky insects and attract beneficials. It makes a nice border for crops such as calendula and echinacea, which are susceptible to diseases carried by insects. Plant spacing is twelve inches between plants within the row, twenty-eight inches between rows.

## Medicinal Uses

Catnip is a wonderful nervine that is safe to use with children and adults. It has a calming, sedative effect and can help ease fussiness in children; it promotes sleep and is good for reducing digestive upset. Catnip has bitter properties and not only calms the stomach but acts as a carminative and helps dispel gas. In addition to its nervine and bitter actions catnip helps promote sweating, brings down fevers, and is commonly used with herbs like boneset and yarrow to treat flu symptoms. Slightly astringent, it is also good for treating children with diarrhea and can be taken as a tea or tincture. Another way we use catnip is to make an herbally infused honey with it and add it to beverages. Catnip is a stimulant for cats and is often found in their toys and treats.

## Harvest Specifications

Large commercial herb farms often produce catnip much like they would a hay crop, using machinery to cut, field dry, and bale the harvest. Field-drying catnip produces low-quality results and is not recommended for medicinal purposes. Catnip is harvested when it goes into full bloom in mid- to late summer. Cut the whole plant, both leaves and flowers, with a field knife or sickle bar cutter. When harvesting be sure to leave at least six to eight inches of growth for regeneration. This will help plants overwinter, and some warmer climates allow for a second harvest.

## Postharvest and Drying Considerations

Dry catnip at temperatures of 95 to 100°F. Do not dry catnip at very high temperatures or the volatile essential oils can be lost. Tops should be dried whole, in partial shade and with good airflow. Under good conditions catnip can dry in two to three days. Leaves and flowers can be garbled after drying by rubbing them over quarter- to half-inch stainless steel mesh

and picking out the stems. Once garbled, mix leaves and blossoms together.

## Pests and Diseases

Catnip has minimal disease or pest pressure due, in part, to its highly aromatic nature.

## Yields

Two hundred pounds of dried catnip leaf and flowers per one-eighth-acre bed (multiple harvests per season). Moisture ratio for catnip leaf flowers is 3.5:1 fresh:dried.

## Pricing

Retail price for one pound organic:
- Dried catnip leaf/flower: $12 to $24
- Fresh catnip leaf/flower: $12 to $14

**Figure 18-17. Catnip.** Photograph courtesy of Kate Clearlight

# Chamomile, German      *Matricaria recutita*

## Life Cycle

Chamomile is a self-seeding annual.

## Plant Description

Chamomile is native to Europe and north and western Asia. Introduced to North America, it grows well in many environments and is a highly sought-after medicinal. Chamomile is a lovely member of the Asteraceae family. It grows two feet tall and is a multistemmed, light green plant that has lacy leaves and a white daisylike flower. The flowers bloom in mid- to late summer and have yellow centers surrounded by white petals.

## Growing Conditions

Even though chamomile looks dainty, it is a pretty tenacious grower and does well in cooler climates. It likes full sun and well-drained, sandy loam with good fertility and organic matter. This annual will readily self-seed; however, self-seeded beds are often interspersed with weeds that chamomile will not outcompete. This can be problematic and make harvesting difficult. For production it is good to start with a clean, weed-free seedbed and replant each season.

## Propagation

Chamomile can be direct-seeded using a broadcast or precision seeder in fall or early spring. Because chamomile will not outcompete or smother weeds, it is important to be proactive with weed control. Broadcast seeds at a rate of approximately four to sixteen ounces per acre. Chamomile can also be direct-seeded with a push seeder (an Earthway or something similar). Chamomile can also be grown from transplanted plugs. The seeds are light-dependent germinators and should be sown on the surface or covered lightly with potting mix.

## Planting Considerations

We like to grow German chamomile because it grows well in our region, is very productive, and is highly medicinal. In our crop rotations we often plant chamomile nearby other annual blossoms, such as calendula and California poppy. They like similar growing

**Figure 18-18.** Chamomile in bloom. Photograph courtesy of Bethany Bond

**Figure 18-19.** Chamomile rake.

conditions and look fabulous together, and it's easier for planning purposes to keep all the annuals together, moving them to new beds each season. Plant spacing for chamomile is eight to ten inches within row, fourteen inches between rows, and three rows per bed. Once established, chamomile will spread to fill in the spaces between plants, making a nice full bed of blossoms.

## Medicinal Uses

Chamomile is a medicinal herb that has widespread popularity and recognition. We can thank Peter Rabbit for cultivating a love of chamomile in generations of parents and children alike. As the story indicates, chamomile is a tasty nervine that has a calming effect on both the nervous and digestive systems. While safe enough for children and people who are in a debilitated or weakened state, chamomile is still highly effective at treating an array of health issues. Chamomile is an anti-inflammatory, a mild sedative, and bitter. It has antiseptic properties and is used topically in washes for skin, eyes, and mouth. The essential oil of chamomile is prized for cosmetic purposes and often found in creams, oils, and salves. When brewed as a tea it is delicious and brings a sense of well-being; those sweet little blossoms lighten the spirit and restore peace. Chamomile can also be formulated with other herbs and taken in extract form as a digestive, a sleep aid, and an overall nerve tonic. A must-have for every herb pantry, this medicinal is highly valued and increasingly in demand.

## Harvest Specifications

Large operations like those found in Eastern Europe and Egypt use specialized combines to harvest chamomile blossoms, allowing for much lower prices, but the quality certainly suffers. Most mid- to small-scale farmers hand-harvest the flowers, creating a much finer product that retains higher quantities of essential oils and medicinal components. While hand harvesting is great for the medicine it is not always easy on the bottom line. Hand harvesting takes a lot of time and is a significant factor to consider when looking at the cost of production as it relates to pricing. Chamomile is harvested during full bloom every seven to ten days for approximately a month. After about four weeks blossom production begins to decrease.

Flowers are picked individually by hand, raked by a blossom comb, or cut from the stem. Each technique has its benefits and detractors. We prefer harvesting with a chamomile rake or comb or by hand. The comb is raked over the plants to pull the blossoms off the stems. Using a comb can be a bit slower but allows for more blossoms and fewer stems to be harvested. We also have had success picking by hand. You get a few more stems, but it is considerably faster than raking.

## Postharvest and Drying Considerations

Blossoms are not difficult to dry, but care should be taken to ensure that the centers are dried completely while at the same time the volatile oils are not lost. To this end dry at lower temperatures with excellent airflow and limited exposure to light. Temperatures of 85 to 95°F work well. To check for dryness select several blossoms and break them apart. When dried completely chamomile comes apart easily but is not overly crumbly or desiccated. Most buyers want their blossoms whole, so there is no need to garble them.

## Pests and Diseases

Chamomile does not attract many pests and diseases. Crop rotations and good cultivation practices are recommended to avoid any pest and disease issues that may arise.

## Yields

Fifty to seventy pounds of dried chamomile flowers per one-eighth-acre bed (multiple harvests per season). Moisture ratio for chamomile flowers is 6:1 fresh:dried.

## Pricing

Retail price for one pound organic:
- Dried chamomile flower: $17 to $81
- Fresh chamomile flower: $15 to $60

## Commonly Imported From

Egypt and Bulgaria

# Codonopsis     *Codonopsis pilosula*

**Figure 18-20. Codonopsis.** Photograph courtesy of Larken Bunce

**Figure 18-21. Codonopsis flower.** Photograph courtesy of Larken Bunce

## Life Cycle

Codonopsis, also known as dang shen, is an herbaceous perennial that is hardy to USDA zones 4 to 11.

## Plant Description

Codonopsis, a member of the Campanulaceae family, is a vine that grows up to six to seven feet tall. It is native to central and east Asia and has been used for thousands of years in traditional Chinese medicine. It has green oval leaves that are slightly toothed and covered with small hairs. Codonopsis produces lovely bell-shaped flowers that are approximately one inch wide and are a white-purplish color. Roots are brown on the outside and fleshy white on the inside. Codonopsis blooms in late summer to early fall.

## Growing Conditions

Codonopsis grows well in cool climates in full sun to partial shade. It does best in well-drained, sandy loam with average fertility and good organic matter. When growing codonopsis for root production, it is beneficial to loosen the subsoil before planting and to create trellising for the vines to grow on. The vines can be left to twine on the ground, but that is not recommended because it makes weeding and cultivation challenging.

## Propagation

Codonopsis seeds are light-dependent germinators and should be sown on the surface or covered lightly with potting mix. Seeds can be direct-seeded into the fields in the fall or spring or can be transplanted out.

## Planting Considerations

Codonopsis is a vigorously growing vine and is fairly weak stemmed. Therefore, it needs to have a very sturdy trellising system. Plant spacing is twelve inches within the row and twenty-eight inches between rows. Codonopsis can be used as an integral part of permaculture design and is a medicinal vine that will grow easily up hawthorn and other trees, using them for support as it matures.

## Medicinal Uses

Codonopsis is known as poor man's ginseng and has many of the same adaptogenic functions in the body. When taken over a long period of time, it builds energy, helps the body cope with stress, and restores vitality. It is somewhat gentler than ginseng and can be beneficial for people who find that herb too strong. In addition to its work as a longevity tonic, codonopsis has been shown to increase hemoglobin levels and nourish the blood. It can also be used in combination with other herbs to improve brain function and also as a decongestant for the respiratory system. The root has a mild, sweet flavor and is good in teas, tinctures, and capsules.

## Harvest Specifications

Codonopsis root is harvested after the fall of the third year. The roots grow up to one foot in depth and can be dug by hand with a spading fork or mechanically using a modified potato digger or bed lifter. Prior to digging the roots it is helpful to mow down the aerial tops and loosen the soil by running a shank or cultivator up the row on either side of the codonopsis.

## Postharvest and Drying Considerations

Codonopsis roots are relatively easy to clean because they do not collect dirt like many root crops that form dense root mats. Simply chop large roots into pieces (either diagonally or in cross sections) and wash. Codonopis roots can be chopped easily with a field knife. After chopping the roots, wash thoroughly, then dry at approximately 100°F. Codonopsis roots dry under optimum conditions in three to four days. Mill or chip roots after they are fully dried.

## Pests and Diseases

Minimal disease pressure reported. Roots are pleasant tasting, so some growers report rodent damage.

## Yields

Harvest yields are not available. Moisture ratio for codonopsis is 4:1 fresh:dried.

## Pricing

Retail price for one pound organic:
• Dried codonopsis root: $29

## Commonly Imported From

China

---

# Comfrey                                        *Symphytum officinale*

## Life Cycle

Comfrey is an herbaceous perennial that is hardy to USDA zones 3 to 9.

## Plant Description

Comfrey is native to Europe and was introduced to North America when it was brought to this country by settlers as animal fodder. It grows prolifically around the United States and is a member of the Boraginaceae family. It has large lance-shaped leaves that are deep green with pronounced veins; the stems and leaves of comfrey are rough and hairy and contain high water content. One of the most striking aspects of this plant is its vivid purple flowers

**Figure 18-22.** Comfrey leaf and flower. Photograph courtesy of Bethany Bond

that hang gracefully in arcs at the apex of the plant. These flowers bloom throughout the summer and are a favorite of pollinators, especially bumblebees. The roots of comfrey are black on the outside and fleshy and white on the inside and grow up to two feet deep.

## Growing Conditions

Comfrey will grow well in many conditions. It thrives in moist, well-drained soil with high fertility and will grow in full sun or partial shade. It grows easily and vigorously, and some people consider it an invasive. In our own experience we find that comfrey will quickly establish itself and does not easily relinquish a beloved growing location. Even after the root is dug, small pieces of root will regenerate. To utilize comfrey's propensity for growth, we have established permanent beds for it in areas where it can grow freely and be easily accessed. Rather than move the beds every three to four years, we add compost and manures each season to feed the plants.

## Propagation

Comfrey likes to grow and can be propagated by vegetative root divisions, which is the preferred method of propagation on our farm.

## Planting Considerations

Comfrey grows easily and is a great medicinal that can also be used as animal fodder and to enrich compost piles. On our farm the chickens eat comfrey, which is really beneficial because the plant is full of vitamins and minerals and is extremely nutritious. Consider planting stands of comfrey near chicken coops, animal pens, and pastures. Letting some grow near compost piles is good, too; layer leaves in piles for extra fertility. Plant spacing for comfrey is eight inches within the rows and twenty-eight inches between rows.

## Medicinal Uses

Comfrey roots and leaves have many medicinal uses. The word "comfrey" comes from the Latin *confervere*, meaning " to grow together," and that is exactly what it does. Containing high levels of allantoin, comfrey is a powerful cell proliferant internally and externally and is used to knit wounds and damaged tissue back together. Comfrey is a common ingredient in salves, poultices, oils, and ointments. When treating deep wounds care should be taken to avoid regenerating surface tissue before the internal wound is healed and chances of infection are gone.

Rich in vitamin B, amino acids, protein, calcium, potassium, and trace minerals, comfrey is a wonderful food source. Some people caution against the internal use of comfrey root because of the possible presence of pyrrolizidine alkaloids, which may have harmful side effects if consumed in large dosages, especially by those with liver disease. Traditionally, comfrey root and leaf were used internally for their demulcent qualities, to sooth irritated mucous membranes in both the respiratory and digestive systems.

## Harvest Specifications

This generous plant will give multiple leaf harvests in one season and abundant root harvests in the early spring and fall following two to three seasons of growth. To harvest leaves throughout the summer, cut the whole plant down to three to four inches above the crown and let it regenerate. Roots should be harvested in the spring or the fall when there is no aerial growth. The root grows up to two feet in depth and can be dug by hand with a spading fork or mechanically using a modified potato digger or bed lifter. Prior to digging roots, it is helpful to mow down the aerial tops and loosen the soil by running a chisel shank or cultivator up the row on either side of the comfrey.

## Postharvest and Drying Considerations

Pull leaves off the stems prior to drying to speed up the drying process. Leaves have a high water content and can easily bruise and turn brown. Browning is often caused by rough handling and by drying too rapidly in high heat. Comfrey leaf can also easily compost if not processed quickly after harvest. To process lay out leaves in a single layer with minimal overlap. Good airflow is very important. Begin drying leaves with fans at lower temperatures of 80 to 90°F. Then gradually raise the heat after the leaves begin to lose their moisture, and finish drying at temperatures of no more than 100°F.

Soil often gets compacted in the central crown of comfrey where the roots begin to branch off. Comfrey likes to grow in rich, moist soils, so proper cleaning takes care and attention. It is helpful to quarter the comfrey roots before washing. Comfrey roots are relatively soft and can be chopped easily with a field knife. After splitting the roots, wash thoroughly before drying in temperatures of 100 to 110°F. Mill roots after they are fully dried. Comfrey root is full of mucilage and will get gummy if chipped when fresh. Roots dry under optimum conditions in four to six days.

## Pests and Diseases

Comfrey has very few pests and minimal disease pressure. Comfrey rust or powdery mildew can sometimes be an issue, but proper crop rotations and good cultivation practices are recommended to avoid any pest and disease issues that may arise.

## Yields

Five to six hundred pounds of dried comfrey leaf per one-eighth-acre bed (multiple cuttings per season). Moisture ratio for comfrey leaf is 6:1 fresh:dried.

One hundred fifty pounds of dried comfrey root per one-eighth-acre bed. Moisture ratio for comfrey root is 3:1 fresh:dried.

## Pricing

Retail price for one pound organic:
- Dried comfrey leaf: $10 to $24
- Fresh comfrey leaf: $13 to $18
- Dried comfrey root: $10 to $20
- Fresh comfrey root: $12 to $13

## Commonly Imported From

Hungary

# Dandelion

## Taraxacum officinale

## Life Cycle

Dandelion is one of the world's most recognizable and useful herbs. It is an herbaceous perennial and grows in USDA zones 3 to 10.

## Plant Description

Dandelion is native to Europe and Asia and is a member of the Asteraceae family. It produces sunny yellow flowers, graceful smooth stems that contain white latex, and dark green leaves that are deeply toothed. The leaves grow in a rosette fashion at the base of the plant, which reaches heights of eight to twenty-four inches tall. Dandelion taproots are brown on the outside and white inside. Mature roots average eight to twelve inches in length.

## Growing Conditions

Dandelion is a tenacious grower and grows well in many different soil types. It thrives in soils with good fertility, adequate moisture, and good sun.

## Propagation

Dandelion can be grown from transplants or sown directly in the fields in late fall or early spring. Plant spacing is eight inches apart within the row. Within a single bed we plant triple rows of dandelion with fourteen-inch spacing between the rows.

## Planting Considerations

Dandelion can be grown as an annual or for two seasons before root harvest. We like to plant dandelion with other "short-time" roots, such as burdock, yellow dock, and angelica. All of these crops we grow for root production, and planting them together helps streamline bed preparation, harvest, and crop rotation. For root beds we do more subsoiling to break up any hardpan, and also raise the beds. This allows for optimum root growth. We also make sure to rotate these crops into fields that have been cover cropped with clover or alfalfa, building fertility that will be needed by these heavy feeders.

## Medicinal Uses

Glorious dandelion! Not only is it a lovely plant, dandelion is highly medicinal and used for many different health needs. First and foremost, dandelion roots and leaves are a fantastic food source. They are highly nutritive and are rich in calcium, potassium, protein, and vitamins A and C. Traditionally, the spring leaves were collected and eaten as greens in salads and soups, or steamed and cooked in casseroles. They are bitter but delicious and are used to stimulate digestion and tonify the body.

The roots and leaves are also used as tonics for the liver and the kidneys. Dandelion helps the body produce bile and is a safe, effective diuretic for the kidneys. A common and delicious way to take dandelion is as an aperitif or bitter. Dandelion blends well with other herbs such as yellow dock and angelica to make a tasty combination that will aid in digestion and tone the liver.

## Harvest Specifications

Dandelion leaves can be harvested starting in spring through early fall. The leaves are most tender in the spring and are the best for cooking because they are more tender and palatable. Leaves that are harvested in summer and fall, although more bitter, make very good medicine and can be tinctured fresh or dried. To harvest dandelion leaves in the spring and summer cut the leaves with a field knife to about an inch above the crown. There is one caveat: When harvesting leaves be sure to leave some green growth on the plant if you are also growing for root harvest. We do this by taking some leaves from each plant but leaving some intact to allow the plant to grow larger and more robustly. Harvest leaves when they are clean and dry. In the fall you can also get a leaf harvest after digging the roots. Once the roots are dug, you can simply chop the leaves off while processing the roots.

Root can be grown as an annual and harvested in the fall of the first year. In one season they size up well and are highly medicinal. Many growers prefer to wait until the spring or fall of the second year to harvest dandelion root because the yields are greater. Dandelion roots are easy to harvest by hand with a spading fork or mechanically using a modified potato digger or bed lifter. Prior to digging roots it is helpful to loosen the soil by running a shank or cultivator up the row on either side of the dandelions.

## Postharvest and Drying Considerations

Dandelion leaves are one of the few aerial parts that we wash before drying. Because dandelion leaf is low growing, it often contains soil. It is important to get this particulate matter off the plants before drying. We double rinse them in bins of clean water and soak them for five to ten minutes, agitating gently with our hands. Once they are clean we drain the leaves, spread them on our drying racks outside, and rinse them one more time with clean water. The racks drip-dry for ten to fifteen minutes, then are placed in an empty drying shed. Introducing this much moisture could damage other crops if they are in the shed trying to dry, so we initially dry the dandelion leaf alone for one or two hours. Once the majority of the moisture is gone, we can add other crops into the shed.

When transplanted, dandelion roots present as small taproots or, more commonly, root clusters that can be tedious to clean. (Large dandy taproots are most often found in direct-seeded and wildcrafted plants.) We divide the multi-stemmed roots before washing in order to clean them. Taproots can be washed whole. Dandelion roots are relatively soft and can be chopped easily with a field knife. After splitting the roots, wash them thoroughly before drying. Mill roots after they are fully dried. Dandelion roots dry under optimum conditions in three to four days. The recommended drying temperature is 95 to 110°F.

## Pests and Diseases

Minimal pest and disease pressure occurs on dandelion.

## Yields

Fifty pounds of dried dandelion leaf per one-eighth-acre bed. Moisture ratio for dandelion leaf is 4-5:1 fresh:dried.

One hundred pounds of dried dandelion root per one-eighth-acre bed. Moisture ratio for dandelion root is 4:1 fresh:dried.

## Pricing

Retail price for one pound organic:
- Dried dandelion leaf: $12 to $28
- Fresh dandelion leaf: $13 to $14
- Dried dandelion root: $16 to $22
- Fresh dandelion root: $13 to $16

**Figure 18-23. Dandelion.** Photograph courtesy of Larken Bunce

# Echinacea

## *Echinacea purpurea*

## Life Cycle

*Echinacea purpurea* is an herbaceous perennial that is hardy to USDA zones 3 to 8 and is drought tolerant.

## Plant Description

Native to the prairies of the Midwest, *E. purpurea* is one of the showier members of the Asteraceae family. Growing up to five feet tall, this gorgeous perennial is known for both its striking purple coneflowers and its important medicinal attributes. *E. purpurea* flowers grow from several stalks and are daisy shaped, with a central seed head encircled by bright purple petals. Leaves are cauline and grow from the stem more densely at the base of the plant, becoming sparser at the top. Leaves are lance shaped, and like the stems are somewhat coarse and rough. *E. purpurea* roots grow in a spreading, multibranched root mass, whereas the *E. angustifolia* and other echinacea species produce more of a taproot. Echinacea begins flowering in midsummer and continues flowering for many weeks. It is a favorite among pollinators.

**Figure 18-24.** *Echinacea purpurea*. Photograph courtesy of Bethany Bond

## Growing Conditions

*E. purpurea* grows best in full sun and well-drained loamy soil. As with many perennial crops, echinacea responds well to ample fertility and can benefit from top-dressings of fertilizer according to soil tests taken during its several years of growth. A consistent water supply increases growth, but echinacea plants can also tolerate long dry spells.

## Propagation

Echinacea purpurea can be grown by direct-seeding in the fall or early spring or by transplanting plugs or crown divisions. When direct-seeded, plants do best with a fine seedbed and cool temperatures. Although seeds do not require stratification it can help increase germination rates. *E. purpurea* takes approximately four weeks for germination, and seeds should be planted on the surface and covered lightly with soil.

## Planting Considerations

This is one perennial that loves to be showcased. We often joke that *E. purpurea* is the highly photogenic "cover girl" of the herb world—and for good reason. It has deep green foliage and fantastic purple and red flowers. Plant *E. purpurea* where you can see it! We like to plant *E. purpurea* in large blocks, in beds that have been well worked and cover cropped. Because of its susceptibility to aster yellows (a disease carried by leafhoppers) we avoid planting it next to calendula, which also is prone to the disease. *E. purpurea* will need at least three years to grow before root harvest. Therefore, good bed preparation is essential. *E. purpurea* plant spacing is eighteen inches within the row and twenty-eight inches between rows, with two beds per row.

## Medicinal Uses

The entire plant of *E. purpurea* is medicinal. Most people are familiar with the root and seeds; however, the leaves and flower are also healing. *E. purpurea* is a highly effective and safe immune herb that helps to jump-start the immune response in the body. Specifically it increases the macrophage T-cell activity and helps boost the immune system at the onset of infection. It has been used successfully for hundreds of years to treat colds and flus and works well against both viral and bacterial infections. Echinacea can be used as a tea, made into tinctures, powdered and capsulated, or made into mouthwashes and throat sprays. It is safe to use for children and adults and has a consistently solid market value.

## Harvest Specifications

Leaves and flowers can begin to be harvested in the first year if planted early enough when they are in full blossom, usually around midsummer. Because it is important to leave enough aerial growth to feed the root system, we do not cut all the tops back. Instead, we selectively harvest 10 to 20 percent of the tops. *E. purpurea* is a multistemmed plant, so it is easy to use field knives to cut a few stems from each plant. This produces a good harvest and also allows the roots to develop. Heavier leaf and flower harvests can be made from more mature plants without severely compromising root growth.

The roots are ready to harvest in the fall of the third or fourth year. *E. purpurea* roots are easy to harvest by hand with a spading fork or mechanically using a potato digger or bed lifter. Prior to digging roots, mow down or harvest the aerial parts. It is also helpful to loosen the soil by running a shank or cultivator up the row on either side of the *E. purpurea* before digging.

Seeds can be harvested after the second year of growth. For "true" seeds that haven't cross-pollinated, make sure *E. purpurea* is isolated from other echinacea species. Harvest ripe seed heads when the stems have begun to turn brown and the seed head shatters, meaning that it easily releases the seed. Seed heads are not hard to harvest and can be gathered by hand using pruners or a field knife.

## Postharvest and Drying Considerations

Once flowers are harvested from the field, clip all blossoms off the stem and run them through a chipper. Spread the chipped blossoms on drying screens with good airflow at temperatures of 90 to 100°F. Chipping the blossoms before drying helps them dry more uniformly and makes for easier processing.

Leaves can be dried easily on the stalks. Place the aerial parts in a single layer on drying racks, and dry at temperatures of 100 to 110°F. Once dried, garble the plants over quarter-inch stainless steel mesh and pick out the stems. Once both the flowers and leaves are dried, they can be mixed together or sold individually.

Soil can get compacted in the central crown of *E. purpurea* where the roots make a bit of a mat. Therefore it is helpful to quarter the roots before washing. Echinacea roots are relatively soft and can be chopped easily with a field knife and mallet or small hatchet. After splitting the roots, wash thoroughly before drying. Mill roots after they are fully dried. *E. purpurea* roots dry under optimum conditions in three to four days.

It takes a little work to free the seeds from the seed head. Richo Cech, from Horizon Herbs,[2] recommends running the seed heads gently through a hammer mill at low rpm with a one-inch screen. This breaks up the heads but does not damage the seeds. Next, hand-clean seeds by running them through quarter-inch screens and winnowing. This will separate the chaff from the seed. Once seeds are clean lay them on drying screens with good airflow and dry at temperatures of 90 to 100°F.

## Pests and Diseases

*E. purpurea* is susceptible to leafhoppers and aster yellows, sclerotinia stem and root rot, and botrytis. Proper crop rotations and good cultivation practices will help reduce these disease and pest pressures. We also carefully monitor our fields and remove any plants that show signs of aster yellows. This has been effective at preventing the disease from spreading without reducing yields considerably.

## Yields

250 pounds of dried echinacea leaf and flower per one-eighth-acre bed. Moisture ratio for echinacea leaf and flower is 4:1 fresh:dried.

125 pounds of dried echinacea root per one-eighth acre bed. Moisture ratio for echinacea root is 3:1 fresh:dried.

Echinacea seed data not available.

## Pricing

Retail price for one pound organic:
- Dried *Echinacea purpurea* root: $20 to $31
- Fresh *Echinacea purpurea* root: $13 to $21
- Dried *Echinacea purpurea* tops: $8 to $25
- Fresh *Echinacea purpurea* tops: $6 to $13
- Dried *Echinacea purpurea* seed: $70

# Elder                                                *Sambucus nigra*

## Life Cycle

Elder, or elderberry as it is often called, is a perennial deciduous shrub that is hardy to USDA zones 3 to 7.

## Plant Description

Native to Europe, Africa, and Asia, elder is a member of the Adoxaceae family, growing up to ten feet tall and producing beautiful clusters of small, creamy-white flowers that form umbels at the apex of the plant. Elder goes into bloom in July and sets berries in August. The berries are shiny and smooth and are a lovely deep purple. The leaves of elder are also handsome. They are pinnately compound and comprised of groups of five to nine leaflets.

## Growing Conditions

Elder can grow in a fair amount of moisture and likes very rich soil and partial shade. In many places elder can be field-grown in full sun, provided there is adequate moisture. Because it thrives on fertility, consider top-dressing elder with compost or composted manures or growing a living mulch such as alfalfa that will fix nitrogen into the soil, feeding the bushes during their many years of growth.

## Propagation

Elderberry can be propagated from seed or cuttings. Seeds should be sown in outside conditions, allowing exposure to fluctuating temperatures. Elder needs the cycle of stratification to germinate. Sow seeds into flats, keep moist, and leave outside in a shaded, protected area. If sown in the fall, seeds should germinate in the subsequent spring. When working with cuttings there may be a higher mortality rate than with seeds.

To improve success rates make cuttings at least ten inches long and include two or more nodes. Plant cuttings into good potting soil, keep moist, and let plants develop substantial a root system before transplanting. We prefer to propagate elder using dormant hardwood cuttings taken in late March. We place the cuttings—in gallon pots with a lightweight, well-draining potting soil—on the floor of the greenhouse and generally achieve more than 75 percent success rate at producing viable plants using this method. This also allows us to select cuttings from our most productive bushes to perpetuate good genetic traits.

## Planting Considerations

In climates like Vermont's, elder grows very well in full sun; however in hotter, sunnier places elder may

**Figure 18-25.** Elder in berry.

**Figure 18-26. Elder flower.** Photograph courtesy of Larken Bunce

**Figure 18-27. Elderberry being destemmed.** Photograph courtesy of Bethany Bond

grow better in partial shade. Consider planting elder as an understory shrub on the edge of fields beneath the higher canopies of maples and other hardwoods. Another consideration when planting elder groves is that it takes three to four seasons to establish plants that will produce berries in significant quantities. It is important to keep weed pressure down during early growth and to maintain fertility. Consider planting red or white clover or alfalfa beneath shrubs as a dual-purpose cover crop. These legumes will feed the soil and also reduce weeds. Elderberry plant spacing is six feet between the plants and six feet between the rows. Growers may wish to space rows wider for mechanical mowing between rows which can help reduce weed pressure, promote good airflow and pollination, and allow easier access for harvest and maintenance.

## Medicinal Uses

Elderberries and elder flowers are extremely delicious and effective medicinals. Both leaves and berries are diaphoretic, help promote sweating, and bring down fevers. They stimulate the immune system, are antiviral, and are commonly used to treat flus, colds, and respiratory infections. In addition to these medicinal attributes, elderberry is highly nutritious, is rich in bioflavonoids and vitamin C, and is a strong antioxidant. One of the reasons elder is so beloved is that it is delicious! Elder flowers and berries can be made into teas and are often used in combination with peppermint, yarrow, and boneset for flus and colds. The berries are cooked and made into syrups, jams, juices, concentrates, and tinctures.

*Cautionary note:* The black elder (*Sambucus nigra*) is edible; red elder, however, is not. Berries from the black elder should be cooked before using, and the seeds should be strained out.

## Harvest Specifications

The umbels are harvested in early July when elder begins to flower. The small flowers that make up the umbels stagger their blossoms; some open earlier than others. As a result umbels should be picked when most of the flowers are beginning to open, but some may still be in bud. The key is to not let the blossoms become overly mature or the flowers will lose their vitality and the prized pollen will also be lost. To harvest simply snap the stem at the base of the cluster, taking the whole lacy flowering top.

Elder sets berries in mid-August and will allow for a substantial harvest after three to four years of growth. The berries are dark bluish-black and shiny when ripe. Avoid picking the berries when they are still pinkish. Like the flowers, berries are picked by snapping the stem at the base of the cluster. Berries are then seperated from clusters by hand.

## Postharvest and Drying Considerations

Elder flowers are dried by placing the whole flowers onto drying screens in a single layer. For optimum color and retention of medicinal compounds, dry in complete darkness at temperatures of 95 to 100°F. Good airflow is also key; make sure fans are blowing directly over the racks. Elder flowers are small and light and should dry within two days. It is helpful to process the flowers when they are completely dry but when the stems are still slightly pliable. Run the blossoms gently over quarter-inch stainless steel mesh to remove stems.

We sell the majority of our berries fresh or frozen. To dry elderberry the best course of action is to rack berries in a single layer and position them directly in front of fans and a heat source. To keep the vibrancy of the berries it is a good idea to limit the exposure to light because the berries can take a week or two to dry thoroughly. To test for dryness use a sharp blade to cut open the berries. They should be dry and hard, not pliable or mushy.

## Pests and Diseases

Elder has very minimal insect or disease pressure. The main pest issue facing elderberry producers is harvesting the berries before birds and other animals eat them. Netting is expensive and time consuming. The best course of action is to monitor berries carefully in August and pick immediately when ripe.

## Yields

Elder flower yield data per one-eighth-acre bed is not available. Moisture ratio for elder flower is 5–6:1 fresh:dried.

Five hundred pounds of fresh elderberry per one-eighth-acre bed. Moisture ratio for elderberry is 4:1 fresh:dried.

## Pricing

Retail price for one pound organic:
- Dried elderberries: $12 to $37
- Fresh elderberries: $21
- Dried elder flower: $15 to $60
- Fresh elder flower: $20 to $36

## Commonly Imported From

Hungary, Poland, and Bulgaria

# Elecampane                                    *Inula helenium*

## Life Cycle

Elecampane is an herbaceous perennial that is hardy to USDA zones 3 to 8.

## Plant Description

Native to Europe and Asia, elecampane is a gorgeous member of the Asteraceae family. It produces bright yellow, daisylike flowers that are approximately three inches wide. There are multiple flowers on each stem, and our daughter likes to call this plant "the little sunflower plant." The leaves of elecampane are large, basal, and slightly toothed, with undersides that are white and hairy. The roots grow in branching clusters that can be quite large, going two to three feet deep. Elecampane is a statuesque plant that grows five to seven feet tall.

## Growing Conditions

Elecampane grows best in full sun to partial shade, in loamy, well-drained soil. It thrives in fertile soils and can respond well to top-dressing with bands of compost. To increase root harvest it is worth the time to break up any compaction by either double-digging beds, creating raised beds, or using cultivation tools to loosen subsoil structure.

## Propagation

Elecampane is a light-dependent germinator. Sow seeds in early spring on the surface or cover with a thin layer of planting medium. Keep well watered. Elecampane can be direct-seeded in the fields or indirect-seeded for transplanting.

## Planting Considerations

Elecampane is lovely and large and needs at least two years to mature before harvesting. Consider planting elecampane next to other tall perennials such as blue vervain or echinacea. Not only will they visually complement each other, but they also grow in similar conditions. As with any plant raised for its roots it is recommended to use a subsoiler or other devices to loosen the soil and to form raised beds for better yields. Plant spacing is eighteen inches within the row and twenty-eight inches between rows.

## Medicinal Uses

The yellow flowers of elecampane are a good example of the doctrine of signatures. The bright gold color points to its traditional use to fight infection and clear dense, sticky mucus from the bronchial system. Elecampane is warming and is indicated for wet, persistent infections. Paired with pleurisy root

**Figure 18-28.** Elecampane.

and ginger, it is often made into cough syrups, teas, and tinctures. Our family is prone to bronchial infections, and we wouldn't dream of going through a winter without this incredible plant.

## Harvest Specifications

Elecampane is harvested after the fall of the second or third year. Waiting until the second year allows the roots to develop more fully and reach maximum size. Preferably, the roots are harvested while the plants are still dormant. Mature roots are fleshy and large, branching out from a thick centralized stalk, and can grow up to two feet deep. As a result, it is helpful to mechanically loosen the surrounding soil before digging the roots. This can be accomplished by running a shank or cultivator up the row on either side of the elecampane. Once the soil is loosened roots can be dug out by hand with spading forks or mechanically, using a bed lifter or modified potato digger.

## Postharvest and Drying Considerations

Soil often gets compacted in the central crown of elecampane where the roots begin to branch off.

Therefore, it is helpful to quarter the roots before washing. Elecampane roots are relatively soft and can be chopped easily with a field knife. After splitting the roots, wash thoroughly before drying. Mill roots, after they are fully dried. Elecampane roots dry under optimum conditions, at temperatures of between 100 and 110°F, in three to four days.

## Pests and Diseases

Elecampane has minimal pest and disease pressures.

## Yields

Three hundred dried pounds of elecampane roots per one-eighth-acre bed. Moisture ratio for elecampane root is 4:1 fresh:dried.

## Pricing

Retail price for one pound organic:
- Dried elecampane root: $10 to $20
- Fresh elecampane root: $9 to $13

## Commonly Imported From

Bulgaria

# Garlic (hardneck)                                    *Allium* spp.

## Life Cycle

Garlic is a perennial that is hardy to USDA zones 3 to 9.

## Plant Description

Garlic is one of the most common and well-known medicinal herbs that we grow. It is native to Asia but is commonly used for medicinal and culinary purposes all over the world. A member of the Amaryllidaceae family, garlic is a bulbous plant that comes in both hard- and softneck varieties. We prefer to grow hardneck, which is the preferred medicinal and

culinary variety—our favorite variety being the German Porcelain White. This variety is extremely hardy, medicinal, and flavorful. It grows well in our area and stores easily. Hardneck varieties evolved directly from wild garlic and are close relatives to onions and leeks. Elephant garlic, however, is not a true garlic, rather it is a type of shallot, and although often used as a culinary, it is not recommended for medicinal use.

Garlic puts out leaf growth in early spring. The leaves are long and slender, and grow and look similar to grass but are thicker and larger. In the summer garlic will set a flowering stalk arising from the main

**Figure 18-29.** Garlic curing in the greenhouse.

## What Are Scapes?

Scapes are the centralized stalks of garlic that make graceful loops and produce a small seed head called a bulbil capsule. This seed head is not to be confused with "seed garlic." Seed garlic is the mature bulbs that are broken apart into individual cloves for planting. Seed garlic vegetatively produces a mature garlic bulb in one season. The bulbil capsules when ripe contain upward of a hundred bulbils that can also be planted. These bulbils will eventually grow into larger, good-size bulbs after a two- to three-year cycle of growing, harvesting and replanting.

Besides producing planting stock, scapes are wonderful because they can be harvested and used as a culinary vegetable. Scapes are delicious steamed, pickled, and sautéed and are also wonderful in pestos. In our area there is a small market for scapes in local restaurants and co-ops. The scapes are also lovely in the way they grow, and we have been known to put a few in a vase for an unusual but pretty flower arrangement.

leaf stem that can grow to between one and five feet. This is known as a scape, and it will grow in loops that are initially succulent and green. As the plant matures, these scapes straighten out and become woody, eventually producing a flowering head called a bulbil capsule.

The roots of garlic are a bulb composed of (on average) four to ten cloves that are encased in skins or wrappers with a fibrous basal root system located at the bottom of the bulb. When cured properly these wrappers serve to protect the garlic, helping to maintain freshness and shelf life.

### Growing Conditions

Garlic does well in a variety of soils. To get maximum yields it is recommended to plant garlic in well-drained, loamy soil with ample moisture, good fertility, and plenty of organic matter. It is important to keep garlic well weeded to prevent nutrient competition, promote good airflow, and provide access to sunlight. This helps maintain bulb health and reduces the chances of disease. Good weed maintenance also helps ease with harvesting. Some agricultural publications report that poor weeding can result in up to a 50 percent reduction in yields.[3] Keeping up with weeding is well worth the time and effort. Garlic is also a heavy feeder. It can be beneficial to top-dress it in the spring with composted manures. However, adding additional nutrients later in the summer is discouraged because it can damage bulbs.

### Propagation

It is best to plant garlic from cloves originating from good seed stock. Plant it in the fall early enough for roots to develop but not too early to allow the plant

**Figure 18-30. Garlic scapes.** Photograph courtesy of Bethany Bond

to set too much green growth and sap energy needed for winter protection and spring growth. Before planting prepare garlic by "cracking" the bulbs, which means breaking the bulbs apart into individual cloves. Be careful not to damage the individual cloves by nicking the fleshy parts or disrupting the base where the roots will grow. It is recommended to do the cracking no more than forty-eight hours before planting. This prevents cloves from drying out too much before they go into the ground and reduces exposure to diseases.

## Planting Considerations

Garlic does well when planted in triple rows with eight-inch spacing in the row and fourteen inches between the rows. Cloves should be planted with tips up and roots down at a depth of approximately three inches below the soil surface. After planting many farmers like to mulch with leaves or straw to maintain moisture, discourage weeds, and prevent deep freezing in cold climates.

If planted in the fall, garlic will quickly produce vegetative growth the following spring and by early July will begin producing graceful curling seed stalks—the scapes. It is beneficial for many varieties of garlic to remove the scapes to allow more nutrients and energy to be focused in the root bulbs. Removing scapes is best done with a sharp field knife after they start growing into loops. Scapes can be marketed to resturaunts and individuals for eating or processing into pestos and other products.

In crop rotations wait at least four years before planting on the same soil to help minimize any soil-borne disease or pest pressure. Use good, strong seed stock. When saving your own seed choose the largest, healthiest bulbs for planting. Proper drainage is essential for good garlic production, so choose beds wisely.

## Medicinal Uses

Garlic has a rich history of medicinal use dating back over seven thousand years. It is truly one of the most revered herbs in our herbal pharmacopoeia. Garlic is both antimicrobial and antiseptic and is highly effective at stimulating the immune system and fighting infection. It is wonderful for treating bronchial infections and the common cold and is also good for lowering cholesterol and blood pressure when taken over a long period of time. It is delicious and can be prepared a number of ways: sautéed, pickled, added to vinegars and oils, eaten raw, tinctured with other immune herbs such as echinacea and goldenseal, encapsulated, and used topically.

## Harvest Specifications

Harvest garlic when some (40 to 60 percent) but not all of the leaves are beginning to go brown. At harvest time an examination of bulbs should show fully developed cloves with skins still intact. Do not wait until the skins are starting to pull away. This can make the bulbs susceptible to disease. In Vermont we harvest around late July or early August when the soils are dry and there is less moisture. Wet soils around harvest time can be problematic because excess moisture can increase the chance of fungal diseases forming during the curing process. Garlic can easily be pulled out of the ground by hand and the soil gently brushed off while still in the field. It is important to handle garlic carefully, as it is easily bruised. Do not bang garlic bulbs against one another. Place them gently on tarps in bins, wheelbarrows, trucks, or tractor buckets to transport, and avoid damaging the tender bulbs.

## Postharvest and Drying Considerations

Proper curing is essential for long-term storage and maintaining garlic's medicinal properties and flavor. Cure garlic out of direct sun in a place that has good airflow. We dry garlic in our greenhouse on benches under shade cloth, keeping the temperatures between 80 and 90°F. When curing leave the aerial tops on. This allows nutrients and many of the medicinal compounds to continue to transfer to the bulbs. In good conditions curing takes three to four weeks. The garlic is fully cured when the tops have gone from green to brown and all vitality has left the leaves. The outer skins are papery and dry but still tight around the bulb.

To process garlic after curing we cut off the tops and roots using a sharp pruner. When topping the bulbs, trim the tops half an inch above the cloves. Remove soil by gently rubbing off the outermost skins while leaving the rest of the wrappers intact to protect the cloves and maintain freshness. Store garlic in temperatures between 55 and 65°F with low humidity and good airflow. For long-term storage garlic can be kept in netted onion bags, or in some cases loosely closed paper bags can work.

## Pests and Diseases

Soilborne diseases and pests can be an issue with garlic. To help prevent the spread of disease proper crop rotation and management is important. If disease does occur, immediately (and carefully) remove and dispose of any deformed plants. Do not use planting stock that comes from diseased garlic, and do not replant garlic in the same spot two years in a row. Mites, thrips, nematodes, and onion maggots can sometimes be a problem. Again, proper sanitation and good cultivation techniques can help reduce the impact of harmful insects. Planting insectaries and encouraging populations of beneficial insects can also be useful.

## Yields

375 pounds of cured garlic per one-eighth-acre bed.

## Pricing

Retail price for one pound organic:
• Cured garlic bulbs: $8 to $20

## Commonly Imported From

China

# Ginkgo

*Ginkgo biloba*

## Life Cycle

Ginkgo is a beautiful, long-lived deciduous tree that is hardy to USDA zones 3 to 8.

## Plant Description

Native to China, ginkgo, also known as maidenhair tree, is the oldest living fossil, related to trees that lived over 270 million years ago. Ginkgo grows to be very old; some trees in China are reported to be close to four thousand years old.[7] Strikingly beautiful, ginkgo produces two-lobed, fan-shaped leaves that are two to four inches wide and crowns that are statuesque, growing sixty to one hundred feet tall. Ginkgo is a dioecious species, meaning that it has separate male and female plants. The male plants produce pollen, and the female plants are known for their malodorous fruit.

## Growing Conditions

Ginkgo prefers to grow in partial shade or full sun and in soils that are moist but well drained. For leaf production, arborists plant trees approximately eight

**Figure 18-31. Ginkgo.** Photograph courtesy of Kate Clearlight

feet apart and fertilize them well with composted manure. Trees for leaf production are also pruned to shorter heights, focusing on lateral rather than vertical growth. For fruit and seed production, trees are planted farther apart and are allowed to grow to full height to promote good pollination.

## Propagation

Ginkgo can be grown from seed. First, the seeds need to be cleaned of the smelly fruit that encases them. Then the seeds have to stratify first in warm temperatures for two months, followed by two months of cold. Once stratified, seeds can be sown in the spring. It is beneficial to grow seedlings in small containers for two to three years before planting out to more permanent locations.

Ginkgo can also be grown from softwood cuttings. The benefit of propagating ginkgo vegetatively is that germination of seeds can be spotty. Also, propagation from cuttings allows growers to select for sex. In some cases people prefer to grow only male plants to avoid stinky fruit. Trees from seeds cannot be sexed until they mature, which can take up to twenty years.

## Planting Considerations

One wonderful thing about ginkgo is that it is very tolerant to air pollution and can grow well in urban settings. It does not mind growing in confined spaces and is often a favorite for city parks. The trees require little maintenance and are very lovely.

The fruit of ginkgo tends to cause a stir because of its offensive, odiferous nature. Some people when growing ginkgo for leaf production grow only the male plants to avoid stinky fruit. However, when ginkgo is grown for seed production, both sexes are needed, and the fruit is collected and processed. For fruit, trees should be planted farther apart, at ten- to fifteen-foot spacing. Trees need to be at least twenty years old before good fruit is set and seven to eight years old for good leaf production. So whether it is grown for leaves, fruit, wood, or simply for its beauty, ginkgo can be a great component of permaculture guilds on a farm and should be planted for future generations.

## Medicinal Uses

Ginkgo seeds are considered a delicacy in Asian cooking and are used traditionally as seasoning in soups and egg dishes. While ginkgo seeds are a specialty food and are used less commonly, ginkgo leaves are a well-known medicinal, having been used for hundreds of years as a memory tonic and to increase brain function. Ginkgo leaves act as a vasodilator and help to increase circulation throughout the body, including the brain. The leaves are also rich in antioxidants and help the body stay healthy during stressful times. Some herbalists blend ginkgo with other nervines to help strengthen the nervous system. In addition to helping the nervous system, ginkgo can be paired with herbs like hawthorn and linden to tonify the heart and circulatory system.

## Harvest Specifications

There has been debate in the herbal community about whether to harvest ginkgo leaves when they are green or after they have turned yellow. Because the leaves contain different chemical constituents at different stages, ginkgo leaves should be harvested when the leaves are just beginning to turn yellow at the end of August or in early September. This allows harvesters to gather leaves that contain important properties that are present when the leaves are green (ginkgolides and bilobalides) and yellow (flavonoids/antioxidants).[8] Leaves can be harvested by hand using a ladder or the bucket of a tractor. After harvesting as much as possible by hand, additional leaves can be captured by tarps spread out beneath the trees. Ginkgo drops all its leaves within a two-week window. The leaves are bright yellow when they drop, but after combining them with the harvest of green leaves the result is a product that contains the desired ginkgolides, bilobalides, flavonoids, and other antioxidants.

## Postharvest and Drying Considerations

Dry ginkgo leaves on drying racks with good airflow, little light, and at temperatures around 100°F. Under good drying conditions leaves will dry in one or two days. Leaves will crumble and lose pliability when dry. Ginkgo leaves can be sold whole or can be run over a garbling rack if a finer grade is desired.

## Pests and Diseases

The wood of ginkgo is insect resistant, and it has very few pest or disease problems.

## Yields

250 pounds of dried ginkgo leaves per one-eighth-acre bed. Moisture ratio for ginkgo leaf is 4:1 fresh:dried.

## Pricing

Retail price for one pound organic:
- Dried ginkgo leaf: $10 to $20
- Fresh ginkgo leaf: $12 to $13

## Commonly Imported From

China

# Ginseng                                    *Panax quinquefolius*

## Life Cycle

Ginseng is a long-lived perennial that grows in USDA zones 3 to 8.

## Plant Description

Ginseng is a member of the Araliaceae family and grows in the deciduous forests in the mountains of the eastern United States and Canada. In the United States its range extends from northern Vermont to Georgia and as far west as Minnesota. A beautiful woodland plant, ginseng grows one to one a and half feet tall. The leaves grow off a simple stem that branches off into two-, three-, or sometimes five-pronged plants as the plant ages from year to year. The more branching that occurs, the older the ginseng plants are. The leaves are palmate, primarily composed of five leaflets that are finely toothed with pronounced tips. The ginseng flowers are small and yellowish-green and produce lovely clusters of dark red berries. The roots, sometimes said to look like a little man, are pale yellowish-brown and at maturity measure four to six inches in length.

## Growing Conditions

Ginseng requires very specific growing conditions and needs a cold winter for dormancy. Ginseng also needs at least 70 percent shade to grow and does well in humusy soil that is well drained and has high calcium and phosphorus content. The plants tend to favor growing in places where there is plenty of space and airflow. In the wild, ginseng is found growing in mixed hardwood forests, its ideal habitat. Ginseng requires a good amount of water. It does not tolerate prolonged drought, nor does it grow well in overly wet environments where the roots will tend to rot.

## Propagation

Ginseng can be grown from fresh seeds or rootstock. Plant seeds in late summer or early fall approximately half an inch deep. Seeds will go through a required cold dormancy of winter, stratifying naturally. Be patient! Ginseng takes a long time to germinate, up to one to two years. This slow germination rate is one of the reasons wild populations are at great risk of being harvested out of existence.[4] Recommended seeding rates are approximately ten to fifteen pounds of seed per acre. To plant, rake away leaves and remove major weeds. Plant seeds every three inches in rows that are eighteen inches apart, top-dress with amendments as needed, then restore leaf mulch. After planting leave the bed alone and let nature take her course. Low germination rates and natural factors (such as pests, disease, and weed pressure) will naturally thin the bed. The key to wild simulation is that, once planted, the ginseng is left to the forces of nature to develop and grow like

**Figure 18-32.** American ginseng. Photograph courtesy of Bethany Bond

**Figure 18-33.**

its wild counterparts, creating strong medicine and good roots. Top-dress every two to three years with gypsum and rock phosphate if needed as directed by soil test results, but other than that, leave the tending to Mother Nature.

Ginseng can also be transplanted by rootstock, although it is more expensive to procure rootstock than it is to get seed. Plant roots in the fall with one-foot spacing. Do not till beds as this can damage the delicate fungal hyphae that make up good, fungally dominant forest soils and also damage the roots of surrounding trees. Instead, rake leaves and debris out of beds, plant roots, top-dress with amendments as needed, and replace leaf mulch. Then leave the plants alone until harvest.

## Planting Considerations

Because of highly specific growing conditions, farmers cultivate ginseng under artificial shade (i.e., in shade houses) or in the woods using wild-simulated techniques. Research and the experience of farmers in the United States, Canada, and China show

that growing ginseng under artificial shade is more expensive and more labor intensive and has increased disease pressure that often results in a greater dependence on chemical interventions.[5] Another detractor is that roots grown in shade houses tend to fetch lower prices on the market because they are perceived as less desirable than woods-grown or wild. The soil in shade house conditions is primarily bacterially dominant, which is different than the fungally dominant soils found in wild and wild-simulated ginseng habitats. These bacterially dominant soils are more prone to harboring fungal diseases. As a result more farmers are turning away from shade houses and considering transforming the understory of hardwood forests into woodland beds.

Site selection for a woodland bed is key. The ideal site is northeastern facing, sloped, and located under a canopy of 70 to 75 percent shade. Thick leaf mulch and good drainage are also important. It is highly recommended to companion-plant ginseng with other plants from the woodland communities. Consider interplanting ginseng with goldenseal,

Solomon's seal, wild ginger, and trillium/bethroot. They grow well together, help prevent the spread of disease within the beds, and are all valuable medicinals. In addition to the ecological aspects of site selection, plant woodland beds containing ginseng in secluded, protected areas where poaching (a serious concern; see chapter 12) is less likely. Some bigger ginseng operations employ security measures, but this can be expensive and tricky. For smaller farms careful site selection and discretion are often enough.

When preparing wild-simulated beds the soil is not tilled or double-dug. Instead the roots grow in the relatively undisturbed soil, as they would in nature. Leaving the soil as undisturbed as possible reduces the spread of disease and replicates more closely wild conditions. That said, it is beneficial to top-dress plants with gypsum and rock phosphate to increase the calcium and phosphorus levels. This helps promote plant growth, root development, and disease resistance. Top-dressing can be done every couple of years. Use soil tests to determine the amount of amendments needed for beds, as soil composition in woodland beds can vary greatly. In general, it is recommended to use five pounds gypsum per one hundred square feet of bed. When using rock phosphate, try to achieve soils that contain approximately ninety-five pounds of phosphorus per acre.[6]

## Medicinal Uses

Ginseng root has been used for centuries as a prized adaptogenic herb. Taken as a tonic, ginseng helps the body deal with stress and restores overall health and vitality. Used to treat depletion and exhaustion, ginseng improves physical stamina and athletic performance. It is also used frequently as a reproductive tonic to restore and build sexual chi. Most commonly used in elixirs and extracts, ginseng can also be taken in capsule form or drank in teas.

## Harvest Specifications

The roots of ginseng are harvested after at least five seasons of growth. Generally speaking, the more mature the roots are, the higher the price they can be marketed at. Roots should be harvested in the fall *after* the ginseng plant has produced its berries. Roots are easily dug with a spading fork. Take care not to damage the roots, and keep roots intact for best marketability. Ginseng can grow for a long time (some wild ginseng has been reported to be eighty-plus years old), and it is a good practice to plant and harvest in succession, allowing some plants to continue to grow to a ripe old age and collect seed stock from them. Determining the age of ginseng roots can be performed by counting growth rings on the root crown where the stem emerges.

## Postharvest and Drying Considerations

Ginseng is easy to clean and should be kept whole for marketing purposes. Rinse roots free of dirt, but avoid prolonged soaking or tumbling in root washers. This can leach important chemical constituents or break off roots. After washing let the roots drain, then dry whole roots slowly at temperatures of 85 to 95°F. It can take up to two weeks to dry whole roots completely.

## Pests and Diseases

Because ginseng roots are pleasant tasting and nutritious, they are a favorite food of rodents and deer. Fungal disease and slugs can also impact ginseng. This is all a part of the natural ecosystem, and when growing wild-simulated ginseng, growers tend to not intervene but instead let natural selection take its course. The key is to set the stage for success through good site selection, companion planting with other woodland herbs, and planting strong, viable ginseng seed or rootstock.

## Yields

Forty dried pounds of ginseng roots per one-eighth-acre bed. Moisture ratio for ginseng root is 3:1 fresh:dried.

## Pricing

Retail price for one pound organic:
- Dried American ginseng root (cultivated): $89 to $500+
- Fresh American ginseng root (cultivated): $25 to $150+

# Goldenseal                    *Hydrastis canadensis*

## Life Cycle

Goldenseal is an herbaceous woodland perennial and is hardy to USDA zones 3 to 8.

## Plant Description

Native to the central and eastern hardwood forests of the United States and southern Canada, goldenseal is one of the most prized members of the Ranunculaceae family. It grows six to twelve inches tall and has a single stem that produces two palmate leaves composed of five to seven large lobes. Leaves can be five to eight inches wide and are deep green. Flowering in early summer, goldenseal blossoms are diminutive but lovely. They grow out from between the leaves about two inches above the foliage and are greenish-white with little rays of stamens circling a central point. The berries by comparison are bright red and resemble the shape and size of a plump raspberry. The roots are also vivid and full of energy. They grow horizontally and are bright yellow and bumpy. Goldenseal roots, on average, are one-half to three-quarters of an inch thick, with many rootlets branching off the main rhizome.

## Growing Conditions

This woodland plant likes to grow in the understory of hardwood forests that have approximately 70 to 75 percent shade, are rich with humusy soils composed of leaf mulch, and are well drained but with good moisture content. Growers either utilize hardwood forests to create woodland beds (wild simulated) or grow goldenseal in shade houses. Unlike ginseng, goldenseal tends to fare better in shade houses and is less susceptible to diseases. That said, we grow goldenseal using wild-simulated techniques in the woods. It has less expense associated with it and requires fewer resources than building and maintaining shade houses. The end product is beautiful and highly medicinal roots.

## Propagation

Goldenseal can be grown from seed or rootlets. On our farm we plant rootlets one to two inches deep with the small nascent buds (the buds located at the apex where stem growth originates) oriented up. Goldenseal needs cold winter dormancy, so it is recommended to plant rootlets in the fall or very early spring before they begin to break dormancy at about one-foot spacing.

## Planting Considerations

Goldenseal beds in the woods are prepared in a manner similar to that of ginseng beds. When laying out the beds, we find it helpful to make them one or two arms' lengths wide to allow us to reach in easily from either side to plant, weed, and harvest without having to walk in them. Goldenseal often forms into an intertwined mat of rootlets once established, and it is good to minimize disruption and foot traffic. Before planting, rake away surface leaf matter and top dress with soil amendments if needed based on soil test results. Goldenseal does well in soils that have a pH of 5.5 to 6.5.[9] After planting be sure to restore leaf mulch; this will help retain moisture and add nutrients to the soil as leaves decompose. Avoid using straw to mulch, as it can inadvertently introduce fungal diseases. It is important that the rootlets do not dry out and should be irrigated as needed.

Habitat loss from mountaintop-removal mining and overharvesting from the wild are two main factors putting this beautiful woodland plant at risk of disappearing in its natural environment. It is currently on the United Plant Savers' "At-Risk List"[10] and is a plant that farmers should consider cultivating. It is relatively easy to grow in the understories

**Figure 18-34.** Goldenseal in flower.

**Figure 18-35.** Goldenseal root.

**Figure 18-36.** Goldenseal berries.

of hardwood forests or in shade houses. Like its woodland friend ginseng, goldenseal does well when companion-planted with other species such as trillium, wild ginger, and Solomon's seal.

## Medicinal Uses

Goldenseal is a very important herb for fighting infections. It is both antimicrobial and antiviral, and it stimulates mucous membranes by triggering the body's immune response. Goldenseal can be used internally in moderation or topically and helps the body rid itself of foreign bodies and infection. Typically it is recommended to use goldenseal at the onset of illness and discontinue use after two to three weeks. Prolonged use of this strong medicine can be contraindicated and irritating to the system. Goldenseal is also a wonderful bitter and excellent at stimulating sluggish digestion. It can be used alone or

in combination with other herbs to make tinctures, bitters, mouthwashes, and topical liniments.

## Harvest Specifications

Goldenseal roots can be harvested after three years and are easily dug with a spading fork. The leaves are also medicinal and can be harvested every season. It is best to harvest both the roots and the leaves of goldenseal after fruit has matured to allow the plant to set seed. These seeds can be harvested and replanted to help repopulate stands.

## Postharvest and Drying Considerations

Goldenseal is easily washed with a hose or barrel root washer. Because the roots are small, there is no need to chop them before washing. Soil can get tangled in the small rootlet hairs, so it is helpful to pull root mats apart to ensure proper cleaning. Do not overwash or soak the roots, as this can leach important medicinal compounds from them. After washing, dry on racks with good airflow at temperatures of 100 to 110°F. Mill roots after they are fully dried. Goldenseal roots dry under optimum conditions in three to four days.

Like other aerial plant parts, leaves can be dried on racks as well and should be spread out in a single layer with access to good airflow. Goldenseal leaves will dry in approximately two days and can be garbled by running the leaves and stems over half-inch stainless steel mesh.

## Pests and Diseases

Goldenseal can be susceptible to slugs, deer, leaf spots, and some fungal diseases. The best course of action is to create good woodland beds and let nature take its course. In the spirit of wild-simulated practices, let the environmental factors run their course. Some plants may die, but the ones that survive will be strong, potent, and full of wild vitality.

## Yields

125 pounds of dried goldenseal root (three-year-old roots) per one-eighth-acre bed. Moisture ratio for goldenseal root is 4:1 fresh:dried.

## Pricing

Retail price for one pound organic:
- Dried goldenseal root (cultivated): $78 to $200
- Fresh goldenseal root (cultivated): $70
- Dried goldenseal leaf (cultivated): $21

# Hawthorn                                    *Crataegus* spp.

## Life Cycle

Hawthorn is a long-lived tree that is hardy to USDA zones 3 to 9.

## Plant Description

Also known as may-tree or thornapple, hawthorn has hundreds of species and is native to the northern regions of Europe, North America, and Asia. Hawthorn, like apple, is a member of the Rosaceae family. It is a dense, thorny tree that is often used as hedgerows in Europe. It can be short and shrubby or reach heights of up to twenty-five feet. Hawthorn is a lovely, magical tree. It has deep green leaves that are serrated and lobed, growing two to four inches long. The flowers look very much like apple blossoms, beautiful clusters of white flowers composed of five petals and beloved by pollinators. In the fall the tree sets bunches of small, quarter-inch red fruit called pomes. Earning the title *crataegus* (the

Greek word for strong), these trees are resilient and tough, growing well in harsh conditions. One elder hawthorn in England is reported to be over seven hundred years old.

## Growing Conditions

Hawthorn likes to grow in well-drained, slightly acidic soils. It can do well in full sun to partial shade. It does not have an extensive root system and does not need a great deal of nutrients. It is often found growing on hardscrabble land at the tops of fields where soils are thinner.

## Propagation

Hawthorn can be grown from seed, but it is much easier and quicker to grow it from cuttings. To vegetatively propagate, simply take softwood stem cuttings with a sharp knife or pruner, wet the cuttings, and plant in potting soil. Keep cuttings moist and warm.

**Figure 18-37. Hawthorn flower.** Photograph courtesy of Larken Bunce

Roots will begin to develop on the cuttings in two to three months. Continue to let the cuttings grow in pots until they are well established. Once well rooted, the saplings can be planted outdoors.

## Planting Considerations

Hawthorn is called thornapple for a reason. Its thorns can be two to three inches long and are very sharp. When planting hawthorn, be mindful of where you locate it so it doesn't accidentally poke you or the ones you love. Thorns can be useful, however, when planted as a hedge or a border to keep deer or other pests out.

## Medicinal Uses

The leaves, flowers, and berries of hawthorn are all used medicinally. Hawthorn is one of the best tonics for the cardiovascular and nervous systems and is safe to take over a long period of time. Full of vitamins and bioflavonoids, hawthorn is highly nutritive to the body and totally delicious. Herbalists use hawthorn in a number of ways, including jams, teas, tinctures, syrups, and elixirs. One of the great benefits of hawthorn is that it normalizes the heart's action and is good for both hyper- and hypotensive conditions. When taken regularly it lowers blood pressure and cholesterol, tonifies the nervous system, and reduces anxiety. It also strengthens and improves the elasticity of arterial walls of the circulatory system and is the herb of choice for people with arteriosclerosis. Looking at the energetics of hawthorn, it can be used to help heal injuries of the heart, strengthen a person's feeling of well-being, and restore a sense of beauty and awe with the world.

## Harvest Specifications

The leaves and flowers of hawthorn are harvested in early summer when the trees are just starting to bloom and the buds are beginning to open. It is a wonderful harvest; it feels like being a bee flitting from flower to flower. When the tree first starts to bloom, the flavonoid levels in the leaves are at their peak, so pick the buds and the first cluster of leaves surrounding them. Be mindful not to overharvest, as blossoms are an important food source for pollinators and are the sources of berries. Depending on the density of your trees or hedge, long sleeves and gloves can be helpful to avoid hawthorn's sharp thorns. The berries are harvested when ripe in late summer to early fall. In Vermont our trees are ready to harvest in September, and we access them using ladders and the bucket loader of our tractor.

## Postharvest and Drying Considerations

Hawthorn leaves and flowers dry easily on racks with good airflow, minimal light, and temperatures between 100 and 110°F. Under good drying conditions aerial parts usually dry in two days. Leaves and flowers can be sold whole or run over a half-inch stainless steel mesh screen if a finer grade is desired.

The berries are a different drying experience altogether. Because they are harvested in the fall and they are fleshy fruits, the berries take much longer to dry. The best course of action is to rack berries in a single layer and position them directly in front of fans and a heat source. To keep the vibrancy of the berries it is a good idea to limit the exposure to light because the berries can take approximately two weeks to dry thoroughly. To test for dryness, use a pruner to cut or open the berries. They should be

**Figure 18-38.** Hawthorn berries.

dry and hard, not moist and pliable. Once they get really dry they are difficult to cut open. At this point the berries are most likely dry. However, it is hard to be 100 percent sure. To safeguard against spoilage we store and ship in breathable paper bags and check the state of the berries frequently, encouraging our customers to do the same. Another good way to "process" berries is to sell them fresh or make them immediately into products such as tinctures, jams, syrups, or elixirs. There is a good market for fresh berries, and we sell more than half of our berry harvest as fresh medicine.

## Pests and Diseases

Like many other plants in the Rosaceae family, hawthorn is susceptible to a number of pests and diseases, including leaf spots, rusts, scales, and numerous insects. For preventive practices and useful treatments, see chapter 11. Also very helpful is the information on orchard management and pest control and prevention given by Michael Phillips in his book *The Holistic Orchard*.

## Yields

Information not available.

## Pricing

Retail price for one pound organic:
- Dried hawthorn berry: $8 to $51
- Fresh hawthorn berry: $10 to $36
- Dried hawthorn leaf/flower: $10

## Commonly Imported From

Chile and Bulgaria

# Hops      *Humulus lupulus*

## Life Cycle

Hops is an herbaceous perennial vine that is hardy to USDA zones 4 to 8.

## Plant Description

Hops is a fragrant and popular member of the Cannabinaceae family that is native to temperate regions in the Northern Hemisphere. It has deeply lobed leaves and square stems that are covered with sticky hairs that help it climb as it grows. Hops are a dioecious species with both male and female plants producing yellowish-green coneflowers called strobili. The strobili of both sexes are highly fragrant. The male strobili produce copious amounts of pollen, whereas the female cones have bright yellow lupulin glands, which contain the resins and oils that are prized by herbalists and beer brewers alike. Besides being a relatively easy and tenacious plant to grow, hops grow quickly. In the first year hops can easily grow six to seven feet tall and in subsequent seasons reach heights of twenty to thirty feet, depending on the variety.

## Growing Conditions

Because it is such an avid grower, hops likes fertile soil and responds well to green manures and top-dressings of compost. When choosing a site, it is best to grow hops in well-drained land, in full sun, and with some sort of trellising. As it grows, hops requires a significant amount of water. Drip irrigation is preferred to aerial watering systems because excess water on leaves can sometimes make for weak plants that are susceptible to wilts or mildews. Also important to healthy hops cultivation is good airflow and plenty of space. Within beds, space hops vines four to five feet apart and allow up to eight feet between beds.

## Propagation

One of the best ways to propagate hops is from rhizome cuttings and layering. Rhizomes are small roots that grow out of the main root systems of mature plants. Rhizome cuttings can be taken from well-established plants and easily rooted in potting medium. Some farmers prefer to grow female plants exclusively because their cones contain a higher concentration of resin glands. Vegetative propagation allows growers to select for the sex of the plants. As with all cuttings, it is important to keep the plants moist as they are beginning to root and to wait until the plants are well established before moving them into the fields.

Another way to vegetatively propagate is through layering, done by bending the vines onto their sides, removing the leaves, and covering the bare stems with soil. It's best to leave six to twelve inches of the top part of the plant exposed. Water the layered stems in well, then let the layering develop for a few months. The buried stems will begin to form roots from the leaf nodes. Once a solid set of roots is produced, sections of the stems can be divided from the parent plant into individual new plants and used for transplanting.

**Figure 18-39.** Hops strobili.

## Planting Considerations

One of the major considerations surrounding hops is its need for a trellising system. Depending on the variety, hops can grow as much as thirty feet tall and will twine and vine around almost anything. On our farm we grow three to four hops plants against the south-facing side of our house. The hops do well there in the full sun and use our porch railings and roof as a trellis. Each individual vine, once mature, can produce two to three pounds of dried hops. Using a ladder and the bucket of our tractor, we can easily pick the cones off the vine.

Growers who want to produce higher volumes of hops will need to explore a more advanced trellising system. There are many different ways to trellis. One simple but elegant system is the single-pole trellis that consists of a centralized post that the hops are planted around. Using a pulley and cable system, the hops are raised to grow and lowered to harvest. Whatever system is used, the minimum height of the trellis should be twelve feet, with fifteen to eighteen feet recommended.

Another important consideration when constructing a hops yard is that, once established, hops will most likely be located in a field for many years. It can be beneficial to use mulching and cover cropping to reduce weeds and build fertility. Side-dressing plants with an application of manure or compost in the fall is also highly beneficial for the plants. One great thing about a hops yard is the shade it provides and the opportunities for polyculture. Consider planting shade-loving plants beneath hops. One of our favorites is red clover, which can be both a medicinal crop and a great green manure.

## Medicinal Uses

Pure and simple, we love hops. We are big fans of bitters and nervines, and hops is one herb that finds a solid place in our apothecary. We use it primarily in tincture form, but it can also be brewed into beer and formulated into tea blends. Hops has a somniferous effect and can be excellent in treating insomnia. It relaxes the nervous system and calms the body. It is often paired with valerian to treat sleep issues and to ease anxiety and tension. The bitter attributes of hops are also highly recommended for the digestive system.

## Harvest Specifications

Hops strobili are harvested in late summer when the cones are ripe. The cones should be plump, fragrant, and sticky. It is important that the cones are beginning to open and fluff out. They should not stay compressed when squeezed together (too young) or be completely open or overly dry (too old). When they become too mature the hops will crumble apart when dry, and the medicinal value will be diminished. To determine when hops are ready to harvest, collect cones from different parts of the vine. Some places may ripen more quickly than others, so it is a good idea to monitor strobili carefully. Gently squeezing them is a good way to tell if they are ready to harvest. Your hands should get sticky, the lupulin glands and pollen (if it's a male cone) should be bright yellow, and the cones should have some density to them but also feel as though they are beginning to open and "lighten up" and not stick together.

To gather the hops, small growers like us pick the cones directly off the vines on the trellis where they grow. Larger operations, by comparison, will release the plants from the trellis, cut down the vines, then strip the strobili off the vine by hand or with mechanical hops pickers. Either way works, but it is easier, if harvesting large amounts of hops, to bring the harvest to ground level rather than spend time on ladders or scaffolds.

## Postharvest and Drying Considerations

To dry hops, spread the cones in a single layer on racks where there is good airflow and a consistent heat of 80 to 100°F. Hops will degrade quickly if exposed to direct light, so it is best to dry in the shade. With good conditions hops takes three to four days to dry. When dry the strobili will maintain their bright green color and the lupulin glands and pollen will be yellow. At the correct level of dryness the central stem will no longer be pliable but will break when bent. The entire cone should still be springy and not crumble apart when handled. For beginning growers, it can be tricky to tell if the hops are dry, so it is a good idea to monitor them closely before storing.

## Pests and Diseases

Hops is for the most part resistant to pests and diseases. With good site selection, crop rotations, and field management most wilts and mildew issues can be avoided.

## Yields

180 pounds of dried hops strobili per one-eighth-acre bed. Moisture ratio for hops strobili is 5:1 fresh:dried.

## Pricing

Retail price for one pound organic:
- Dried hops strobili: $30 to $36
- Fresh hops strobili: $14

# Lavender                                    *Lavandula* spp.

## Life Cycle

Lavender is a woody perennial that is hardy to USDA zones 4 to 9.

## Plant Description

Native to the mountainous zones of the Mediterranean, lavender is an extremely popular member of the Lamiaceae family. It is a short, bushy shrub that has rough woody branches and grows two to three feet tall. The leaves of lavender can be broad or narrow depending on the variety; most are lance shaped and greyish-blue and grow directly off the stem. Lavender flowers are very fragrant and grow on slender stalks that reach up above the leaves. Flowers consist

**Figure 18-40.** Bee in lavender blossoms.

of whorls of bright purple flowers that are small and spiky at the top of the stalk.

## Growing Conditions

True to its Mediterranean roots, lavender likes full sun and well-drained, sandy soil. It takes a couple of seasons to become established and to produce a solid blossom crop. While farms in Canada not too far from us have had success growing lavender in large volumes, we have not. Perhaps we are just too cold or it's not quite sunny enough in our area. We are able to produce lovely nursery stock and therefore have focused on live plant sales of lavender rather than on row crops.

## Propagation

Lavender can be grown from seeds or vegetative cuttings. On our farm we grow lavender from seed and also from cuttings that are made from soft (not woody) stems that are planted in potting medium

and kept well watered. We grow the cuttings until roots begin to form, then transplant them out into our gardens when they are well established.

## Planting Considerations

When lavender is in full bloom it is quite striking: rows of purple blossoms with a delightful fragrance. Because of its sweet appearance and aromatics, we plant lavender where we can see and interact with it daily. Here in Vermont where we farm in USDA zone 4b, lavender is marginally hardy and is prone to winter kill, especially if it isn't heavily mulched. Therefore, we grow varieties of lavender such as *Lavandula angustifolia* var. Munstead that are very hardy yet not ideal for commercial production due to their relatively low yields and low essential-oil content. However, we continue to grow lavender because we love this plant. We are also experimenting with growing different varieties of lavender and ways of marketing that may be more commercially viable in the future. Wherever you plant your lavender fields, they will remain for a while. Therefore, site selection is important. Lavender plants take approximately two growing seasons to set good blossoms and will continue to produce for five or six years before yields begin to drop off. Lavender does not require rich soils but after subsequent seasons may benefit from a top-dressing of compost. Plant spacing is sixteen inches within the row and twenty-eight inches between rows. Pruning mature plants is also important to help them overwinter and to increase yields. After the second season prune plants, leaving one to two inches of green growth about the woody section of the plant.

## Medicinal Uses

Lavender is a popular herb for herbalists, foodies, and artisans. At any given health food store or farmers' market you can easily find lavender in soaps, sachets, essential oils, lotions, salves, extracts, teas, decorative weavings, baked goods, flavorings, body powders, and bath salts. Lavender is familiar and beloved—as well it should be. Lavender is a powerful nervine, helps to reduce anxiety, promotes relaxation, and restores a sense of well-being to the frazzled. The essential oil is safe to use directly on the skin (dilute in a carrier oil for those with sensitive skin) and can be used topically to relieve headaches and insomnia. Lavender is a favorite for soaps and is antimicrobial and soothing. The essential oil of lavender is also good to use directly on insect bites and stings to reduce inflammation and pain.

## Harvest Specifications

Lavender produces good blossoms after the second season and should be harvested for medicinal purposes when it is beginning to bloom but before all the flowers are fully open. When growing for oil production, wait to harvest until the flowers are fully open. In Vermont our lavender harvest occurs at the end of July or early August. To harvest lavender use a sharp knife to cut off the flowering stalks a couple of inches above the leaves. It is faster to cut the stems and garble off the flowers once dried, rather than pick the small blossoms off the plant. It should be noted that the stems also contain essential oils, and some distillers will utilize both the flowers and the stem. If you are working with artisans or selling bundles, keep the stems and flowers intact, as it is better for weaving and crafts. Some growers harvest and bundle lavender in the field and dry the bundles by hanging them. We harvest lavender but do not bundle. We use and sell lavender in its loose form, so we dry our lavender in racks in a drying shed, then garble off the blossoms. It is helpful to keep all the lavender oriented the same way when harvesting onto tarps. This makes racking and garbling the lavender easier.

## Postharvest and Drying Considerations

To dry lavender lay out the flower stalks in a single, compact layer on drying racks and dry at temperatures of 100 to 110°F. It is best to keep lavender out of direct light to help maintain vibrant color and medicinal quality. In good drying conditions lavender should dry in a couple of days. When dry the blossoms will rub off the stems easily and the stems will snap and no longer be pliable. Rub flower stalks over quarter-inch stainless steel mesh, and separate the stem from the blossoms.

## Pests and Diseases

In general, lavender is extremely disease resistant and not susceptible to many pests. Large commercial growers do report having issues with leaf spot, some fungal diseases, and pests such as spittlebugs and caterpillars. Good site selection, crop rotation, and management are the best course of action for growing strong crops and avoiding difficulty.

## Yields

Forty pounds of dried lavender flowers per one-eighth-acre bed. Moisture ratio for lavender flowers is 5:1 fresh:dried.

## Pricing

Retail price for one pound organic:
• Dried lavender flower: $18 to $26

## Commonly Imported From

France

# Lemon Balm                                    *Melissa officinalis*

## Life Cycle

Lemon balm is an herbaceous perennial hardy to USDA zones 4 to 9.

## Plant Description

Lemon balm is native to the Mediterranean and Eurasia and is a lovely mint, belonging to the Lamiaceae

**Figure 18-41. Lemon balm.** Photograph courtesy of Bethany Bond

family. It has square stems and opposite leaves that are slightly heart shaped and toothed. They grow two to three inches long and are very fragrant and lemony tasting. The flowers of lemon balm grow in whorls around the stem and are composed of small white blossoms. The plants are upright and branching, growing three to three and a half feet tall.

## Growing Conditions

Lemon balm grows well in full sun to partial shade. Many herbalists report growing large plants well in the shade, but we like to grow lemon balm in full sun because the plants seem much more vibrant and medicinally potent, with higher concentrations of essential oil. Regardless of whether the plant grows in full or partial sun, it needs good, fertile but not excessively fertile, well-drained, loamy soil, with ample soil moisture. Lemon balm can grow in one spot for many seasons and may be banded with compost or manure in early spring. We find that after three to four seasons our lemon balm stands get tired from multiple harvests, so we plant new beds in different locations.

## Propagation

Lemon balm is easily propagated from seeds and vegetative cuttings. Sow seeds indirectly in early spring for transplants. The seeds are light-dependent germinators and should be planted on the surface, covered lightly with potting medium. Some growers scarify the hard seed coats of lemon balm seed but we have had good germination results without scarifying our seed. To make softwood cuttings take slips from new green growing tips, strip off the lower leaves, and plant in moist potting mixture. Keep cuttings moist and warm. They are usually ready to transplant into the fields in three to five weeks.

## Planting Considerations

Lemon balm is an easy-to-grow, highly medicinal, and extremely yummy member of the mint family. It can grow well in lots of different soils and is a very generous plant, giving multiple harvests in one season. We often grow our lemon balm stands for at least three years in one location. Recommended plant spacing is twelve inches within the rows and twenty-eight inches between rows, with two rows per bed. Lemon balm grows fast when it becomes established and will form a dense hedge that will help to outcompete weeds.

When growing lemon balm, it can be a good idea to mulch. This helps plants overwinter in colder climates and also decreases soil splash that can not only make the leaves dirty but can also introduce foliar diseases and leaf spots. Other things to consider when planting lemon balm are that it is a favorite of bees and is antimicrobial and antiseptic. This makes it an excellent choice to plant around beehives. The lemony smell has also been known to repel many types of harmful insects and can act as a protective border or companion plant.

## Medicinal Uses

Lemon balm is a very special member of the mint family. It has nervine properties that can reduce nervous tension, relieve agitation, and help the body relax. It is also very uplifting and soothing to the spirit. Lemon balm is highly aromatic and pleasant tasting and is a favorite for teas, tinctures, and cordials. Besides being totally delightful, lemon balm has antiviral and antimicrobial properties and is a good source of vitamin C. It is often used in healing protocols to treat cold sores and fever blisters and to fight infection. Underneath its wonderful lemony taste, lemon balm also has slightly bitter attributes and is antispasmodic, making it an excellent carminative and treatment for gastric upset.

## Harvest Specifications

It is best to harvest lemon balm when the plant begins to bloom. This is when the volatile oils and medicinal attributes are at their peak. Harvest during hot, dry weather when the plants are clean and vibrant. To harvest use a field knife or sickle bar cutter to cut the entire aerial part of the plants down, leaving six to eight inches of green growth on the plant to regenerate. In a good season it is possible to get two to three harvests from lemon balm. Lemon balm tends to heat up quickly in the field and should be processed immediately after harvest to maintain high quality.

## Postharvest and Drying Considerations

Lemon balm can be tricky to dry because it can easily degrade and lose important chemical constituents. For this reason some herbal resources recommend using only fresh lemon balm, stating that the dried plant is not as medicinal. While this may be true for a lot of the mass-marketed lemon balm, we find that dried lemon balm can be exceptionally healing if it has been processed carefully. The key to maintaining quality is to dry lemon balm at lower temperatures with good airflow and minimal exposure to light. Airflow and lower drying temperatures are essential to minimizing the volatilization of key essential oils. It also prevents lemon balm from turning brown. We harvest lemon balm in the morning after the dew has dried and rack it in the drying shed before it gets too hot. It is a good idea to place lemon balm on drying racks in a single layer to maximize airflow and prevent composting. We keep the drying shed sides rolled up and the fans on and put lemon balm on our lower racks, which tend to be cooler than the ones at the top of the shed. Gradually, as the plants start to lose their moisture and begin to dry, we close up the shed and allow temperatures to increase to no more than 95°F. By monitoring closely, our goal is to dry lemon balm in two to three days. As soon as the leaves lose their pliability and start to get crunchy, but before the stems are completely brittle, we remove the lemon balm from the shed and garble it by running the plants over quarter- to half-inch stainless steel mesh. This breaks the dried leaves off the stem and produces a desirable grade of herb.

## Pests and Diseases

Lemon balm has minimal pest and disease issues. The strong citrusy scent works to repel many harmful insects. The mildews or leaf spots that can affect lemon balm are easily avoided with proper crop rotation and soil management. When choosing a site for lemon balm make sure it is well drained and has good airflow. Lemon balm is one of those plants whose leaves seem to attract dirt, so it can be helpful to mulch lemon balm to reduce soil splash when it rains. An added benefit to mulch is that it will reduce weeds that can rob the soil of nutrients and weaken the plants.

## Yields

150 to 200 pounds of dried lemon balm leaf per one-eighth-acre bed (multiple harvests per season). Moisture ratio for lemon balm leaf is 5:1 fresh:dried.

## Pricing

Retail price for one pound organic:
- Dried lemon balm leaf: $11 to $25
- Fresh lemon balm leaf: $12

# Lobelia     *Lobelia inflata*

## Life Cycle

Lobelia is an annual that is native to southeastern Canada and the United States.

## Plant Description

Lobelia is a member of the Lobeliaceae family. The plant grows one to two feet tall. It has hairy stems and lance-shaped leaves. Lobelia has small purplish-white flowers, and the seedpods that form along the stem look like little inflated bladders or balloons. If you take a tiny nibble of fresh leaves, the taste is highly acrid and reminiscent of tobacco smoke.

## Growing Conditions

Lobelia grows well in full sun to partial shade and does well in poor to average soil. It will not outcompete weeds easily, so it does best when planted in a clean bed.

## Propagation

We find it most useful to indirect-seed lobelia into plugs that we transplant in early spring. Lobelia is a light-dependent germinator and should be sown on the surface and pressed lightly into the soil. If desired, a fine dusting of potting mix can be applied over the seeds. Be careful not to wash the tiny seeds out of flats when watering. We use a fine mister to water. Because the seeds are so tiny, we find it helpful to seed lobelia into twenty-row seed flats. Once the seedlings develop, the plants are pricked out and grown on in larger plug cells before being transplanted into the fields.

## Planting Considerations

Because lobelia is an annual, we plant it with our other annuals and rotate it to new beds each season. Plant spacing for lobelia is twelve inches within the row and fourteen inches between rows with three rows per bed.

## Medicinal Uses

Lobelia is a powerful antispasmodic and expectorant. It is used traditionally to treat asthma and bronchitis and to alleviate spasmodic, unproductive

**Figure 18-42.** Lobelia flower. Photograph courtesy of Larken Bunce

**Figure 18-43.** Lobelia seed pods. Photograph courtesy of Larken Bunce

coughing. The native peoples often included lobelia in smoking mixtures. It has been used successfully in some cessation protocols to help curb nicotine cravings due to the fact that the primary alkaloid, lobeline, is almost molecularly identical to nicotine yet it has no addictive properties. Because of the powerful effect of lobelia it should be used in small dosages. When used in greater amounts lobelia can cause nausea and have emetic effects. Lobelia is also a good herb to use topically in liniment preparations for bruises and sore muscles. Lobelia helps to move and release energy. Whether it is used as an emetic, as an expectorant, topically, as a tincture, or in smoking mixtures, lobelia moves congestion and stagnation out of the body. It relaxes the smooth muscles and eases tension. It is a powerful and important herb but should be used with care.

## Harvest Specifications

Lobelia is harvested when the plant is in bloom and beginning to create seedpods. This is when lobeline, one of the desired chemical constituents, is at high levels. Using a field knife or a sickle bar mower, cut the aerial parts of the plant down to within an inch of the ground. Harvest lobelia when the plant is clean and dry and after the dew has evaporated.

Keeping with the high energetics of this plant, we like to harvest it on bright, hot days.

## Postharvest and Drying Considerations

Lobelia is an easy plant to dry. After harvesting, place lobelia on drying racks in a single layer with good airflow. Dry at temperatures of 100 to 110°F. Under good conditions lobelia will dry in one to two days. When dry, lobelia leaves snap from the stem and can be processed by running the plants over quarter- to half-inch stainless steel mesh. This makes it easy to separate the leaves from the stem and creates a marketable grade of herb.

## Pests and Diseases

Lobelia has very minimal pest and disease issues. Good crop rotation and field management produce healthy plant stands that are pretty much resistant to leaf spots and other issues that can impact lobelia.

## Yields

126 pounds of dried lobelia leaf per one-eighth-acre bed. Moisture ratio for lobelia leaf is 4–5:1 fresh:dried.

## Pricing

Retail price for one pound organic:
• Dried lobelia leaf/flower: $19 to $31

# Marshmallow                    *Althaea officinalis*

## Life Cycle

Marshmallow is an herbaceous perennial that is hardy to USDA zones 3 to 9.

## Plant Description

Native to Europe, marshmallow is a gracious member of the Malvaceae family, grows four to four and a half feet tall, and has lovely soft leaves and stems. The downy leaves are a silvery green color and grow two to three inches long. The flowers are white and composed of five petals that often have a purple hue. Marshmallow produces large, fibrous roots that are whitish-yellow and full of rich, gooey, medicinal mucilage. Marshmallow does not produce a single taproot; instead roots form in deep growing clumps and are sometimes as much as two feet long.

## Growing Conditions

Marshmallow can grow in many different soils but prefers moist, fertile soils that are rich in nutrients. We grow marshmallow in full sun, and it thrives. It can also do well in partial shade and in areas that are not well drained. Although it can make harvesting difficult, marshmallow can be planted in swampy areas that may not be suited for other crops. Just be sure that you can still get people and equipment into these areas to cultivate and dig without damaging delicate wetland habitat.

## Propagation

Marshmallow can be grown by root divisions, cuttings, or indirect seeding. Before seeding, it can be helpful to scarify the seeds to help increase germination. To scarify run the seeds over medium-grit sandpaper until the seed coat is nicked but the inside germ is not compromised. We have found that scarification is not absolutely necessary and we sow the seed a bit heavier when we don't scarify it to make up for slightly lower germination rates. We do find that cold-conditioning (stratification) seems to greatly increase germination rates.

## Planting Considerations

Marshmallow needs to grow for two to three years before a root harvest. It should be planted in a place that has good moisture content. Other plants that can share moist beds with marshmallow are boneset, valerian, and angelica. We do not interplant within the rows but often companion plant within the bed. Plant marshmallow sixteen inches within the row and twenty-eight inches between rows, with two rows per bed.

## Medicinal Uses

Marshmallow is soothing in every sense of the word. The leaves, flowers, and roots are medicinal, and full of gooey mucilage that is good for healing irritated mucosa. Traditionally marshmallow was used to treat inflammation of the urinary tract and respiratory systems. The whole plant has anti-inflammatory properties and can be used alone or in combination with other herbs. For treating respiratory problems, marshmallow leaf is often combined with other bronchial herbs such as coltsfoot and mullein. When used for urinary infections marshmallow can work very effectively with other herbs, such as cornsilk and uva ursi. Because of its demulcent properties and pleasant flavor, marshmallow makes excellent teas, extracts, and throat lozenges. It can also be used topically to soothe irritation by making an herbal wash, a poultice, or a balm.

## Harvest Specifications

Harvest the leaves of marshmallow as the plants begin to flower, when some blossoms are open and others remain in bud stage The aerial parts of the plant can be cut with a field knife or sickle bar mower. They should be harvested when dry and clean and after the dew has evaporated. In the second and third year roots can be harvested (see details below). During those growing seasons it is better to harvest the leaves with knives, cutting a few stems from each plant rather than mowing down all the aerial parts. This allows for a good leaf harvest while maintaining some aerial parts to photosynthesize and provide nutrients for the plants throughout the season. This is essential for root development and overall health of the plants.

Roots are harvested in the fall of the second year or in the spring or fall of the third season. Dig before aerial growth has started in the spring, or if digging in the fall, mow down the tops of the plants before root harvest. To loosen the soil run a shank or cultivator sweep parallel to plants between the rows. Be careful not to go too close to the plants or you may break off some of the lateral roots. Roots can be mechanically dug with a bed lifter or modified potato digger. They also can be lifted out of the bed with a spading fork. The roots are big and often long, so mechanical digging can really save in labor costs and wear and tear on the body.

## Postharvest and Drying Considerations

To dry marshmallow leaf and flowers, place the stems in a single layer on drying racks with good airflow. Marshmallow leaf should be dried in temperatures of 100 to 110°F in partial shade. Marshmallow does

**Figure 18-44.** Marshmallow flower.

not need to be dried in complete darkness but maintains higher quality if out of direct light. With good drying conditions marshmallow leaf and flowers should dry in approximately three days. To process the tops run dried material over quarter- to half-inch stainless steel mesh and separate out the stems. The leaves and flowers are sold together and when dry should be light and fluffy but still somewhat pliable.

Marshmallow root must be washed thoroughly before drying, but take care not to soak the roots excessively, as many of its medicinal compounds are water soluble and can be leached out during the washing process. Soil often gets compacted in the central crown of marshmallow, where the roots begin to branch off. Therefore it is helpful to quarter the roots before washing. Marshmallow roots are relatively soft and can easily be chopped with a field knife. After splitting the roots, wash in a barrel washer or with a hose before drying. Mill roots after they are fully dried to avoid gumming up your chipper. Marshmallow roots dry in temperatures of 100 to 110°F with good airflow, in four to five days.

## Pests and Diseases

Marshmallow has minimal pests and diseases. Use good crop rotation and field management to produce healthy plant stands.

## Yields

200 pounds of dried marshmallow leaf per one-eighth-acre bed. Moisture ratio for marshmallow leaf is 5:1 fresh:dried.

100 to 150 pounds of dried marshmallow root per one-eighth-acre bed. Moisture ratio for marshmallow root is 3:1 fresh:dried.

## Pricing

Retail price for one pound organic:
- Dried marshmallow leaf: $15 to $26
- Fresh marshmallow leaf: $13
- Dried marshmallow root: $10 to $20
- Fresh marshmallow root: $9 to $13

## Commonly Imported From

Albania and Bulgaria

# Meadowsweet　　　　　　*Filipendula ulmaria*

## Life Cycle

Meadowsweet is an herbaceous perennial that is hardy to USDA zones 3 to 9.

## Plant Description

Known as Queen of the Meadow, meadowsweet is a majestic member of the Rosaceae family that grows more than five feet tall and two to two and a half feet wide. It is native to Europe and western Asia and has striking leaves that are deep green on top and silvery on the underside. The leaves are pinnate and deeply divided into three to five lobes. The flowers of meadowsweet are also quite lovely. They are small white clusters that grow at the top of flowering stalks. Their fragrance is sweet and strong, and flowers come into full bloom in midsummer.

## Growing Conditions

Meadowsweet grows well in rich, humusy soil that has plenty of fertility and moisture. It likes full sun to partial shade, and it's quite adaptable. Naturalized meadowsweet is often found in meadows, swampy areas, ditches, and riverbeds. It doesn't require a lot of care, but it responds well to top-dressing annually with compost.

**Figure 18-45. Meadowsweet in full bloom.** Photograph courtesy of Larken Bunce

**Figure 18-46.** Photograph courtesy of Larken Bunce

## Propagation

Meadowsweet can be grown from root divisions, and we have had good success growing from seed. The seeds are light-dependent germinators and should be sown on the surface, pressed in lightly, and covered with a fine layer of potting medium. Keep the seeds moist.

## Planting Considerations

We transplant plugs into well-prepared fields in spring with plant spacing within the row of sixteen inches and twenty-eight inches between rows, with two rows per bed. Plants grow very tall, so it is beneficial to perform cultivation and mechanical weeding when the plants are smaller to avoid breaking off aerial growth as they reach their full height. Once well weeded and established, the plants tend to shade out other weedy species.

Because meadowsweet grows very tall, it often does well sharing a bed with other statuesque moisture lovers. We grow meadowsweet in beds with blue vervain and valerian because they like similar growing conditions. A little herbal lore: Both meadowsweet and vervain were considered sacred plants of the Druids, and this is another sweet reason to grow them together—there is an ancient affinity between them.

## Medicinal Uses

Meadowsweet has a rich medicinal history. It was one of the precursors to aspirin due to its salicylic acid content, which helps to reduce inflammation and is analgesic. The leaves and flowers of meadowsweet are used in teas and tinctures to reduce pain, bring down fevers, and treat diarrhea. It also has diaphoretic and diuretic properties and is good for soothing the stomach. It's a fantastic plant to have in your herbal pantry because it is great for both children and adults and treats an array of ailments. It is pleasant tasting, even though it is slightly astringent and really helps the body when it is in pain, suffering from colds and the flu, and feeling inflamed.

## Harvest Specifications

We harvest meadowsweet when it is in full bloom. Here in Vermont that is in mid-July. To harvest the plants pick a nice, dry, sunny day and use either a field knife or a sickle bar mower to cut all the aerial parts. The tops are very tall, and we find it easiest to lay out the stems intact in a single row. Sometimes we have to bend the flowering stalks back on themselves to get them to fit on drying racks. This works just fine and does not damage the quality of the final product.

## Postharvest and Drying Considerations

It would be wonderful if everything dried as easily as meadowsweet. It will dry beautifully in one to two days with little effort. Dry at 100 to 110°F with good airflow. The leaves do not have a tendency to brown easily, and the blossoms are light and dry completely in a short amount of time. The trick is to watch meadowsweet carefully so it doesn't overdry. It also is helpful if you can time it so the leaves and blossoms are dry but the stems are still slightly pliable. This makes garbling easier. To garble, run the tops over quarter- to half-inch stainless steel mesh, picking out the stems. The leaves and blossoms are sold together.

## Pests and Diseases

Meadowsweet has minimal pest and disease issues. It is also a favorite of many pollinators.

## Yields

One hundred pounds of dried meadowsweet leaf per one-eighth-acre bed. Moisture ratio for meadowsweet leaf is 4:1 fresh:dried.

## Pricing

Retail price for one pound organic:
- Dried meadowsweet leaf/flower: $12 to $20
- Fresh meadowsweet leaf/flower: $6

## Commonly Imported From

Hungary

# Motherwort                    *Leonurus cardiaca*

## Life Cycle

Motherwort is an herbaceous perennial hardy to USDA zones 3 to 8.

## Plant Description

Motherwort is native to Europe and central Asia and grows very well in temperate areas of the United States. It is a striking member of the Lamiaceae/Labiatae family and has deeply divided leaves of three to five lobes that are dark green with silvery-grey undersides. The stems are hairy, causing the plant to be somewhat prickly when mature. The flowers form in multiple whorls of pinkish white blossoms that grow on the spiky tops of the plant. Motherwort is extremely bitter and has a strong, elemental earthy smell. It is so strong that if you lick your lips while harvesting, you can often taste the bitter properties on your skin!

## Growing Conditions

Motherwort likes well-drained, loamy soil and full sun to partial shade. While it is a versatile plant and can do fine in poor soil, it thrives with fertility, and it is recommended to top-dress crops with compost when growing for production.

## Propagation

Motherwort readily self-seeds and can also be grown from divisions. We grow motherwort by indirect seeding and start seeds in the early spring. They germinate easily in two to three weeks.

## Planting Considerations

Motherwort can create quite a hedgerow and will reach heights of up to seven feet tall. It is lovely but not necessarily showy. Consider planting it on the edge of fields as a windbreak or buffer. Pollinators love it, so it is also nice to plant next to hives. Motherwort can be transplanted into fields using plant spacing of two feet between plants and twenty-eight inches between rows, with two rows per bed. Motherwort tends to be a very aggressive self-seeder, so care should be taken to be sure motherwort seedlings aren't growing where you don't want them to grow.

## Medicinal Uses

Motherwort is very effective at tonifing and strengthening the heart and circulatory system. Energetically, it helps people to feel lionhearted (note its Latin name *Leonurus cardiaca*), hopeful, and full of life. Motherwort is also a strong nervine and a potent bitter. When used for the nervous system, it can alleviate melancholy and reduce nervous tension. Motherwort also relaxes the muscles and can ease menstrual discomfort and cramping. These strengthening, restorative, and relaxing properties make it a fabulous plant to use to help women transition during postpartum and menopausal changes. Motherwort is a common treatment for hot flashes and emotional upheaval; it is uplifting yet grounding. As a bitter it aids in digestion and tonifies the liver. Because of the flavor, it is highly recommended to take motherwort as an extract, formulated in an elixir, or in capsules/herb balls.

## Harvest Specifications

Motherwort is harvested when it is in full bloom in mid- to late July. Harvest motherwort on a hot, dry summer day using a knife or a sickle bar mower to cut the entire top of the plant. It is possible to get two cuttings, even in Vermont. The first cutting is the mother lode, and the second cutting is somewhat smaller. When handling motherwort, it is recommended to use gloves, as motherwort can be prickly to work with at times and the hairs and flowering tops are a bit sharp.

## Postharvest and Drying Considerations

To dry motherwort spread the aerial parts out in a single layer on drying racks with good airflow. Dry at temperatures of 100 to 110°F out of direct light. Under good drying conditions motherwort should be dry in two to three days. After drying rub the tops over quarter- to half-inch stainless steel mesh and pick out the stems. Flowers and leaves are sold together.

## Pests and Diseases

Motherwort has minimal pest and disease pressures. Utilizing good crop rotations and cultivation practices will create strong plants that are resistant to most challenges.

## Yields

100 to 150 pounds of dried motherwort leaf per one-eighth-acre bed. Moisture ratio for motherwort leaf is 4:1 fresh:dried.

## Pricing

Retail price for one pound organic:
- Dried motherwort leaf/flower: $9 to $24
- Fresh motherwort leaf/flower: $13

## Commonly Imported From

Poland

**Figure 18-47.** Motherwort in flower. Photograph courtesy of Bethany Bond

# Nettle (stinging)                    *Urtica dioica*

## Life Cycle

Nettle is an herbaceous perennial, hardy to USDA zones 2 to 10.

## Plant Description

A native to Europe and North America, nettle is a part of the Urticaceae family and is an amazing medicinal plant, a superfood, and a source of natural cordage that can be used much like hemp for fiber. The plant has square stems and lance-shaped leaves that are finely toothed. The stems and leaves are covered with fine hairs that contain formic acid. When fresh these hairs can produce an itchy sting if they come in contact with the skin. Therefore, it is recommended to wear gloves and long sleeves for harvest. The plants grow up to eight feet tall in fertile soil and produce bunches of small, cream-colored flowers ripening to round seed clusters that form along the stem at the top of the plant. The roots of nettle are fibrous and deep yellow, growing in a dense mat.

## Growing Conditions

Nettles grow well in full sun to partial shade and thrive in soils rich with organic matter. The plants prefer a good supply of moisture but can tolerate dry conditions. In natural habitats nettle is found growing on edges of streams and meadows. On farms nettles can often be found happily growing in old manure and compost piles and pretty much anywhere else they can find a fertile niche to occupy.

## Propagation

Nettle can be propagated from seeds or vegetative cuttings. When seeding use fresh seed and sow thinly on top of the soil or cover lightly with soil. Nettle germinates somewhat sporadically in two to four weeks. When propagating from cuttings take vegetative cuttings or root divisions in early spring, plant in potting medium, and keep well watered. Let the cuttings become fully established before transplanting into fields.

## Planting Considerations

Nettle can grow quite large and bushy. It also stings when people brush against it, so it is recommended

**Figure 18-48. Stinging nettle.** Photo courtesy of Larken Bunce

to plant in a low-traffic area that is out of the way. For cultivation plant nettle twelve inches spacing within the row and twenty-eight inches between the rows, with two rows per bed. Mature nettle beds will form a dense hedgerow that can be effective at outcompeting weeds.

## Medicinal Uses

Nettle is an incredible tonic herb that is fortifying to the body, full of chlorophyll, vitamins, minerals, micronutrients, and protein. It is wonderful for promoting liver and kidney health and feeds the entire body. It is a superfood that restores health and vitality. Nettle can be used in many ways: in teas, extracts, and powders. For those "foodies" in the know, nettle is also an amazing spring green to eat in soups and casseroles and is delicious steamed, sautéed, or braised. Once dried or cooked, nettle loses its sting, and what is left is green goodness of the highest order. When we first began going to farmers' markets twelve years ago we would cut bags of fresh, tender nettles in hopes of enticing locals to eat this spring tonic. Not many people knew what they were good for, and we spent lots of nights after market dining on leftover piles of nettle. As the weeks went on we talked with loads of people about this delicious superfood, and slowly folks began looking past the "sting" to the deep healing. Our efforts were not in vain. One day an elderly gentleman stopped by with his grandson and became completely animated when he saw our stand. He said that in his Russian homeland it was a common practice to cook with nettle. He had a beloved soup recipe that he wanted to share with his grandson but had been struggling to procure the ingredients. Delighted, he bought up all our fresh bags, and off they went. Every week after that the pair would return and scoop up our nettles. As it turned out, the grandson loved the soup, and this sweet man passed on the important tradition of using food as medicine. A great reminder!

Besides being a delicious spring edible, nettle has astringent, anti-inflammatory properties and can be used successfully to treat allergies, urinary tract infections, and liver imbalances. The roots and seeds are also medicinal and work well to reduce prostate inflammation.

## Rejuvenating Nettle Soup by Lisa Masé

This nettle soup recipe comes from our friend Lisa Masé and is very similar to the one that the Russian grandfather from the farmers' market made. Lisa Masé is a whole foods cooking educator, food writer, translator, and herbalist living in Vermont. She grew up in Italy, where she learned that healing comes from eating local foods, as close to the source as possible, and attuned to the foods of our ancestors. In her teaching and consultation she draws from the wisdom traditions of European herbalism and ayurveda.

### Ingredients

2 tablespoons olive oil
2 large shallots
1 inch fresh ginger root
½ teaspoon salt
1 teaspoon garam masala
½ teaspoon turmeric
2 large zucchini
1 tablespoon stone-ground brown mustard
1 cup water or vegetable stock
4 cups freshly harvested young nettle tops

Peel and dice shallots. Mince ginger root. Place oil in a soup pot, warm it to medium heat, and sauté shallots and ginger for 5 minutes. Add spices. Sauté for a few more minutes. Dice zucchini, and add to the pot. Add mustard. Sauté for 5 minutes, stirring occasionally. Add the water or stock and nettles. Bring everything to a boil, reduce heat to medium, and simmer with a lid on for 20 minutes. You can add marinated tempeh or roasted chicken to the soup for a delicious meal.

## Harvest Specifications

When harvesting nettle, it is really helpful to wear long sleeves and gloves to avoid being stung by the hairs on the nettle leaves and stems. Nettle leaf can be harvested at its early stages of growth all the way to the onset of flowering but before the plant sets seeds. Harvest aerial portions of the plants, using field knives or a sickle bar mower to cut the stems twelve inches above the base of the plant. This encourages the plant to produce more growth for a second and sometimes even a third harvest.

Before harvesting nettle roots it can be helpful to use a chisel plow, disc harrows, or a shank to loosen the soil surrounding the nettle plants and break up the plant's fibrous root mass. It's best to dig roots when the plant is dormant, in either the fall or the spring. If harvesting in the fall, mow down the aerial parts of the plant before digging the roots. This will make it much easier to work the soil. Using a digging fork or bed lifter to lift the root clumps, and bang out as much of the soil from the roots into the field as possible. Sometimes it can be helpful to use a flat-head shovel to slice through big clumps of roots after the bed lifter has pulled them up. It takes work to dig the nettle roots, but it produces beautiful medicine.

For seed harvest wait until the plant sets seeds; then, using gloves, strip the seeds off the plants while they remain growing in the fields. You may get some leaf matter, which is fine because it can be removed after drying by winnowing.

## Postharvest and Drying Considerations

Nettle leaves dry easily on the stem and should be placed whole in a single layer on drying racks. Dry at temperatures of 95 to 100°F, out of direct light in good airflow. The aerial parts are ready to garble when the leaves are dry and break easily off the main stems but before the stems lose all pliability. This allows the leaves to be removed from the stem, while leaving the main stems intact so they can be picked out. To garble, rub the dried tops over quarter- to half-inch stainless steel mesh. Leaves will pass through, and the stems will remain on top. While there may be specialized markets for stems for cordage and fiber production, most buyers want the leaves.

Nettle root should be washed thoroughly before being dried. Because the roots grow in a mat they often collect dirt and mud. It is helpful to pull the mats apart before washing or use pruning shears to cut them apart. Field knives don't work as well because the roots are very tough. Once they are washed lay the roots out in a thin layer, and dry at 100 to 110°F with good airflow. It takes approximately a week to dry roots under good conditions. Once dry, the roots can be chipped to the proper grade and size.

Nettle seed is dried the same way as the leaf. After drying, any leaf matter can be removed by pouring the seeds through a fine mesh screen, winnowing, or using a Clipper seed cleaner.

## Pests and Diseases

Red Admiral butterfly larvae are the major pests we see on nettle. The Red Admiral larvae can skeletonize the plants and can seriously impact growth and minimize harvest yields. The larvae can be managed by monitoring the crop, handpicking, and applying raw neem oil preparations. See chapter 11 for details. Besides neem applications we try to plant extra nettle so there is enough to withstand some larvae pressure and still be enough for harvest.

## Yields

Five hundred pounds of dried nettle leaf per one-eighth-acre bed. Moisture ratio for nettle leaf is 4:1 fresh:dried.

Root and seed data are not available.

## Pricing

Retail price for one pound organic:
- Dried nettle leaf: $10 to $24
- Fresh nettle leaf: $9
- Dried nettle root: $20 to $37
- Dried nettle seed: $30 to $37

## Commonly Imported From

Poland and Bulgaria

# Oats                                                    *Avena sativa*

## Life Cycle

Oats are an annual grass that grows prolifically in most temperate regions.

## Plant Description

Oats can trace their origins to the Fertile Crescent in the Middle East and have over time migrated successfully to Europe, India, and the United States. A member of the Poaceae family, oats are a common commodity crop grown for cereal grains and livestock feed. They are also very much beloved by herbalists as a supreme nervine and can be grown and harvested with minimal equipment. Oats grow three to four feet tall and consist of green leaf sheaths and fruiting stalks on which the oat seeds grow. The seeds are cylindrical and light green when they first emerge, then become plump and full of white milky latex as they mature. Their color also changes with maturation from green to a light yellowish color. A self-fertile plant, oats are wind pollinated.

## Growing Conditions

Oats grow best in well-drained loamy soil with good fertility. However, they can grow in a variety of conditions, including sandy soil and clay if necessary. Oats like full sun and consistent but not excessive soil moisture. They can tolerate droughty conditions, but yields are highest with adequate water. Oats

**Figure 18-49.** Milky oats ready to harvest. Photograph courtesy of Bethany Bond

require substantial nitrogen, so consider planting with nitrogen-fixing legumes or in fields that have previously had them as cover crops.

## Propagation

Direct-seed with a broadcast spreader or seed drill in the spring or early summer. Sow at a rate of twenty pounds per one-eighth-acre bed. Oats can be planted throughout the season in successive plantings to allow for multiple harvests. After broadcast sowing it is helpful to lightly disc harrow the field to help cover the seeds, and it is important to maintain adequate soil moisture during germination.

## Planting Considerations

Oats not only are a wonderful medicinal plant, they are a beneficial cover crop, improving tilth when tilled into the soil. A great way to fully utilize oats is to plant it on land that is lying fallow as part of a crop rotation. Interplanting oats with other cover crops such as red clover helps to do a number of things. Initially as they grow, the oats provide shade for the red clover, helping it to become established. The red clover in turn matures and in subsequent seasons will fix nitrogen from the air into the soil. While working their magic as cover crops, red clover and oats also produce medicinal crops that can be harvested and sold. Once the crops are harvested the oats and clover can be tilled under, adding green matter and biomass to the soil and leaving the land ready to plant.

## Medicinal Uses

Oats, ahhh oats! Just thinking about them brings feelings of restoration and rejuvenation! Oat are one of the most soothing and nutritive nervines to be found and in this harried world should be included in almost everyone's daily routine. Milky oat tops (and oat straw to a lesser extent) are mucilaginous and help to reduce inflammation and irritation. Specifically they help to soothe the myelin sheath that encases the axons of the nervous system. The myelin sheath is essential to proper nervous system function, and when it gets frayed or damaged oats

can help restore and heal. Oats also provide much-needed nutrients for people who are feeling depleted and run down. Oats contain soluble fiber, a protein known as avenalin, calcium, and silica, all very useful for nerve, adrenal, and cardiac health. Not only do they work directly on these systems, oats also help revitalize and restore the whole body, providing nourishment, and have been used in formulation to treat depression, exhaustion, and extreme anxiety. Oats are very effective yet safe to use for most people. They are excellent for children, elders, and people recovering from debilitating illness. Versatile oats can be used in oatmeal baths (wonderfully healing to the skin and topical irritations), teas, and tinctures and can be eaten as good old oatmeal porridge.

## Harvest Specifications

Both the oat straw (the leafy part of the plant) and the milky seed heads are medicinal and can be harvested for market. That said, the ripening seed heads contain the highest concentration of the medicinal milky latex and are what we harvest on the farm. Oats should be harvested when the seed heads are mature, plump, and full of latex but still a light yellowish-green color. As the seed heads begin to mature, the stalks that they grow on will begin to "stretch out" of the leaf shaft. The seeds at the top of the stalk tend to mature more quickly than those at the base. It is easy to think that the plant is ready to harvest when the top heads are beginning to get plump. Be patient! It is important to wait until most of the seeds have come out of the leaf sheath and are maturing. To check readiness pick many seed heads from several different plants. When ready the heads will easily strip off the stem and when squeezed will squirt out a nice offering of white oat milk.

Harvesting oats is a real joy and is reminiscent of how we used to pop seed heads off grasses as kids. It's the same motion, albeit not quite as carefree as that of children playing in unmowed fields. To harvest oats it is helpful to wear tight-fitting gloves or use Vet Wrap or duct tape to protect fingers. Even the most calloused hands will get raw from stripping oats. We have found that wrapping our fingers and thumbs with duct tape provides a layer of protection to the skin, while allowing flexibility for the fingers to harvest. Sometimes gloves are too smooth, and the tape provides the friction needed to strip the oats without abrading the skin. Using harvesting baskets tied around our waists, we walk through the oat field like human combines, stripping the mature seed heads. We make sure to harvest in overlapping swaths to capture all the heads. It is also best to harvest the oats on dry, sunny days when the oats are upright. Sometimes oats will be knocked down by the wind. They can still be harvested, but it is more challenging.

When you are harvesting, some of the oat grass will come off as you strip the heads, and that's fine. It all goes into the basket and is winnowed off later. This is much faster than picking out grass as you go. Once collecting baskets are full, we dump them on clean, dry tarps in the tractor bucket or wheelbarrow to transport to the processing shed. Oats can heat up and begin to compost quickly, so be sure to limit how long they sit before transport. It is best to keep the tractor or wheelbarrow in the shade if possible.

## Postharvest and Drying Considerations

Winnowing is the best way to separate the oats from the chaff (the unwanted oat grass). To winnow we set up a ten-foot stepladder and attach two box fans to the side. Then we place the winnowing ladder on a large, clean tarp. Next we slowly pour baskets full of the oat harvest from the top of the ladder in front of the fans. The dense oats drop down into a pile, and the light chaff is blown by the fans and drifts into another, separate pile. Once done winnowing, you are left with a pile of clean oat heads and another pile of oat grass. Depending on your needs, both can be dried and processed. We process the oat heads and usually compost the grass. To dry oat heads, lay them out on a thin layer on drying racks. Oats can be tricky to dry, so it's important to keep the layer thin and to have very good airflow. Dry oats out of direct light in temperatures of 100 to 110°F. It takes about five days to a week, sometimes more, to completely dry the oats.

The key to successful processing is to be sure that the oats are thoroughly dry all the way through the seed head. Because the oats are harvested in their milky stage, it takes a while to remove all the moisture. Tricky drying is also compounded by the very dense nature of the seed. It takes time and consistent heat to dry each layer of the head. To test for dryness take oat samples from several racks around your drying shed. It is common that some places in a drying shed dry faster than others. Using sharp fingernails or scissors, cut open the seed heads. The milk inside should be crumbly dry and white. If there is any moisture present in any of your oat samples, the whole batch needs to dry more. It can't be overstated how important it is to check oats thoroughly. It only takes one small pocket of moisture to ruin the whole batch while in storage. Once the oats are dried, there is no need to garble or process further. The herbs can go directly into storage. When you bag up the herbs, they should sound "dry" and feel pretty light.

## Pests and Diseases

Oats are susceptible to some rusts and ergot. We sometimes see rust forming during dry spells. We do not treat our oats at all. Instead we employ good crop management techniques and rotations and also companion plant with red clover. We plant many crops of oats in different beds throughout a season. If crops develop rusts or show signs of ergot, we do not harvest the milky heads from those plants, and we till them under. We find it helpful to use good sanitation techniques with equipment to avoid spreading diseases to other beds.

## Yields

150 pounds of dried oat milky seed head per one-eighth-acre bed. Moisture ratio for oat milky seed head is 4:1 fresh:dried.

## Pricing

Retail price for one pound organic:
- Dried oat tops: $15 to $22
- Fresh oat tops: $9

# Passionflower     *Passiflora incarnata*

## Life Cycle

Passionflower is an herbaceous vine that becomes woody as it matures and is hardy to USDA zones 5 to 10.

## Plant Description

Passionflower is a truly magnificent member of the Passifloraceae family! When it is in full bloom, there is nothing quite like its resplendent beauty. Passionflower is native to the southern and eastern United States, and it is possible to overwinter passionflower in zone 4 if it is heavily mulched and planted in a protected location. The flowers bloom in late summer and are purple and white, growing up to two to three inches in diameter. The flower is a complex layer of riotous symmetry and design. The base of the passionflower is composed of two layers of finely fringed sepals and petals. The anthers and styles have a lovely geometry and rich history that is the genesis of the flower's name. The name, *passiflora*, alludes to the Passion of Christ. In the allusion the three styles and the five anthers are likened to the nails and hammers that were used to crucify Jesus. To continue the floral metaphor further, some think the vibrant petals represent the halo of God and the deep green, three-lobed leaves symbolize the Holy Trinity. However one views the plant, most are in agreement that passionflower is a strikingly beautiful and highly medicinal herb. In Vermont passionflower vines can grow ten to fifteen feet tall. Growers in warmer climates may have plants reaching twenty feet or more.

## Growing Conditions

Despite its exotic looks passionflower is relatively low maintenance and easy to cultivate. It can grow in weak, sandy soil but responds well to fertile, well-drained areas and likes to grow in full sun. Because it is a vining perennial, it requires some sort of trellising system and should be planted in a location where it can climb.

## Propagation

Passionflower can be grown from seeds, vegetative cuttings, or layering. When seeding passionflower, sow seeds in regular potting medium (which can be amended with sand if desired) and keep moist. Passionflower can be somewhat sporadic in its germination, so don't throw in the towel too quickly. It can sometimes take many weeks for plants to emerge.

To propagate from cuttings take six- to eight-inch cuttings from the stems, plant cuttings three to four inches deep in a light potting mix, and water well. Keep in a warm, sunny environment, and let the plants fully root before transplanting out. Another way to vegetatively propagate is through layering, done by bending the vines onto their sides, removing the leaves, and covering the bare stems with soil. It's best to leave six to twelve inches of the top part of the plant exposed. Water the layered stems in well, then let the layering develop for a few months. The buried stems will begin to form roots from the leaf nodes. Once a solid set of roots is produced, sections of the stems can be divided from the parent plant into individual new plants and used for transplanting.

## Planting Considerations

As previously mentioned, passionflower is a climber and requires a trellis of some sort, either naturally occurring or man-made. It can be trained to grow against the side of a building, on wooden trellises, fence posts, or the like. If you are creating a permaculture design, consider planting passionflower on the south-facing side of large trees (such as sugar maple, black locust, and slippery elm) to help ensure that the plant gets enough light and is not shaded out.

## Medicinal Uses

Passionflower is a wonderful nervine for calming the body, easing nervous tension, and helping people sleep. It has sedative qualities without leaving the body feeling sluggish or drugged. Passionflower has the ability to ease anxiety and reduce wired energy in both children and adults. Clinically, it has been shown to help treat panic attacks when taken over long periods of time. It is simultaneously relaxing and uplifting. Passionflower also has antispasmodic qualities and can help to reduce pain. It can be taken alone or in formulation with other nervines and makes pleasant-tasting teas and extracts. The flower essence of passionflower is also beneficial to restore a sense of wonder and hope. *Passiflora* produces delicious fruit, commonly known as maypop or passionfruit. Plants can produce anywhere from five to twenty fruits per mature vine.

## Harvest Specifications

Both the leaves and flowers are used medicinally. The aerial parts of the plant are harvested when the plant is in full bloom; in Vermont it flowers in early to mid-August. To harvest, cut the vines down to about a foot above the ground, leaving a few leaves to help the plant regenerate in subsequent seasons. Vines are easily cut with a sharp field knife or pruner and can be transported on clean tarps.

## Postharvest and Drying Considerations

Dry the vines intact on drying racks with good airflow and minimal light. This helps maintain the vibrancy of the blossoms, which can easily degrade in direct light. Dry at temperatures of 100 to 110°F. In optimum conditions passionflower will dry in three to four days. Once they are dry strip leaves and flowers off the vine and run the leaves and flowers over quarter-inch stainless steel mesh. Leaves and flowers are commonly sold together.

## Pests and Diseases

Passionflower has minimal pest and disease issues.

## Yields

Information not available.

## Pricing

Retail price for one pound organic:
* Dried passionflower leaf/flower: $15

## Commonly Imported From

Italy and France

**Figure 18-50.** Passionflower.

# Peppermint

*Mentha piperita*

## Life Cycle

Peppermint is an herbaceous perennial that is hardy to USDA zones 3 to 11.

## Plant Description

Peppermint's origin is hard to pinpoint, as there is evidence of early use of the plant in many regions. Peppermint is a hybrid of spearmint (*Mentha spicata*) and watermint (*Mentha aquatica*) and as a result is sterile and needs to be vegetatively propagated. Other mint varieties produce seed that is often true to type, but this one does not. Peppermint belongs to the Lamiaceae family and is a fragrant, easy-growing plant. Peppermint produces whorls of purple flowers and has smooth square stems. Its lance-shaped leaves are lovely—dark green, finely toothed, and with a pronounced central vein that is reddish-purple. Peppermint rhizomes grow readily, spreading under the surface of the soil as runners. It is an aggressive, fast-growing plant that can quickly take over an area if not kept in check. Peppermint can grow two to three feet tall and spread far and wide.

## Growing Conditions

Peppermint likes moist soils, warm days, and cool nights. It will grow in full sun to partial shade and thrives in well-drained humusy soil with good fertility. Once well established, peppermint outcompetes most perennial grasses and weeds. However, spot weeding is needed occasionally, and after several years of harvesting, beds can start to lose their vitality and should be replanted.

## Propagation

Peppermint is a hybrid and does not produce seeds. It's easiest to propagate from rhizome cuttings. To make these take six-inch cuttings from the rootlets, being sure to keep a bit of green growth (one to two leaves) per cutting. Plant the cutting three to four inches deep in a light potting mix, leaving some of the green growth above the surface, and water well. Keep in a warm environment out of direct sunlight, and let the plants fully root before transplanting out.

Another way to vegetatively propagate is through layering. Layering is done by bending the peppermint plant onto its side, removing the leaves from the parts of the stem that are touching the ground, and covering these denuded stems with soil. Leave six to twelve inches of the top part of the plant exposed, with leaves still intact. Water the layered stems in well, and let the layering develop for several weeks. The buried stems will begin to form roots from the leaf nodes. Once a solid set of roots is produced, sections of the stems can be divided from the parent plant into individual new plants and used for transplanting.

## Planting Considerations

Peppermint produces many rootlets and will take over the space that it is growing in through rhizomatous spreading. A good bed of peppermint will last three to five years and will be highly productive, allowing for multiple harvests per season. When planting it's best to give peppermint its own space and let it take over and flourish. Some people worry that peppermint will become invasive. We find that our beds do not outgrow their boundaries. With regular cultivation practices the peppermint stays in place and is fine planted next to other perennials. Plant spacing for peppermint is twelve inches between plants within the row and twenty-eight inches between rows, with two rows per bed. Beds planted with this spacing quickly mature to form a dense hedgerow of peppermint that helps to outcompete weeds.

**Figure 18-51. Peppermint.** Photograph courtesy of Bethany Bond

## Medicinal Uses

Peppermint leaves and flowers are used medicinally in teas, extracts, essential oils, and food. Aromatic and very flavorful, peppermint is an herb that is wonderful for the digestive system, acting as a carminative to reduce gas, ease nausea, and sooth cramping. It also is very uplifting to the nervous system and cooling to the body. It can help bring down fever and is pleasant tasting. Peppermint is a favorite of children and adults alike and is safe to use with most people.

## Harvest Specifications

The aerial parts of peppermint are harvested when the plant begins to bloom. Harvest on a hot, sunny day when the volatile oil levels are high and the dew has evaporated off the plants. Cut the tops with a field knife or a sickle bar mower, leaving six inches of growth for the plant to regenerate. You can get multiple harvests in one season. Peppermint should be processed quickly after harvest. If left in the direct sun, it will heat up quickly and start to decompose.

## Postharvest and Drying Considerations

To prevent browning and to maintain the volatile oils in the plants, start drying peppermint slowly. We like to rack our peppermint on the lower, cooler levels of the drying shed. Spread out the peppermint in a single layer, and be sure to maximize airflow. Begin drying at lower temperatures of 90 to 95°F to allow the moisture to leave the plant slowly. Once the plants have fully wilted, the heat can be gradually brought up to 100 to 110°F to finish the drying process. Peppermint dries in three to four days and is ready to process when the leaves come off the stem easily and can be crumbled in your hand. It is helpful to time drying so the leaves are dry but the main stems are slightly pliable. This helps with the garbling process. The leaves will come through the stainless steel mesh but the moist stems will not break off easily. This helps produce a better, less stemy product.

## Pests and Diseases

Peppermint has minimal pests or diseases. Some larger operations report issues with rust and fungal diseases, but we find that using good crop rotations and field management produces strong, healthy plants, and we have not had any issues with disease.

## Yields

150 to 200 pounds of dried peppermint leaf per one-eighth-acre bed (multiple harvests per season). Moisture ratio for peppermint leaf is 5:1 fresh:dried.

## Pricing

Retail price for one pound organic:
- Dried peppermint leaf: $10 to $22
- Fresh peppermint leaf: $12

## Commonly Imported From

India

---

# Pleurisy Root                    *Asclepias tuberosa*

## Life Cycle

Pleurisy root is an herbaceous perennial that is hardy to USDA zones 3 to 9.

## Plant Description

Pleurisy is native to North America and historically grew in the wide-open prairies. A member of the milkweed family, Asclepiadaceae, pleurisy root is a beautiful plant. It grows around two feet tall and has striking umbels of orange flowers. Each umbel is made up of a cluster of small individual flowers that consist of five petals and look quite literally like starbursts of bright color. Pleurisy has alternate lance-shaped leaves, a slightly fuzzy stem, and long, light-colored taproots. Unlike most members of the family, pleurisy does not have milky white latex in its stems.

## Growing Conditions

Pleurisy likes to grow in full sun and well-drained, sandy soil. It is not a fast grower, and in cultivation, care should be given to site selection and weed control. In New England we have found that pleurisy root will not outcomplete perennial weeds and can become quickly overrun, making harvest difficult. The plants emerge later in the spring than most other plants and often when the weeds have already started to become established; therefore early pre-emergent weed control is necessary. When found in its native soils pleurisy grows vigorously and does just fine intercropped with native grasses and prairie beauties such as the echinaceas, compass plant, and other asclepia species.

## Propagation

Pleurisy root can be propagated from rootlets and seeds. When sowing seeds it is helpful to stratify them to improve germination. Sow pleurisy seeds on the surface, and cover with a light layer of potting medium. Keep well watered, and seeds should germinate within two to three weeks. If using rootstock, plant in the fall or early spring. Be sure to cover roots with two to three inches of soil and to orient the roots so the nodes that produce the aerial parts of the plants are facing up.

## Planting Considerations

Pleurisy root is also called butterfly weed and is a wonderful plant for attracting pollinators. The blossoms are quite striking and should be planted where they can be admired. Pleurisy can easily share a bed with other plants such as *Echinacea purpurea*. For cultivation purposes use plant spacing of twelve inches within the row and fourteen inches between rows, with three rows per bed.

## Medicinal Uses

As the name indicates, it is the root of pleurisy root that is used medicinally to treat deep lung infections and to help clear airways. Pleurisy acts as an expectorant and helps the body expel mucus. It also has antispasmodic actions and is used to treat unproductive coughing. Pleurisy is great to use in combination with other herbs such as elecampane and mullein to treat bronchitis and the flu. It can be taken as a tea or tincture and is used for acute infections, as opposed to using a long-term tonic.

## Harvest Specifications

We harvest pleurisy root after three seasons of growth. The roots can be harvested in the fall after the aerial tops have died back or in the spring before green growth has begun. Pleurisy roots are relatively easy to dig with a spading fork or mechanical digger such as a bed lifter or modified potato digger. It can be helpful to run a cultivator or shank parallel to the row before digging to loosen the soil. Once the roots are dug, knock out as much soil into the field as possible, then process the roots before drying.

## Postharvest and Drying Considerations

Soil often gets compacted in the central crown of pleurisy roots. Therefore, it is helpful to halve or quarter the roots before washing. Pleurisy roots are relatively soft and can be chopped with a field

**Figure 18-52.** Pleurisy root blooming.

knife. After splitting the roots wash them thoroughly before drying. Mill roots after they are fully dried. Pleurisy roots dry under optimum conditions in four to five days.

## Pests and Diseases

Pleurisy root has minimal pest and disease issues.

## Yields

One hundred pounds of dried pleurisy root per one-eighth-acre bed. Moisture ratio for pleurisy root is 3:1 fresh:dried.

## Pricing

Retail price for one pound organic:
• Dried pleurisy root: $15 to $20

# Red Clover                              *Trifolium pratense*

## Life Cycle

Red clover is a short-lived herbaceous perennial that is hardy to USDA zones 4 to 9.

## Plant Description

Red clover is a lovely plant that is native to Europe, Asia, and areas of Africa. It has multiple clusters of three leaves on each stem (hence the name *trifolium*) and purplish-pink flowers. The leaves of red clover have white variegation in the shape of an inverted V on the top side, whereas other clovers (such as white clover) do not. Herbal lore credits the fairies with painting these light little brushstrokes to help distinguish red clover from all others. Red clover has hollow stems that grow one to two feet tall and large taproots that can reach several feet deep. They begin blooming in early summer and will continue to produce flushes of blossoms throughout the summer.

**Figure 18-53.** Red clover leaf and blossom.

## Growing Conditions

Red clover grows well in many soil types, actually improving the soil that it grows in by fixing gaseous nitrogen from the air into a solid form on its root nodules and making it a wonderful cover crop. It does well in cool, moist conditions and can grow in full sun to partial shade. Red clover uses significant calcium, phosphate, and potash when growing and grows best in soils that have good fertility.

## Propagation

Red clover can be broadcast-seeded or drilled in the early spring through midsummer before it gets too hot. Seeding rates are seven to eighteen pounds per acre. It is also extremely beneficial to inoculate red clover with rhizobia bacteria before planting; this helps increase the plant's vigor and its ability to fix nitrogen into the soil.

## Planting Considerations

Red clover is not only a wonderful medicinal plant, it is a beneficial cover crop, increasing nitrogen levels, suppressing weeds, and improving soil structure and biomass. A great way to utilize red clover is to plant it on land that is lying fallow as part of a crop rotation. Interplanting red clover with other cover crops such as oats does a number of things. Initially, as the annual oats grow, they provide shade for the red clover, helping it to become established. The red clover in turn matures and in subsequent seasons will fix nitrogen from the air into the soil. Red clover also produces medicinal crops of blossoms that can be harvested and sold even as it works to improve the soil. Once the crops are harvested the clover can be tilled under, adding biomass as a green manure and leaving the land ready to plant with future crops. Red clover is tolerant of shade and is very useful to plant as a living mulch beneath orchards, in hops yards, and under elderberry stands.

## Medicinal Uses

First and foremost, red clover blossoms are very nutritious and are rich with calcium, iron, and nitrogen. They are also one of the premier herbs used to help clean and detoxify the blood and are often found in formulas to treat skin problems such as eczema and psoriasis. Red clover is also an excellent expectorant for the lungs and is used to treat bronchitis and deep, persistent coughs. Red clover also contains phytoestrogens (estrogen-like substances from plants) and can help ease difficulties associated with menopausal changes. Red clover makes an excellent, mildly sweet tasting tea and can also be used in extract form.

## Harvest Specifications

Blossoms and the top leaf set that grows directly below the blossom should be hand-picked early in the morning when the dew is still on the flowers. This is when they are the most succulent and full of medicine. Be sure to pick blossoms that are completely purple and do not have any browning. Red clover can heat up quickly and begin composting, so keep harvest baskets shaded and process blossoms as quickly as possible. The flowers will continue to flourish for several weeks, so you can return to your patch to harvest multiple times.

## Postharvest and Drying Considerations

Drying red clover blossoms can be tricky. The key is to start cool, dry slowly, and finish with higher temperatures to ensure complete dryness. Begin by laying blossoms out in a single layer in an area with good airflow and out of direct light. On our farm we rack blossoms on the lower racks of our drying shed because they tend to be cooler. Start drying at 90 to 95°F and wait approximately a day before increasing temperatures to 100 to 110°F. This allows the flowers to lose moisture gradually and decreases the possibility of their turning brown. The blossoms are dense in the middle and take several days to dry completely. Red clover has finished drying when the centers of the blossoms are free from moisture. To check for dryness take several blossoms from different racks (because some racks may dry more quickly) and crush them gently in your hand. Blossoms should feel and sound crunchy and break apart when rubbed between your fingers. Then try to snap the stem at the base of the flower. If completely dry, the stem will snap off easily. If there is still moisture present, the stem will be pliable and will bend, indicating that more drying time is needed. It is important to fully dry the blossom but also to take care not to overdry them because they will lose their vivid color and medicinal qualities.

The last precaution we take is to check our red clover in storage frequently, especially when we initially put them away. Often you can feel if there is any moisture present in the bag. When you stick your hand in the bag, the blossoms should be crispy and dry. If they are beginning to feel and sound "soft," you should consider reracking and drying. If you catch this quickly enough, red clover can be redried with no noticeable loss in quality.

## Pests and Diseases

Red clover has minimal pest and disease issues when grown for blossom harvest. Occasionally red clover will have powdery mildew on leaves, and sometimes there is insect pressure. However, with good crop rotation and field management pests and disease have not been an issue.

## Yields

Two hundred pounds of dried red clover blossoms per one-eighth-acre bed (multiple harvests per season). Moisture ratio for red clover blossoms is 5:1 fresh:dried.

## Pricing

Retail price for one pound organic:
• Dried red clover blossoms: $33 to $51
• Fresh red clover blossoms: $41
• Dried red clover leaf: $9 to $25
• Fresh red clover leaf: $13

## Commonly Imported From

Chile

# Red Raspberry                                    *Rubus* spp.

## Life Cycle

Red raspberry is a perennial herb that is considered a "cane fruit" or a "bramble" and is hardy to USDA zones 3 to 9.

## Plant Description

Red raspberry is native to Asia and North America, grows wild, and is a favorite for cultivation. Red raspberry is a member of the Rosaceae family and produces highly medicinal leaves as well as delicious fruit. When left to its own devices, red raspberry can get quite large, growing to five feet tall. The roots and crowns are perennial, but the individual raspberry canes have a two-year cycle. The first year canes are green and produce vegetative growth. In the second year canes become woody and brown. The second-year growth sets light-pink flowers and produces fruit, then dies at the end of the season. The leaves of raspberry are dark green on the top and a beautiful silver on the underside. This is an easy way to distinguish red raspberry from the other *Rubus* species. The leaves are compound, made up of three to five leaflets that are finely toothed. The stems of red raspberry are covered with small thorns that can scratch but aren't nearly as painful as the large thorns of blackberry.

## Growing Conditions

Red raspberry likes to grow in full sun or partial shade in well-drained, loamy soil. When cultivating raspberry, it is beneficial to build organic matter and fertility by preparing the soil with compost, manure, and cover crops before planting. Red raspberry also needs an adequate water supply. Besides being cultivated,

**Figure 18-54.** Red raspberry leaf.

red raspberry also likes to grow in wild places that have been disturbed or cleared. On our farm we have patches of wild red raspberry growing on old logging roads and at the edges of fields. We harvest from these patches and let Mother Nature do her thing.

## Propagation

The most common way to propagate red raspberry is through vegetative cuttings or root cuttings. To take leaf cuttings cut five to six inches of green stem and remove leaves from the lower three inches. Plant the stem in potting medium, leaving the top of the stem that has leaves above the soil. Water well and keep out of direct light until rooting begins. Roots will grow from the leaf nodes that are located beneath the soil.

Root cuttings are another great way to grow red raspberry. Take six-inch cuttings from the roots of the parent plant and plant them two to three inches deep. Keep well watered, and they will begin to produce shoots after a couple of months.

## Planting Considerations

Red raspberry is called a bramble for a reason. It grows well and spreads out, taking over an area. Because of the thorns red raspberry is best planted out of the main thoroughfare. All of the red raspberry leaf we harvest at ZWHF is wild-harvested from our farm as well as from the land we lease from our neighbors where it grows abundantly in clearings in the forest.

## Medicinal Uses

Red raspberry leaf is a wonderful herb for both men and women. It is extremely nutritious and contains vitamins and minerals such as iron, niacin, and magnesium that strengthen the body, build the blood, and help restore energy. For women it is specifically used to tonify the reproductive system. For menstruation it can help ease heavy bleeding and is wonderful to give to teens as they begin their cycles, giving them extra nutrients and support. A favorite for pregnant women for much the same reason, raspberry is a powerhouse of green energy, safe enough to use every day, and delicious. It also has astringent properties and can work with the body to "tighten" things up, tonify the tissues, and treat ailments such as diarrhea and leukorrhea. Because of its great flavor red raspberry makes a wonderful tea. It can also be tinctured or added to elixirs.

## Harvest Specifications

Harvest the leaves from the green, first-year canes. The easiest way to harvest the leaves is to use a sharp field knife to cut the canes down and dry the leaves with the canes intact. We take many canes from each plant but are sure to leave some so the plant can grow and regenerate. It is also important to leave some first-year canes in place to go through the life cycle to produce fruit in the following year. Because of the thorny brambles, it's highly recommended to wear long sleeves and pants, as well as gloves, when harvesting.

## Postharvest and Drying Considerations

Red raspberry is a cinch to dry. Lay the canes out on drying racks in a single layer with good airflow. Dry at temperatures of 100 to 110°F, and red raspberry should be dry in one to two days. Take dried canes and run them over quarter- to half-inch stainless steel mesh to separate the leaves from the stems. The only tricky thing with red raspberry is it can be stemy. To remove these stems run the raspberry over the garbling racks several times. It's not uncommon to have some small pieces of stem, but too much can make it less marketable.

## Pests and Diseases

Red raspberry has minimal pests and diseases.

## Yields

Information not available.

## Pricing

Retail price for one pound organic:
- Dried red raspberry leaf: $9 to $25
- Fresh red raspberry leaf: $13

## Commonly Imported From

Poland and Albania

# Rhodiola

## *Rhodiola rosea*

## Life Cycle

Rhodiola is a succulent perennial that is hardy to USDA zones 2 to 7.

## Plant Description

Rhodiola is one of the hardiest perennials we grow. It is native to Russia and Siberia and can be found in the subarctic areas of the Northern Hemisphere. Rhodiola is an extremely beautiful member of the Crassulaceae family that has strong vitality and is capable of growing and thriving in the harshest environments. At maturity rhodiola grows up to a foot tall and has light green, fleshy, succulent leaves with bright flowers ranging from yellow to orange.

## Growing Conditions

In its natural environment rhodiola grows at high altitudes in rocky, sandy soils. Roots from these natural stands tend to be small. However, when cultivated rhodiola can produce substantially larger roots with very good medicinal components. In cultivation rhodiola grows well in full sun and well-drained soil with good fertility. To get maximum root production plant rhodiola in deeply worked soil. Rhodiola is a tenacious plant and can tolerate a variety of soils and also extremely cold temperatures. However, it does not do well in excessively hot climates.

## Propagation

Rhodiola can be grown from direct seed. It is highly recommended to stratify the seeds before planting, then sow them on the surface and cover lightly with potting medium. Water well, and be patient; it can take one to two months for germination. Take care not to let the seedlings dry out, but do not overwater. Rhodiola is slow to grow, especially during its first year. We wait until a substantial number of true leaves have formed, before transplanting out.

Rhodiola can also be successfully propagated from crown divisions and that is our preferred way of increasing our plant stands.

## Planting Considerations

It is important to prepare beds carefully before transplanting rhodiola. It does a poor job of trying to outcompete weeds, so it is time well spent to utilize cover crops to smother perennial grasses and the like before transplanting. Some plants, such as nettle, peppermint, and lemon balm, we put in as soon as roots begin to form, and they thrive in the fields, quickly establishing their place. Rhodiola is not like that; it matures slowly, so we find it helpful to let it mature in the greenhouse for a couple of months before transplanting. When planting use eight- to ten-inch spacing between plants and fourteen inches between the rows, with three rows per bed.

## Medicinal Uses

The roots of rhodiola are a prized adaptogenic and antioxidant herb. They are excellent for the nervous system and are indicated to help improve brain function and strengthen the body's ability to deal with stress. In addition to improving nervous system function, rhodiola is a favorite among endurance athletes, as it increases energy, eliminates fatigue, and helps reduce the effects of altitude sickness. The roots are used primarily in teas and extracts and are safe to use tonically over a long period of time. Because of the growing popularity of this herb, it is being wildcrafted in extremely large volumes, putting naturally occurring plant populations at risk. Therefore, it is best to use rhodiola that has been cultivated.

## Harvest Specifications

The roots of rhodiola are used medicinally and are harvested after the third growing season. While some growers report harvesting the five-year-old roots, we like to harvest our roots in years three to four. After year five the roots can start to deteriorate and have a pithy center. We harvest roots in the fall, once the tops have died back fully. Roots can be dug by hand or with a bed lifter or potato digger.

## Postharvest and Drying Considerations

Soil often gets compacted in the central crown of rhodiola where the roots begin to branch off. Therefore it is helpful to quarter the roots before washing. Rhodiola can be chopped easily with a field knife. After splitting the roots wash them thoroughly before drying in temperatures of 100 to 110°F. Mill roots after they are fully dried. Rhodiola roots dry under optimum conditions in three to four days.

## Pests and Diseases

Rhodiola has minimal pest and disease issues.

## Yields

Four to five hundred pounds of dried rhodiola root (four- to five-year-old root) per one-eighth-acre bed. Moisture ratio for rhodiola roots is 5:1 fresh:dried.

**Figure 18-55.** Rhodiola in flower.

## Pricing

Retail price for one pound organic:
• Dried rhodiola root: $13 to $16

## Commonly Imported From

China and Russia

# Schisandra                    *Schisandra chinensis*

## Life Cycle

Schisandra is a perennial woody vine that is hardy to USDA zones 4 to 9.

## Plant Description

Schisandra is native to China, Korea, and Japan but grows well in the northeastern regions of the United States and Canada. It is a member of the Magnoliaceae family and produces beautiful bunches of ruby-red berries that are delicious, in addition to being medicinal. The schisandra flowers are small

and white, and its leaves are deep green and lance-shaped. Schisandra is dioecious, meaning it has male and female flowers that grow on separate plants. Therefore, you must have both types of plants to produce fruit. Because schisandra usually doesn't flower until its second year, it is hard to determine which plants you have if you grow them from seed. Consider buying vines that have been sexed or vegetatively propagate from vines that have been identified as male or female. Once established and in the right environment, schisandra grows quickly, up

**Figure 18-56.** Shisandra berry. Photograph courtesy of Leonora Enking

to three feet in a single season. Mature plants can be as tall as twenty to twenty-five feet in height.

## Growing Conditions

Schisandra likes to grow in full sun to partial shade and does best in well-drained, sandy-loamy soil. The vines need consistent moisture, especially when young, but do not like to grow in overly wet areas or in clay soils. Being a vine that produces substantial sets of berries, schisandra needs a sturdy trellising system or support to grow on.

## Propagation

Schisandra can be propagated by direct seeding, vegetative cuttings, and layering. We have had good luck propagating schisandra from seed, but it takes a while and requires patience. Schisandra seeds can be stratified to help germination. Soak seeds overnight, then plant in a light potting soil, water well, and wait. Spotty germination usually begins after four to six weeks, but don't give up too soon. We often leave trays of schisandra all summer, pricking out and transplanting the babies as they come up. Patience has paid off, and we have been pleasantly surprised by how many schisandra continue to

germinate given the extra time and TLC. Schisandra can be transplanted out once the plants have become sturdy and produced several sets of true leaves.

If you don't have the patience for seeding, schisandra plants can also be vegetatively propagated by cuttings and layering. Schisandra produces lots of runners that can be easily used to create new plants. Simply cut five to six inches of root runners, and plant in soil. Keep well watered until the plants begin to sprout. Layering is also an easy way to get new vines. Layering is done by bending some of the vines onto their sides, removing the leaves from the parts of the stem that are touching the ground, and covering these denuded stems with soil. Leave six to twelve inches of the top part of the plant exposed, with leaves still intact. Water the layered stems in well, then let the layering develop for a few months. The buried stems will begin to form roots from the leaf nodes. Once a solid set of roots is produced, sections of the stems can be divided from the parent plant into individual new plants and used for transplanting. These vegetative propagation techniques can be especially beneficial, because if you know the sex of the parent vines, you can grow known male or female progeny.

## Planting Considerations

Schisandra bears fruit after three to four years and needs to have something substantial to grow on. Vines can be trained to grow against the side of a building, on wooden trellises, or on fence posts or the like. If creating a permaculture design consider planting schisandra on the south-facing side of large trees (such as sugar maple, black locust, and slippery elm) to help ensure that the schisandra gets enough light and is not shaded out.

## Medicinal Uses

The berries of schisandra are used medicinally and are known as the "five-flavored fruit" in traditional Chinese medicine. Berries are complex tasting and have all five flavors: sweet, sour, bitter, salty, and pungent. Schisandra is an adaptogenic herb that is full of vitamins and minerals. It also has anti-inflammatory properties, builds energy, and aids in digestion. When used in tonics over an extended period of time, schisandra helps to restore balance and calm back to the body. With its highly unique flavor, schisandra is delicious in teas, tinctures, elixirs, and syrups.

## Harvest Specifications

Schisandra produces berries in late summer and should be harvested when fully ripe and juicy. For most growers vines begin producing berries after the third or fourth season. The first crops of berries are often lower yielding. Once well established and mature, schisandra vines will produce bountiful harvests.

## Postharvest and Drying Considerations

Schisandra berries can be used fresh or dried. The berries are not difficult to dry but do take a little special attention. Because they are harvested in the fall and they are fleshy fruits, the berries take much longer to dry. The best course of action is to rack berries in a single layer and position them directly in front of fans and a heat source. To keep the vibrancy of the berries it is a good idea to limit the exposure to light, because the berries can take around two weeks to dry thoroughly. To test for dryness use a sharp blade to cut open the berries. They should be dry and hard, not pliable. Once they become really dry, they will be very difficult to cut open. At this point the berries are most likely dry. However, it is hard to be 100 percent sure. To safeguard against spoilage we store and ship in breathable paper bags and frequently check the berries, encouraging our customers to do the same. Another good way to "process" berries is to sell them fresh or make them directly into such products as tinctures, jams, syrups, or elixirs. There is a good market for fresh berries.

## Pests and Diseases

Schisandra has minimal pests and diseases. However, birds and other animals will eat the harvest if not carefully monitored.

## Yields

Information not available.

## Pricing

Retail price for one pound organic:
- Dried schisandra berries: $16 to $23

## Commonly Imported From

China

# Siberian Ginseng *Eleutherococcus senticosus*

## Life Cycle

Siberian ginseng is a woody perennial that is hardy to USDA zones 3 to 8.

## Plant Description

Siberian ginseng is native to Siberia, northern China, Russia, and Korea. It is a member of the Araliaceae family; however; it is not a true ginseng. Rather it is a member of a different genus. Siberian ginseng is a shrubby bush that can eventually grow ten to fifteen feet tall. It has deep green, oval-shaped leaves that are finely toothed. The plant is self-pollinating, producing male and female flowers on the same round globe of blossoms. The flowers of the female are yellow and of the male are purple.

## Growing Conditions

Siberian ginseng grows well in partial shade. Unlike the true ginsengs, it does not need the deep shade of a hardwood forest and does really well as a hedgerow

**Figure 18-57.** Siberian ginseng in flower. Photograph courtesy of Larken Bunce

at the edges of fields. It can grow in a variety of soils but does best in moist, loamy soil and tends to be a heavy feeder, responding well to top-dressing and added fertility.

## Propagation

Siberian ginseng can be tricky to propagate from seed. When you are working with dried seeds, it can take up to two years to germinate, whereas fresh seed usually will germinate sporadically after one year. The key to good germination is to stratify the seeds and keep them moist. We have had the best luck with sowing fresh seeds in a woodland bed in the spring and letting it go through the natural stratification of a Vermont summer followed by winter. This naturally occurring warm and cold fluctuation breaks dormancy, and the seeds germinate the following spring.

Once you have an established plant, softwood cuttings are the easiest ways to increase your planting stock. In early spring cut eight to ten inches of green-wood cuttings, and plant into moist potting medium. Keep well watered in the shade, and allow the plants to develop a good root system; this will take most of the summer.

## Planting Considerations

Siberian ginseng will grow very tall once it gets established. It likes to grow in partial shade and can easily be added to a woodland bed. That said, if you don't have large stands of hardwoods, consider planting Siberian ginseng on the shady sides of fields along a tree line. Siberian ginseng will grow up like a hedgerow and help make shade for other plants. It also has lovely umbels of purple and yellow blossoms and can be a lovely addition to shade gardens—just be sure to give it room to grow. Transplant with four- to six-foot spacing between plants and six-foot spacing between rows.

## Medicinal Uses

Siberian ginseng is a highly sought-after and beloved adaptogenic herb. Like the ginsengs, it builds energy in the body. It differs, however, in that instead of drastically stimulating the body or giving a buzzed, revved-up feeling like some of the stronger ginsengs do, Siberian ginseng tonifies and fortifies the body. It builds grounded, root energy. In addition to building vitality, Siberian ginseng also helps the body deal with stress, strengthens brain function, and has anti-inflammatory properties. It is wonderful to take over long periods of time and is commonly used in tincture or tea form.

## Harvest Specifications

Harvest Siberian roots after a minimum of three years of growth. It should be harvested in the fall, when the energy of the plant is going back into the root. Before digging, cut tops back, then dig with a spading fork or use a bed lifter or the bucket of the tractor to loosen the roots.

## Postharvest and Drying Considerations

Soil often gets compacted around the central crown of Siberian ginseng where the roots begin to branch off. Therefore, it is helpful to chop the roots before washing. Siberian ginseng can easily be chopped with a field knife. After chopping the roots, wash them thoroughly before drying. Mill roots after they are fully dried. Siberian ginseng roots dry under optimum conditions in three to four days.

## Pests and Diseases
Minimal pests and diseases have been reported.

## Yields
Information not available.

## Pricing
Retail price for one pound organic:
• Dried Siberian ginseng root: $9 to $25

## Commonly Imported From
China

# Skullcap — *Scutellaria lateriflora*

## Life Cycle
Skullcap is an herbaceous perennial hardy to USDA zones 4 to 10.

## Plant Description
Skullcap is from the Lamiaceae family and is a mint native to North America. Here in the Northeast, plants flower in late June to early July and go to seed shortly after flowering. Skullcap plants are erect and multistemmed and grow one to three feet tall from dense rhizomatous mats. Leaves are oval or lance shaped, opposite, and toothed and grow from petioles attached to square stems. Flowers range from white to violet-blue and grow on racemes attached to leaf axils along the length of the stems. We think the flowers resemble dolphins playfully jumping off the leaves. Others refer to the flowers as having "hooded protuberances" or "helmetlike sheaths." The name skullcap may refer to the fact that some have thought that the flowers resemble skulls with little caps. Others say that the name skullcap was given in reference to the herb's ability to calm a headache.

## Growing Conditions
Skullcap prefers growing in full sun or partial shade in moist soil with a high percentage of organic matter. "Muck" is a term often used to describe the soil where we find skullcap growing in the wild. This type of soil is prone to extended periods of being sodden, but unlike anaerobic swamps, muck soils dry out enough periodically to allow for oxygen to promote more aerobic conditions. Here in Vermont we often find this type of soil at the edge of wetlands and in moist depressions known as vernal pools where rich organic matter collects in woodlands. Although skullcap loves rich soil, it does not perform well in soils with excessive amounts of readily available nitrogen. Too much nitrogen can cause the plants to develop more leaf mass than the somewhat delicate stalks can support, and plants tend to grow top heavy and fall over. Excessive nitrogen also greatly increases skullcap's susceptibility to fungal diseases, such as powdery mildew.

## Propagation
Skullcap can be direct-seeded, transplanted, or grown from vegetative cuttings. Direct seeding requires that the seeds be shallowly sown in early spring and kept evenly moist and free from competition. In the greenhouse, seeds can be sown in flats in early spring and transplanted out after two to three sets of true leaves emerge and the root system becomes

**Figure 18-58. Skullcap flower.** Photograph courtesy of Larken Bunce

**Figure 18-59. Skullcap leaf and flower.**

well established in the planting medium. Vegetative cuttings taken from the bright, fleshy, cream-colored rhizomes are easy to grow into healthy plants, and this has become our preferred and most successful method of propagating skullcap.

## Planting Considerations

Since skullcap prefers to grow in rich, moist soil in partial shade, we like to grow our skullcap plants on the eastern side of wooded hedgerows for protection from the hot afternoon sun. We provide them with ample amounts of compost and irrigate them with drip tape to attempt to replicate the types of conditions in which we find them growing in the wild. Skullcap can also be grown in sandy or gravelly soils as long as there is ample organic matter and consistent moisture is provided. We transplant skullcap at a twelve-inch spacing in rows spaced fourteen inches apart, three

rows to the bed. This three-row bed allows us to cultivate in between the rows once or twice during early growth. During this cultivation we set the sweeps deeply so that soil is hilled up around the base of the skullcap plants. This method seems to provide better rooting, and sturdier plants are less likely to tip over from having weak stem or crown unions, which can be common with skullcap transplants. The three rows eventually become one wide bed of skullcap that usually lasts for two to three years before the weeds come in and become too labor intensive to remove during harvest. At this point we take cuttings from the old skullcap bed and start the process over again.

## Medicinal Uses

Skullcap is premier nervine. It is used in tincture and tea form and has a wonderful way of calming and centering the nervous system. Skullcap acts as a mild yet effective sedative and helps relieve nervous tension and anxiety. It also has antispasmodic effects and is wonderful to use for premenstrual tension and cramping. Skullcap can be taken over long periods of time to strengthen and reinvigorate the nervous system, making it more resilient and strong. It also can be taken in acute situations of extreme stress, such as panic attacks and trauma. In these situations skullcap works quickly to bring the nervous system back to a place of homeostasis and calm. It is a versatile herb that is safe to use with children and adults and is also recommended for people who are recovering from long-term debilitating illness.

## Harvest Specifications

Skullcap is best when harvested during the early stages of flowering. We harvest skullcap in the early hours of the day before the plants heat up from solar radiation. Cooler plants hold up better in transit to the drying shed and tend to dry to a higher quality than those harvested with excessive field heat. To harvest cut the plants six to eight inches aboveground with a field knife, taking care not to break the stem/root crown union, which can be fragile in first-year plants. For more mature plants a scythe or sickle bar cutter can be used to harvest skullcap more efficiently. Cutting plants six to eight inches aboveground is recommended to ensure that

the plants don't go into early dormancy and suffer from having too much of their mass harvested.

## Postharvest and Drying Considerations

Skullcap plants contain up to 80 percent water by weight and need to begin drying at lower temperatures. We dry skullcap on the lower racks of our drying shed because they tend to be cooler and more shaded. As the moisture begins to leave the plant and we see it wilting, it is safe to increase the heat to no more than 100°F.

## Pests and Diseases

The major issue we have had with skullcap is powdery mildew. We've had some years when no mildew is present and other seasons when we have lost the whole crop. The powdery mildew seems to thrive in dry years when humidity is high and rains aren't regularly rinsing mildew spores away from leaf surfaces.

## Yields

200 pounds of dried skullcap leaf per one-eighth-acre bed. Moisture ratio for skullcap leaf is 5:1 fresh:dried.

## Pricing

Retail price for one pound organic:
- Dried skullcap leaf/flower: $17 to $26
- Fresh skullcap leaf/flower: $10

# Spilanthes                                     *Spilanthes* spp.

## Life Cycle

Spilanthes is an herbaceous annual in temperate regions and a perennial in tropical regions.

## Plant Description

Spilanthes is native to Africa and South America and is from the Asteraceae family. Here in the north, seeds

sown in flats in the greenhouse in late March or early April are transplanted out into the field in mid-May and flower within four to six weeks of being transplanted. These plants start going to seed in August, as the days grow shorter and nights become cooler. The plants are extremely cold sensitive and succumb to the first frost they experience. Someone visiting our farm and seeing spilanthes growing for the first time remarked that the plants look like something out of a Dr. Seuss book. Others call them "buzz buttons" or "psychedelic gumdrops." They are certainly unusual looking, and when casting a downward gaze at their flowers, you might momentarily wonder whether the plants are staring right back up at you because of the blossoms' resemblance to brightly colored eyeballs. Spilanthes plants reach six to twelve inches in height and grow from a rhizomatous, spreading root system. Stalks are bronze, purple, and green in color; are erect; and bear opposite, glossy, toothed leaves attached via petioles. Flowers are conical and pom-pom shaped, starting out yellow and gradually forming red eyeball-like tips.

## Growing Conditions

Spilanthes grows best in the full sun in rich, well-drained soil with ample moisture. In general, the richer the soil, the larger the plants get. However, there seems to be a diminution of potency when plants are grown in soils with excessive nitrogen.

## Propagation

Spilanthes is best started from seed indoors and transplanted outside after the threat of frost in temperate climates. In tropical climates seeds can be direct-sown into the field in rich, weed-free soil and kept moist until germination. Seeds are light-dependent germinators and should be sown on the surface or lightly covered. Expect them to take seven

**Figure 18-60.** Spilanthes leaf and flower.

to fourteen days for germination. Spilanthes can also be propagated from vegetative cuttings, which easily root within ten to fourteen days. Transplant plugs after two sets of true leaves have appeared, after the threat of frost has passed, and before plants become root-bound for best results.

## Planting Considerations

Spilanthes grows as a dense mounding groundcover that will outcompete most weeds once the plant becomes established. We transplant plugs at twelve-inch spacing in the row, two rows per bed spaced fourteen inches apart. In hot, dry weather in rich soil the plants will quickly grow together to form a dense, single row.

## Medicinal Uses

Once in a while we come across a skeptic, someone who doubts that the plants we grow and harvest are really effective medicines. Spilanthes is a great choice for a quick demonstration of the powerful medicinal properties found in plants. "Try this," we tell him, as we hand him a small flower bud to chew on. The expression on his face speaks volumes before he is even able to open his mouth, which often takes some time because of the powerful and at times even overwhelming tingling sensation spilanthes produces immediately. "Wow, that stuff is potent!" he usually says once he regains his ability to speak . . . case in point!

Although relatively obscure, spilanthes is starting to gain acceptance as one of the most potent yet safe antiviral and antibacterial herbs in our materia medica. Commonly known as "toothache plant" for its oral antiseptic, anesthetic, and sialagogue (saliva-producing) properties, spilanthes contains powerful bioactive compounds such as spilanthol and alkylamide that offer anti-inflammatory, analgesic, and immune-enhancing properties. Spilanthes is also being widely touted and researched for its anti-parasitic properties and is reported to be effective for treating blood-borne parasites such as malaria and Lyme disease. One of the most interesting bits of information we have recently come across regarding

spilanthes is that recent laboratory tests have demonstrated that spilanthes holds promise as a remedy for enhancing sexual function, especially in men. Herbal Viagra, perhaps?

## Harvest Specifications

Although all parts of spilanthes are medicinally active, we harvest the aerial portions by cutting the entire aboveground portion of the plant with field knives. Two or more harvests of the aboveground plant parts are possible in areas with long growing seasons or if plants are started early indoors. Flower buds, which are the most potent parts of the plant, can be harvested every five to seven days throughout the growing season.

## Postharvest and Drying Considerations

Spilanthes plants have a high moisture content and therefore dry slowly. We dry the aerial portions whole after harvesting by spreading them out in a single layer on drying racks. We begin drying at temperatures of 85 to 95°F, then finish the drying at temperatures of 100 to 110°F. This allows the moisture to leave the plant without turning it brown. To test for dryness rub blossoms between your fingers. They should break apart easily, and the stems beneath the blossom should snap. Pass dried leaf and flower over quarter- to half-inch stainless steel mesh to process. Leaf, flower, and stem can be sold together.

## Pests and Diseases

There are no known pest or disease issues that we have experienced or come across in our research.

## Yields

Fifty to sixty pounds of dried spilanthes tops per one-eighth-acre bed. Moisture ratio for spilanthes tops is 4:1 fresh:dried.

## Pricing

Retail price for one pound organic:
• Dried spilanthes leaf/flower: $16 to $24
• Fresh spilanthes leaf/flower: $9 to $10

# Saint John's Wort         *Hypericum perforatum*

## Life Cycle

Saint John's wort is an herbaceous perennial that is hardy to USDA zones 3 to 9.

## Plant Description

Saint John's wort is native to Europe, Asia, and northern Africa. While it is beloved by herbalists, it is considered a noxious weed to others because of its ability to cause photosensitivity in livestock (if consumed in mass quantities). Because of this some states (California, Colorado, Hawaii, Idaho, Montana, Nevada, Oregon, Utah, Washington, and Wyoming) discourage the cultivation of Saint John's wort and prohibit the sale of seeds. Saint John's wort is a member of the Hypericaceae family. It is an erect perennial that grows two to three feet tall and has clusters of bright yellow flowers, each composed of five petals in the shape of a sunburst or star. When blossoming, flowers contain bright red oil that can be seen when the flowers are squeezed. The presence of the red oil is an indication that the tops are ready to harvest. The leaves of Saint John's wort are also unique. They are small and lance shaped and have what look like pinprick holes all over the surface of the leaf. These are actually oil ducts that can be seen when held up to the light. As the plant matures, these clear "holes" will turn a dark purple and will become filled with the medicinal oil of the plant.

## Growing Conditions

Saint John's wort likes to grow in full sun in well-drained, poor soil. Wild stands of Saint John's wort are often found on roadsides, old railroad beds, and parking lots. They are not picky, pampered plants but are scrappy survivors! When cultivating, it is best to plant Saint John's wort in beds that are well drained and not overly rich in nitrogen.

## Propagation

Saint John's wort can be propagated by direct or indirect seeding. The seeds of Saint John's wort are incredibly small and are light-dependent germinators. They should be sown on the surface and planted in sandy, well-drained soil. If indirect-seeding, consider amending potting medium with some perlite or sand to help with drainage. On our farm we prefer to indirect-seed. Indirect seeding is beneficial because it helps the Saint John's wort become established before weed pressure becomes an issue.

## Planting Considerations

Care should be given when planting Saint John's wort, taking into consideration its noxious weed status. In Vermont where we live, Saint John's wort grows well but does not become invasive. However, this is not true in other climates. One Saint John's wort plant can produce thousands of seeds and in the right environment can spread and become problematic. The western states are areas where Saint John's wort has become a significant challenge, and it is not recommended to cultivate it in these regions. We plant plugs into the field at twelve-inch spacing within rows and fourteen inches between rows, with three rows per bed.

## Medicinal Uses

Saint John's wort, despite the controversy of its noxious weed status, is a wonderful healing plant. It is used to make medicinal oils and liniments, and topically, nothing compares to its ability to sooth irritated and inflamed nerve endings and damaged tissue. We have included Saint John's wort oil in our healing salves for years and used the liniment on our tired and sore backs after long days of farming. Internally, Saint John's wort is healing and uplifting to the nervous system as well. It helps, when combined

**Figure 18-61.** Saint John's wort flowers. Photograph courtesy of Kate Clearlight

**Figure 18-62.** Medicinal oils made from Saint John's wort flowers. Note the deep red color. Photograph courtesy of Kate Clearlight

with other nervines, to alleviate mild depression, especially depression caused by the change of seasons and lack of sunshine. It is often said that the golden blossoms of Saint John's wort capture the rays of the sun and bring them into its healing medicine.

## Harvest Specifications

Saint John's wort is harvested when it is beginning to bloom and some of the flowers in the cluster are fully open and others are still in closed bud form. This is when the flowers (and leaves) are full of the highly coveted, bright red hypericum oil. To harvest use a field knife to cut the flowering tops four to six inches long from the plant. Some people want only the blossoms, but the leaves also contain the medicinal properties, and we like to include some of them as well.

## Postharvest and Drying Considerations

Lay the Saint John's wort tops out in a thin layer on drying racks. Dry at temperatures of 100 to 110°F in partial shade with good airflow. Saint John's wort is easy to work with and usually is dry in two days under optimum conditions. Once they are dried, gently rub the tops over quarter- to half-inch stainless

steel mesh to separate the stems from the leaves and flowers. The leaves and flowers are sold together unless customers request blossoms only. In that case it's best to harvest and dry the blossoms alone and not try to separate blossoms from the mixture.

## Pests and Diseases

The most common pests and diseases that impact Saint John's wort are the Klamath beetle (Chrysolina beetle) and the disease anthracnose (*Colletotrichum* spp.). The Klamath beetle is a dark, shiny beetle that feeds off the plant and can significantly decrease yields. We find that picking the bugs off is the best course of action. We also plant extra stands of Saint John's wort every year to compensate for the possibility of beetle pressure. The anthracnose shows up on the plant as rusty leaves and stems. Anthracnose is a fungal disease that can survive in seeds and be transferred by insects coming in contact with infected plants. Some commercial farmers recommend fungicides to combat the anthracnose. We do not apply fungicides. Instead, infected plants are carefully removed from the fields and disposed of. We also quarantine areas that have been impacted with anthracnose. Raw neem oil is another method we are experimenting with to help plants defend themselves against disease.

## Yields

One hundred pounds of dried Saint John's wort flowering tops per one-eighth-acre bed. Moisture ratio for Saint John's wort tops is 4:1 fresh:dried.

## Pricing

Retail price for one pound organic:
- Dried Saint John's wort flower: $9 to $28
- Fresh Saint John's wort flower: $12

## Commonly Imported From

Croatia and Chile

# Thyme, German winter    *Thymus vulgaris*

## Life Cycle

Thyme is a woody perennial that is hardy to USDA zones 5 to 9.

## Plant Description

Thyme has a rich history of use and is native to Europe, North Africa, and Asia. Thyme is a mint and a member of the Lamiaceae family. It is a short, bushy shrub that has rough, woody branches and grows six to twelve inches tall. The small leaves of thyme are narrow, lance shaped, and deep green in color. Thyme foliage is extremely fragrant and contains a valuable phenol called *thymol* that has antiseptic properties. Thyme flowers consist of clusters of white and purplish blossoms that are located at the end of the branches.

## Growing Conditions

Thyme is a sturdy plant. It likes to grow in full sun in well-drained, loamy soil. It can also grow at higher elevations and can overwinter in cold climates if mulched well. Overall thyme is a low-maintenance plant, provided it is not grown in wet, weedy areas.

## Propagation

Thyme can be grown by direct seeding, vegetative cuttings, root divisions, or transplants. Because thyme germinates easily we seed it early in the spring into plugs and transplant them out into fields when the plants are well established. Thyme, as a rule, does not outcompete weeds easily. Therefore, it is recommended to transplant thyme into clean, weed-free beds and to use mulch. Recommended plant spacing for thyme is twelve inches between plants, in triple rows with fourteen inches between rows within the bed.

## Planting Considerations

Thyme may be diminutive in size, but it has hearty, delightful aromatics and strong healing properties. We like to plant thyme next to other low-growing perennials such as oregano, sage, and lavender. The plants have a genuine affinity for one another and do not tend to crowd one another out. Thyme is also beloved by bees and other pollinators and is a great selection for gardens located near hives. Not only will it produce tasty honey, but the antimicrobial components of the plant are thought to be beneficial to the health of the insects.

## Medicinal Uses

Thyme has a rich history of use throughout Europe, the Mediterranean, and Asia. *Thumus*, the Greek word from which our word "thyme" was derived, means courage and speaks of a time when thyme was used in bouquets given to soldiers before going

**Figure 18-63. Thyme.** Photograph by Jorge Moura/Ecos de Pedra

into battle. The thyme was thought to strengthen resolve and to ward off evil and sickness. Extremely fragrant and rich in essential oils, thyme is a staple in cooking, bringing a warm and pungent flavor to food. It is also used medicinally in teas, bronchial steams, cough syrups, washes, and extracts. Thyme has antimicrobial and antispasmodic properties and is often used to treat whooping cough, bronchial infections, and sore throats. In addition to its antiseptic qualities, thyme has a carminative effect that can stimulate digestion and help dispel gas.

## Harvest Specifications

We harvest thyme as it begins to flower in late summer. This is when the volatile oils are at their highest levels and the plant has reached the apex of its growth cycle for the season. We have found it easiest to cut the aerial parts of the plant using very sharp pruners, rather than trying to saw through the woody stems with field knives (which can be time consuming and hard on the arms). When harvesting be sure to leave two to three inches of the plant intact to allow for regeneration and to help the plant overwinter.

## Postharvest and Drying Considerations

To dry thyme lay the flower stalks in a single, compact layer on drying racks, and dry at temperatures of 100 to 110°F. It is best to keep thyme out of direct light to help maintain vibrant color and medicinal quality. In good drying conditions thyme should dry in a couple of days. When dry, the leaves will rub off easily from the stems. Run stalks over quarter-inch stainless steel mesh, and separate the leaves from the stems.

## Pests and Diseases

In general, thyme is extremely disease resistant and not susceptible to many pests. Large commercial growers do report having issues with leaf spot and some fungal disease. Good site selection and crop rotation and management are the best course of action for growing healthy crops and avoiding difficulty.

## Yields

Eighty to one hundred pounds of dried thyme leaves per one-eighth-acre bed. Moisture ratio for thyme leaves is 3:1 fresh:dried.

## Pricing

Retail price for one pound organic:
- Dried thyme leaves: $9.50 to $24

## Commonly Imported From

Eygpt

# Tulsi/Holy Basil    *Ocimum tenuiflorum*, syn. *O. sanctum*

## Life Cycle

Tulsi is an annual or short-lived tropical perennial that will often self-seed and produce successive generations of plants when allowed to go to seed and given ample space and good conditions to grow.

## Plant Description

Tulsi is a fragrant and delicious member of the Lamiaceae family and is native to India, growing prolifically throughout Asia. There are several varieties of tulsi, all of which are native to India. Many have naturalized in tropical and subtropical latitudes around the world. Tulsi is an erect, bushy plant that grows about two feet tall. Depending on the variety, the leaves range from green to purple and are lance shaped. The flowers grow on spikes and are a deep purple with stamens that are a vibrant, eye-popping orange. One of the most delightful aspects of tulsi is its wonderful, heady scent.

## Growing Conditions

Tulsi likes growing in rich soils with ample moisture in full sun. In tropical climates plants can benefit from protection from the hot afternoon sun. In temperate zones at least six hours of full sun are recommended. Tulsi is fairly drought tolerant when mature and extremely sensitive to frost.

## Propagation

Tulsi can be propagated by seed or by vegetative cuttings. Our preferred method is by indirect seeding, which is dependable but often requires patience because seeds often take up to three weeks to germinate. Saving seeds from tulsi is easy in that not only are they prolific seeders but the seed is also easy to collect and clean, allowing even the rookie seed collector to feel like a seasoned vet. In Vermont we indirect-seed into flats and transplant out after the last frost and when true leaves and a substantial root system are formed. In warmer, more tropical climates direct seeding is possible, but we prefer to transplant plugs of tulsi to give it a jump on weeds and to extend our shorter season.

## Planting Considerations

We plant tulsi at a twelve-inch spacing in the row with rows spaced twenty-eight inches apart, two rows to the bed. This spacing allows the tulsi to grow into a dense, compact hedge that is attractive to us because of its purple flowers and also attractive to the bees that can't seem to get enough of its sweet nectar and pollen.

## Medicinal Uses

Tulsi is a heavenly adaptogenic herb that makes delicious teas, tinctures, and elixirs. The lore around tulsi is extremely beautiful and speaks to the holy nature of this "sacred basil." Known as the "incomparable one" in Sanskrit, tulsi is used in religious ceremonies to help bring souls to heaven, is believed by some to have sprung from the tears of a deity, and is thought to bring good luck when planted near homes and dwellings. As a tonic tulsi builds energy, is uplifting to the spirit, and brings a sense of wholeness and well-being. It is one of our favorite tea herbs on the farm, and we drink it daily. Tulsi, besides having a delightful flavor, is also good for releasing stress and easing anxiety and acts as an anti-inflammatory. Externally it can be used to make fantastic hydrosols, spritzers, or skin washes. Its aromatics are refreshing and ease feelings of debilitation and exhaustion. Its antimicrobial properties help to tone and heal the skin.

## Harvest Specifications

Harvesting tulsi by hand could hardly be called work. Although the plant's stalks get somewhat woody, cutting them with a field knife is not extremely difficult, and the harvester is rewarded by the clovelike, camphorous aroma exuded from the plants that is both calming to the mind and energetic to the body. Tulsi is surprisingly dense and heavy when compared to its diminutive stature. This is due to the high moisture and essential oil content in its leaves. We harvest the plants when they are flowering and before they have gone to seed. We cut the entire aerial part of the plants, leaving six to eight inches to allow the plant to regenerate. It is possible in our growing season of approximately 150 frost-free days to get three or four cuttings from a single crop. Tulsi is both generous in spirit and in the harvest.

## Postharvest and Drying Considerations

Because of its high moisture and essential oil content, tulsi should be processed immediately after harvest or it will degrade quickly. If you are selling tulsi fresh, it is essential to cool the plant down quickly to avoid composting. This can be done by laying out the fresh harvest on tarps in a shady, cool basement or barn. If you are drying it, lay the tulsi out on

**Figure 18-64.** Tulsi leaf and flower.

drying racks in a single layer with good airflow. Dry tulsi at temperatures of 100 to 110°F. Unlike most other leaf crops, tulsi takes a long time to dry. It is not unusual for tulsi to take over a week to dry completely. Because it has to be in the drying shed for a longer amount of time, we like to rack the tulsi at lower levels to help retain essential oil content that can be volatized in the higher, warmer drying racks.

## Pests and Diseases

Tulsi has minimal pests and diseases and we have not experienced any.

## Yields

Three hundred pounds of dried tulsi tops per one-eighth-acre bed (multiple harvests per season). Moisture ratio for tulsi tops is 4:1 fresh:dried.

## Pricing

Retail price for one pound organic:
- Dried tulsi leaf/flower: $16 to $30
- Fresh tulsi leaf/flower: $11

## Commonly Imported From

India

# Valerian                                         *Valeriana officinalis*

## Life Cycle

Valerian is a short-lived perennial hardy to USDA zones 3 to 9.

## Plant Description

The valerian family was introduced to the United States from Europe and has naturalized in cool, rich, moist places generally in North America's northern latitudes. Plants start flowering during their second year of growth and generally only live for three to five years before life force wanes and they die. Valerian can be a vigorous self-seeder, so progeny can easily replace parent plants if provided a niche in which to grow. Valerian has deeply divided pinnate leaves emanating from a basal rosette during the first year of growth and forms multiple erect flowering stalks up to six feet tall in subsequent years. The flowers are ray shaped and range from pure white to pink on multiple umbrella-shaped inflorescences. Depending on whom you ask, the fragrance of valerian is either loved or despised, and rarely is there indifference expressed about it. We think the smell is interesting and intoxicatingly beautiful; others, not so much. The pollinators

seem to agree with us, as its scent attracts some of the most obscure and interesting-looking species of bees, wasps, moths, butterflies, flies, and beetles.

Valerian rhizomes resemble a small rag mop in their appearance from the numerous and thin white rootlets branching off the main rhizome. Like the valerian flower, the fresh root is known for its intense aroma, which many people find offensive and comparable to the odor of dirty socks or mold. We are in the minority and think it smells sweet, strong, and earthy. Dried roots, on the other hand, definitely push the boundaries of "interesting" smells, and even we plant lovers find them somewhat stinky, if not downright offensive. Reports abound of the origin of the slang phrase "p-u" used to describe something that smells offensive. Apparently the Roman physician Dioscorides called valerian "Phu," which gave rise to the word association. Whether that is true has been debated, but it certainly makes for fascinating herbal lore.

## Growing Conditions

Valerian likes to grow in partial shade but can tolerate growing in the full sun in cooler climates. It

**Figure 18-65. Valerian in flower.** Photograph courtesy of Bethany Bond

loves fertile, moist, nitrogen-rich soil. Droughty conditions often stunt its growth and force plants into early dormancy or worse.

## Propagation

Many sources say that valerian is difficult to start from seed, but we have not found that to be the case. Fresh seed definitely germinates best and is a light-dependent germinator that should either be sown on the surface or very lightly covered. Valerian can be sown in flats and transplanted or direct-seeded into the field. We prefer to transplant it because of our relatively short growing season. This gives us an earlier start that translates into bigger roots come harvest time. The roots can also be divided and replanted during the plant's dormant period in early spring or late fall, which is an easy way to make many plants from a few plants.

## Planting Considerations

Valerian grows well with other species, such as vervain, marshmallow, and boneset. We often interplant them within the same field but in beds separated by species for easier harvest. They have similar soil requirements and are quite beautiful to see growing together. Valerian is planted at twelve inches between plants and fourteen inches between rows, with three rows per bed.

**Figure 18-66.** Valerian roots.

## Medicinal Uses

Valerian root is an effective and safe nervine seda-tive herb for most people. A very small percentage of people who try valerian experience paradoxical effects and find it stimulating instead of sedating. Fortunately those who benefit from its sedating qual-ities do not find it habit forming, as most narcotic sleep medications can be. Valerian is also useful as a mild anxiolytic (reduces anxiety) and antidepressant. It is one of the most popular and well-known herbal remedies in the Western world, which lends credence to speculation that most of us could use more sleep and less anxiety. Valerian is also reported to be an effective analgesic.

## Harvest Specifications

We have found that the most potent roots are har-vested in early spring of the second or third year of growth, before the roots start putting their energy into leaf, stalk, and flower formation. During this time, when the plants are still dormant, the rootlets resemble masses of thick, fleshy linguine pasta. If we wait too long and the root starts sending up aerial growth, the rootlets shrink and become more angel-hair pasta–like and much less potent. Roots dug in the fall also seem less potent and lower yielding than early-spring-dug roots.

## Postharvest and Drying Considerations

Soil often becomes compacted in the central crown of valerian rhizomes where the rootlets begin to branch off. Therefore it is helpful to quarter the rhizomes before washing. Valerian roots are relatively soft and can be chopped easily with a field knife. After splitting the roots, wash them thoroughly before drying. Mill roots after they are fully dried. Valerian roots dry under optimum conditions in three to four days at temperatures between 90 and 110°F.

## Pests and Diseases

Deer lightly browse on our valerian plants but not enough to concern us. They seem to take a nibble-here-nibble-there approach, and we suspect they may be utilizing the medicine the same way humans do. After all, it's gotta be a little stressful being a deer in this day and age dodging cars and people. Other than deer, we have seen no pest damage on valerian plants. There are reports of fungal diseases affecting valerian plants in commercial production, but we have not seen any signs of disease on our plants in fifteen years of growing them.

## Yields

130 pounds of dried valerian root per one-eighth-acre bed. Moisture ratio for valerian root is 4:1 fresh:dried.

## Pricing

Retail price for one pound organic:
- Dried valerian root: $10 to $20
- Fresh valerian root: $14

# Yarrow

## *Achillea millefolium*

## Life Cycle

Yarrow is an herbaceous perennial that is hardy to USDA zones 2 to 9. It often begins flowering in its first year of growth and sets seed in the fall.

## Plant Description

Yarrow is from the Asteraceae family. Yarrow's bipinnate, fernlike leaves are borne from laterally spreading rhizomes that support multiple flower stalks growing one to three feet tall. Individual ray- or disk-shaped flowers bloom profusely on inflorescences from midsummer through fall and range in color from pure white to pale pink and even yellow. The entire plant is pleasantly aromatic and reminds us of the smell of chrysanthemums. There are also many colorful ornamental varieties of yarrow grown and sold as landscape plants. These plants are not as medicinally active as the wild variety. Yarrow is native to Europe and western Asia and has naturalized throughout the earth's temperate zones.

## Growing Conditions

Yarrow is tolerant of and often found thriving in the seemingly poorest soils. It grows best in full sun, is extremely drought tolerant, and will not tolerate growing in poorly drained soils. Yarrow prefers sandy, gravelly loam but will perform well in rich garden soils where higher fertility generally increases leaf and flower yields.

## Propagation

Yarrow is easily grown from transplants or by direct seeding. The seeds are light-dependent germinators that should be sown directly on top of the soil surface or lightly covered and should germinate in seven to fourteen days. Mature plants can also be divided by chopping and removing sections of the rhizomes, then replanting them.

## Planting Considerations

Yarrow is one of the premier insectary plants known and used because of its attractiveness to a wide variety of beneficial insects. Growers can take advantage of this incredible attribute by planting yarrow in close proximity to other species of plants that are prone to damage from herbivorous insects or the diseases they can carry with them. We plant yarrow at twelve-inch spacing in rows fourteen inches apart with three rows per bed. The plants quickly spread to form a single, dense row that requires very low maintenance and is long lived. We mechanically cultivate between the rows, but the plants within the rows themselves require little to no weeding because of their dense-growing rhizomatous nature.

## Medicinal Uses

Yarrow has a rich history of medicinal use dating back at least as far as the Upper Paleolithic period—remains of a Neanderthal man found in Iraq and dated at more than fifty thousand years old were found with traces of yarrow stems and leaves embedded in his teeth. Greek myths tell the tale of yarrow being given to Achilles by the centaur Chiron to use during battle. Yarrow was referred to as "soldiers' woundwort" or "stanchwort" by soldiers on battlefields in ancient armies because of its ability to stanch bleeding. Native Americans were said to have used the flowers of yarrow as a deodorant. Yarrow is a bitter tonic herb containing tannins, up to 1.4% volatile oil, lactones, flavonoids, and up to a dozen different anti-inflammatory compounds. It is one of our "go-to" herbs for cold and flu season because of its strong diaphoretic properties that assist our bodies in reducing fever and eliminating pathogens. Yarrow is also the first choice in our herbal first-aid kit for its amazing effectiveness in stopping bleeding. It is inevitable that one or more

**Figure 18-67. Yarrow flower.** Photograph courtesy of Bethany Bond

**Figure 18-68. Yarrow leaf.** Photograph courtesy of Bethany Bond

members of our farm crew will cut themselves, usually when cutting herbs with a field knife. Yarrow is always around, growing somewhere close by when we need it. By picking a few leaves, applying a styptic poultice to the wound, then wrapping it with a plantain leaf, we can stanch the bleeding until further treatment is available.

## Harvest Specifications

Yarrow leaf and flower is harvested together when the plants are in the early stages of flowering by hand-cutting with a field knife or mechanically with a sickle bar cutter. By hand is our preferred choice of harvesting this incredibly aromatic herb.

## Postharvest and Drying Considerations

Yarrow leaf and flower dries rapidly in as little as twenty-four hours during ideal drying conditions of low ambient humidity and is very easy to process by hand garbling. Large flower stalks are easily removed, and the leaf and flower appearance, aromatic qualities, and overall potency hold up incredibly well during prolonged storage of a year or more. If only every other herb was as easy to harvest, dry, process, and store as yarrow, the world of

herb farming would be a much happier place than it already is.

## Pests and Diseases

We know of no pests that eat yarrow, other than a few spittlebugs who don't seem to do any harm at all, other than leaving their foamy "saliva" for us to wipe off our hands as we harvest the herb. Some farmers report that yarrow is prone to the fungal diseases botrytis and powdery mildew, but in fifteen years of growing yarrow we have seen no sign of infection on either our cultivated or wild-growing plants.

## Yields

One hundred pounds of dried yarrow tops per one-eighth-acre bed. Moisture ratio for yarrow tops is 5:1 fresh:dried.

## Pricing

Retail price for one pound organic:
- Dried yarrow leaf/flower: $11 to $26
- Fresh yarrow leaf/flower: $9 to $13

## Commonly Imported From

Bulgaria

# Yellow Dock                              *Rumex crispus*

## Life Cycle

Yellow dock is an herbaceous perennial hardy to USDA zones 3 to 9.

## Plant Description

Often called "curly dock" because of the "wavy" nature of its leaf edges, yellow dock is a fairly common weed from the buckwheat family, Polygonaceae. It is native to Europe and western Asia and has naturalized throughout the temperate latitudes. Yellow dock is often considered invasive because of its prolific nature and has thus earned the distinction of being listed as an "injurious weed" under the UK Weeds Act of 1959. The plant sends up a two- to three-foot flowering stalk during the first or second year of growth supporting clusters of small, greenish-white, petalless flowers that mature to form copious amounts of shiny brown seeds. These seeds are encased in dried flower calyxes, which allows them to cling easily to passing mammals whereby they are readily dispersed. Wavy, lance-shaped leaves emanate from a basal rosette, and roots are multiple-branching taproots with brown-colored skin and a golden-yellow fleshy interior.

## Growing Conditions

Yellow dock is often found in disturbed soils in the full sun or partial shade with ample moisture and organic matter.

## Propagation

Propagate yellow dock by seed or root division. The hard-shelled seeds are sporadic germinators that can benefit from having their seed coats scarified or nicked to allow them to absorb water faster. Expect germination to occur sporadically in one to four weeks. We sow our yellow dock seed early in the unheated greenhouse and feel that it benefits from cold conditioning (stratification), although we have found no specific recommendations from seed producers or other growers that recommend stratification.

## Planting Considerations

Yellow dock can yield a large, harvestable-size root in as little as one season, if planted early and harvested late. It should be planted approximately twelve inches apart in the row at a fourteen-inch row spacing to avoid competition and yield large roots. Direct seeding is possible but can be challenging because of the sporadic nature of seed germination. Young seedlings can be outcompeted by more aggressive weeds if germination is delayed. Yellow dock's vigorously growing taproots can help break up compacted soils.

## Medicinal Uses

Yellow dock is primarily grown and harvested for its roots, which are bitter, astringent, and beneficial in assisting the human body with assimilating iron. Small amounts of anthraquinone glycosides also found within the roots are considered by some to be a mild laxative. These glycosides, along with bitter compounds, make yellow dock root an excellent tonic for stimulating digestive enzymes and peristalsis of the bowels, clearing stagnation, and helping to support healthy digestion. The roots and leaves are cooling and drying and are often used to help clear skin conditions such as psoriasis and other types of rashes. Yellow dock leaves harvested in early spring while they are still tender are highly nutritive due to their significant amounts of iron, calcium, beta-carotene, protein, and potassium. They are pleasantly tasty when steamed or boiled, albeit a bit tart because of their oxalic acid content. This necessitates being careful to limit consumption because of the potentially mild toxicity of this compound. Rinsing the greens several times before cooking can help to leach and remove some of the oxalic acid. People

**Figure 18-69.** Yellowdock. Photograph by Harry Rose

with a history of oxalate kidney stones should avoid ingesting the leaves altogether. Mature leaves become extremely bitter, tough, and unpalatable.

## Harvest Specifications

Leaves are harvested when young during the first or second season of growth by cutting with a field knife or mechanically with a sickle bar cutter. Roots are best dug while the plant is dormant in the fall of the first year of growth or the spring or fall of the second year. Roots tend to become pithy and less potent after two seasons of growth and therefore should be harvested within two seasons. Digging roots can be done by hand with a digging fork or with a tractor-mounted bed lifter or mechanical root digger. It is possible to harvest a small portion (25 percent or less) of each root and leave the rest of the plant in the ground for a more sustainable harvest without killing the plants.

## Postharvest and Drying Considerations

Because yellow dock roots are long taproots and only branch off in a few places, they do not "collect dirt" the way other root crops do and are very easy to clean and process. Simply chop large roots into pieces (either diagonally or in cross sections) and

wash. Yellow dock roots are relatively soft and can easily be chopped with a field knife. After chopping the roots wash them thoroughly, then dry them at approximately 100°F. Yellow dock roots dry under optimum conditions in three to four days. Mill or chip roots after they are fully dried.

## Pests and Diseases

Flea beetles seem to enjoy munching on the young leaves but do not noticeably affect the vigorous growth and potential yields of this tough plant. We know of no other pests or diseases that challenge yellow dock.

## Yields

Five hundred pounds of dried yellow dock roots per one-eighth-acre bed. Moisture ratio for yellow dock root is 5:1 fresh:dried.

## Pricing

Retail price for one pound organic:
- Dried yellow dock root: $11 to $20
- Fresh yellow dock root: $9 to $13

## Commonly Imported From

Poland

# Appendix A

**Table A-1.** United Plant Savers Herbs and Analogs

| American Ginseng *Panax quinquefolius* | At Risk | Purchase organic roots only—even so-called "woods-grown" is suspect. |
|---|---|---|
| Chinese Ginseng *Panax ginseng* | | For increasing energy, stamina, fortifying immune system. |
| Siberian Ginseng *Eleutherococcus senticosus* | | Adaptagenic, normalizes energy levels, increases productivity, immunopotentiating. |
| Astragalus *Astragalus membranaceus* | | Protective and recuperative immune tonic. |
| Ashwaganda *Withania somnifera* | | Important ayurvedic herb for strengthening sexual energy and replenishing nervous system exhaustion. |
| Oats *Avena sativa* | | Replenishes nervous system. |
| **Arnica *Arnica* spp.** | **To Watch** | **Cultivated sources only; use all aerial parts rather than just flowers; *Arnica montana* is threatened in its entire range in Europe.** |
| Comfrey *Symphytum officinale* | | For musculoskeletal concerns. |
| Yarrow *Achillea millefolium* | | For bruising. |
| Calendula *Calendula officinalis* | | Anti-inflammatory, vulnerary. |
| Saint Johns Wort *Hypericum perforatum* | | Restores damaged nerve tissue, analgesic. |
| Rescue Remedy: Bach flower essence formula | | Topically and internally reduces trauma. |
| **Black Cohosh *Cimicifuga racemosa*, syn. *Actaea racemosa*** | **At Risk** | **Most in commerce is wildcrafted.** |
| Vitex *Vitex agnus-castus* | | Hormone balancing. |
| Motherwort *Leonurus cardiaca* | | Relieves anxiety, lifts spirits. |
| Skullcap *Scutellaria lateriflora* | | Nerve tonic, analgesic, alleviates mood swings, relieves anxiety. |
| **Bloodroot *Sanguinaria canadensis*** | **At Risk** | |
| Celandine *Chelidonium majus* | | Dissolves warts, also contains sanguinarine. |
| Tumeric *Curcuma longa* | | May resolve skin cancer, anti-inflammatory and astringent to gum tissue. |
| Self-Heal *Prunella vulgaris* | | Anti-inflammatory and astringent to gum tissue. |
| Spilanthes *Spilanthes* spp. | | Stimulating, decay fighting mouthwash, tonifies gums. |
| **Blue Cohosh *Caulophyllum thalictroides*** | **At Risk** | |
| Motherwort *Leonurus cardiaca* | | Uterine tonic. |
| Raspberry leaf *Rubus* spp. | | Uterine tonic. |
| Cottonroot bark *Gossypium herbaceum* | | Oxytocic, promotes or accelerates childbirth by stimulating uterine muscles. Caution: Not to be taken during pregnancy. |
| Pennyroyal *Mentha pulegium* | | Emmenagogue. |
| Senna *Cassia hebecarpa* | | Contains anthroquinoines, powerful laxative. |
| Other *Rhamnus* species | | |
| Psyllium and Flax seeds | | Bulk laxatives. |

| **Echinacea** *Echinacea purpurea* | At Risk | Use only cultivated sources, very available. |
|---|---|---|
| Usnea *Usnea* spp. | | Antibacterial, antifungal, antiviral. Collect windblown specimens on forest floor after a good winter storm. |
| Thyme *Thymus vulgaris* | | Antibacterial, antifungal, antiviral, immune enhancing. |
| Spilanthes *Spilanthes* spp. | | Immune tonic, antibacterial, antifungal, antiviral. |
| Astragalus *Astragalus membranaceos* | | Protective and recuperative immune tonic. |
| Boneset *Eupatorium perfoliatum* | | Relieves aches and pains of flu. |
| Marshmallow *Althea officinalis* | | Immune tonic and restorative. |
| **Elephant Tree** *Bursera microphylla* | To Watch | |
| Myrrh *Commiphora myrrha* | | Stimulates the immune system, disinfectant, astringent, anti-inflammatory, used as incense. |
| Yarrow *Achillea millifolium* | | Immune stimulant, disinfectant, astringent. |
| **Eyebright** *Euphrasia* spp. | At Risk | In the United States it is scarce. In Europe there is still ample supply. |
| Chamomile *Matricaria recutita* | | Soothing eye wash. |
| Self-Heal *Prunella vulgaris* | | Soothing to mucous membranes. |
| Mugwort *Artemisia vulgaris* | | Digestive bitter. |
| Yarrow *Achillea millifolium* | | Digestive bitter, febrifuge, antiseptic. |
| Dandelion *Taraxacum officinale* | | Strengthens the digestive system, bitter principles. |
| **Goldenseal** *Hydrastis canadensis* | At Risk | Presently listed with CITES. |
| Barberry *Berberis vulgaris* | | Contains berberine, alterative, antimicrobial. |
| Usnea *Usnea* spp. | | Heals topical infections, eyewash, antibiotic, antiviral. |
| Garlic *Allium* spp. | | Antibiotic, dries mucous membranes. |
| Plantain *Plantago* spp. | | Antiseptic, heals wounds, alterative. |
| **Goldthread** *Coptis* spp. | To Watch | |
| Usnea *Usnea* spp. | | Heals topical infections, eyewash, antibiotic, antiviral. |
| Garlic *Allium sativum* | | Antibiotic, dries mucous membranes. |
| Plantain *Plantago* spp. | | Antiseptic, heals wounds, alterative. |
| **Helonias Root** *Chamailirium luteum* | At Risk | Virtually non-existent at this point. |
| Motherwort *Leonurus cardiaca* | | Reproductive stimulant. |
| Vitex *Vitex agnus-castus* | | Hormone balancing. |
| Raspberry *Rubus* spp. | | Uterine tonic. |
| **Kava, Wild Hawaiian** *Piper methysticum* | To Watch | Use cultivated sources only. |
| Chamomile *Matricaria recutita* | | Gently calming and soothing. |
| Mugwort *Artemisia vulgaris* | | Digestive bitter, induces vivid dream states. |
| Valerian *Valeriana officinalis* | | Sedating nervine. |
| Hops *Humulus lupulus* | | Sedative, digestive bitter. |
| California Poppy *Eschscholzia californica* | | Mood altering nervine. |
| **Lady's Slipper** *Orchid Cypridium* spp. | At Risk | Delicate forest orchid; use analogs, let it grow! |
| Lemon Balm *Melissa officinalis* | | Antispasmodic, nervine. |
| Skullcap *Scutellaria lateriflora* | | Antispasmodic, nervine, sedative, anodyne. |
| Valerian *Valeriana officinalis* | | Sedating nervine. |

| | | |
|---|---|---|
| **Lobelia** *Lobelia inflata* | **To Watch** | Use cultivated; easy to grow. |
| Lobelia *Lobelia cardinalis* | | Milder action, use cultivated. |
| Thyme *Thymus vulgaris* | | Antispasmodic, expectorant. |
| Hyssop *Hyssop officinalis* | | Expectorant. |
| Violet *Viola odorata* | | Expectorant. |
| Skullcap *Scutellaria lateriflora* | | Antispasmodic nervine. |
| **Lomatium** *Lomatium dissectum* | **At Risk** | Limited range. |
| Echinacea (cultivated) *Echinacea purpurea* | | Antiviral. |
| Saint John's Wort *Hypericum perforatum* | | Antiviral. |
| Rosemary *Rosmarinus officinalis* | | Respiratory tonic, antiseptic, diaphoretic, antibacterial. |
| **Oregon Grape** *Mahonia* spp. | **To Watch** | Though prolific in Pacific Northwest, it has a limited range. |
| Garlic *Allium* spp. | | Antibiotic, dries mucous membranes. |
| Dandelion *Taraxacum officinale* | | Hepatic. |
| Yarrow *Achillea millefolium* | | Topical antiseptic. |
| Barberry *Berberis vulgaris* | | Contains berberine, alterative, antimicrobial. |
| **Osha** *Ligusticum porteri, L.* spp. | **At Risk** | Very limited range. |
| Thyme *Thymus vulgaris* | | Anti-inflammatory, antiseptic, antibiotic. |
| Elecampane *Inula helenium* | | Expectorant, respiratory tonic. |
| Marshmallow *Althea officinalis* | | Demulcent, soothing to irritated mucous membranes. |
| Lovage *Levisticum officinalis* | | For respiratory conditions, antiseptic, diaphoretic, antibacterial, antifungal, antispasmodic. |
| Rosemary *Rosmarinus officinalis* | | Clears mucous, astringent. |
| **Partridgeberry** *Mitchella repens* | **To Watch** | Delicate, slow growing; when harvested, roots are often pulled too, thus taking the whole plant. |
| Motherwort *Leonurus cardiaca* | | Uterine tonic. |
| Raspberry *Rubus* spp. | | Uterine tonic. |
| Catnip *Nepeta cataria* | | Antispasmodic. |
| Oats *Avena sativa* | | Promotes fertility. |
| **Pipsissewa** *Chimaphila umbellata* | **To Watch** | |
| Uva Ursi *Arctostaphylos uva-ursi* | | For urinary tract infections (UTI). |
| Goldenrod *Solidago* spp. | | For UTI. |
| Yarrow *Achillea millefolium* | | Antiseptic to UTI. |
| **Slippery Elm** *Ulmus rubra* | **At Risk** | Limit wild harvest to trees struck by natural disaster (storms, etc.). |
| Other *Ulmus* species | | |
| Mullein *Verbascum* spp. | | Demulcent, antitussive, respiratory tonic. |
| Violet *Viola* spp. | | Demulcent, antitussive, respiratory tonic. |
| Marshmallow *Althea officinalis* | | Demulcent, antibacterial, antitussive, normalizes digestion. |
| **Spikenard** *Aralia racemosa, A. californica* | **To Watch** | |
| Ginseng (cultivated) *Panax quinquefolius* | | Adaptogen. |
| Siberian Ginseng *Eleutherococcus senticosus* | | Adaptogen. |
| Staint John's Wort *Hypericum perforatum* | | Antispasmodic. |

| Stoneroot *Collinsonia canadensis* | To Watch | Easy to cultivate. |
|---|---|---|
| European Horse Chestnut *Aesculus hippocastanum* | | Hemorrhoids, varicose veins. |
| Parsley Root *Petroselinum hortense* | | For kidney concerns, diuretic. |
| **Sundew *Drosera* spp.** | **At Risk** | **Very fragile, use analogs.** |
| Spilanthes *Spilanthes* spp. | | For respiratory complaints, antibacterial, antiviral, antifungal. |
| Sage *Salvia officinalis* | | Sore throats, antibacterial and antiviral actions. |
| Thyme *Thymus vulgaris* | | Antibacterial, antiviral, respiratory complaints. |
| **Trillium, Beth Root *Trillium* spp.** | **At Risk** | **Not available for sale anymore, let it grow!** |
| Motherwort *Leonurus cardiaca* | | Uterine tonic. |
| Yarrow *Achillea millefolium* | | Anti-hemorrhage. |
| Shephard's Purse *Capsella bursa-pastoris* | | Anti-hemorrhage, astringent. |
| Raspberry *Rubus* spp. | | Tonifies reproductive system. |
| **True Unicorn *Aletris farinosa*** | **At Risk** | **Let it rest, if given room it could make a comeback.** |
| Mint *Mentha piperita, M. spicata* | | Carminative. |
| Chamomile *Matricaria recutita* | | Carminative. |
| Raspberry leaf *Rubus idaeus* | | Uterine tonic. |
| **Turkey Corn *Dicentra canadensis*** | **To Watch** | |
| Spilanthes *Spilanthes* spp. | | Reduces tooth pain, helps heal mouth traumas. |
| Clove Bud *Syzygium aromaticum* | | Numbs the gums and eases pain. |
| Hops *Humulus lupulus* | | Sedating and calming to the nervous system. |
| Saint John's Wort *Hypericum perforatum* | | Heals and soothes the nerves. |
| **Venus Flytrap *Dionaea muscipula*** | **At Risk** | **Too fragile for wildcrafting.** |
| Echinacea (cultivated) *Echinacea purpurea* | | Immune stimulating. |
| Red Clover *Trifolium pratense* | | Antitumor. |
| **Virginia Snakeroot *Aristolochia serpentaria*** | **At Risk** | |
| Yucca *Yucca* spp. | | Joint conditions. |
| Dill *Anethum graveolens* | | For digestive concerns. |
| Fennel *Foeniculum vulgare* | | For digestive concerns. |
| Ginger *Zingiber officinale* | | For digestive concerns. |
| Echinacea (cultivated) *Echinacea* spp. | | For ingested poisons, poisonous bites, snakebite. |
| **White Sage *Salvia apiana*** | **To Watch** | **Harvest only sprouting tips.** |
| Garden Sage *Salvia officinalis* | | Soothes irritated mucous membranes, eye wash, throat soreness. |
| Mugwort *Artemisia vulgaris* | | Smudge. |
| Sagebrush *Artemisia tridentata* | | Smudge. |
| **Wild Indigo *Baptisia tinctoria*** | **To Watch** | **Not easy to grow.** |
| Echinacea (cultivated) *Echinacea purpurea* | | Immune enhancing, antimicrobial. |
| Spilanthes *Spilanthes* spp. | | Antimicrobial, fever reducing, antiseptic. |
| Cleavers *Galium* spp. | | Lymphatic, alterative. |

| Wild Yam *Dioscorea villosa, D.* spp. | At Risk | Greatly overharvested at this time. |
|---|---|---|
| Ginger *Zingiber officinale* | | Morning sickness. |
| Chamomile *Matricaria recutita* | | Antispasmodic. |
| Dandelion *Taraxacum officinale* | | Liver restorative. |
| Yerba Mansa *Anemopsis californica* | To Watch | Use cultivated supplies only, grows in polluted lowlands of California's agribusiness. |
| Self-Heal *Prunella vulgaris* | | Astringent to mucous membranes, soothing to mouth and throat inflammation. |
| Yarrow *Achillea millefolium* | | Antiseptic, anti-inflammatory, antimicrobial. |
| Usnea *Usnea* spp. | | Antiviral, antimicrobial. |
| Barberry *Berberis vulgaris* | | Contains berberine, alterative, antimicrobial. |

Note: This list is compiled by Jane Bothwell and does not necessarily reflect the opinions of United Plant Savers.

# Notes

## Chapter 1: Why Grow Medicinal Herbs?

1. Terri Hallenbeck, "A Look at Vermont Organic Farming," *Burlington Free Press*, September 6, 2009, 5D.
2. James A. Duke, "Medicinal Plants and the Pharmaceutical Industry," in *New Crops* by J. Janick and J. E. Simon (eds.), (Wiley, Perdue University, 1993), 664–669, www.hort.purdue.edu/newcrop/proceedings1993/v2-664.html.
3. Julie Denis, "2012 International Herb and Botanical Trends," *Nutraceuticals World,* July 1, 2012, www.nutraceuticalsworld.com/issues/2012-07/view_features/2012-international-herb-botanical-trends/.
4. Steven D. Johnson, "2012 Crop Input Costs Increase, Along with Profit Margin Opportunities," Iowa State University Extension and Outreach, www.extension.iastate.edu/agdm/articles/others/JohSept11.html.
5. D. R. Davis et al, "Changes in USDA Food Composition Data for 43 Garden Crops, 1950 to 1999," *Journal of the American College of Nutrition,* 23(6) (2004): 669–682.
6. Wayne Law and Jane Salick, "Human-induced Dwarfing of Himalayan Snow Lotus, *Saussurea laniceps* (Asteraceae)," *Proceedings of the National Academy of Sciences,* 102(29) (2005): www.pnas.org/content/102/29/10218.
7. Hank Schultz, "Herbal Supplement Sales Rose 5.5% in US in 2012, ABC Says," *NUTRA Ingredients-USA.com,* August 19, 2012, http://mobile.nutraingredients-usa.com/Markets/Herbal-supplement-sales-rose-5.5-in-US-in-2012-ABC-says#.VHzyMq5KG3U
8. Julie Denis, "2012 International Herb and Botanical Trends."
9. Ibid.
10. Ibid.
11. Ibid.

## Chapter 5: Field and Crop Considerations and Planning

1. Bill Mollison, *Introduction to Permaculture* (Tasmania: Tagari, 1991).
2. Courtesy of Rural Action Sustainable Forestry Program & Ohio State University. Adapted from Scott Persons, *American Ginseng: Green Gold* (Fairview, N.C.: Bright Mountain Books, 1994).

## Chapter 11: Pest and Disease Prevention and Control

1. C. Lamb and R. A. Dixon, "The Oxidative Burst in Plant Disease Resistance," *Annual Review of Plant Physiology and Plant Molecular Biology,* no. 48 (1997): www.ncbi.nlm.nih.gov/pubmed/15012264.
2. Erich Kombrink and Elmon Schmelzer, "The Hypersensitive Response and its Role in Local and Systemic Disease Resistance," *European Journal of Plant Pathology,* 107 (2001): 69–78, http://link.springer.com/article/10.1023%2FA%3A1008736629717.
3. Dale R. Walters et al., ed., *Induced Resistance for Plant Defense: A Sustainable Approach to Crop Protection* (Oxford: John Wiley & Sons, Ltd., 2014), 6.

## Chapter 12: Harvest

1. Dave Mengel, "Roots, Growth and Nutrient Uptake," Perdue University Department of

Agronomy Publications, no. AGRY-95-08 (1995), www.agry.purdue.edu/ext/pubs /AGRY-95-08.pdf.

2. Linnie Marsh Wolf, *John of the Mountains: The Unpublished Journals of John Muir* (Madison: University of Wisconsin Press, 1979), 317.

3. Rosemary Gladstar and Pamela Hirsch, ed., *Planting the Future: Saving Our Medicinal Herbs* (Rochester, Vt.: Healing Arts Press, 2000), ix.

4. Ibid., 1–2.

5. United Plant Savers Board of Directors, "Mission Statement," United Plant Savers, www.unitedplantsavers.org/content.php/163 -about-ups_1?s=f6189c7f7093232968c5d 51b5f180054.

6. United Plant Savers Board of Directors, "Journal of Medicinal Plant Conservation," United Plant Savers, www.unitedplantsavers.org /content.php/183-membership-resources.

7. Ryan Drum, "Wildcrafting Medicinal Plants," Ryan Drum Island Herbs, www.ryandrum.com /wildcrafting.htm.

8. Merle Zimmermann, "Good Stewardship Harvesting of Wild American Ginseng (*Panax quinquefolius*)," American Herbal Products Association, www.ahpa.org/portals/0/pdfs /ExportRules.pdf.

9. United Plant Savers Board of Directors, "Saving Wild American Ginseng," United Plant Savers, www.unitedplantsavers.org/content.php/326 -Species-at-Risk-American-Ginseng.

10. Merle Zimmermann, "Good Stewardship Harvesting of Wild American Ginseng," American Herbal Products Association, www.ahpa.org /default.aspx?tabid=154.

11. Janet Rock et al., "Harvesting of Medicinal Plants in the Southern Appalachian Mountains," *Journal of Medicinal Plant Conservation*, Winter (2012), www.unitedplantsavers.org /journal/2012/winter/files/images/pages /page12.swf.

12. Andy Hankins, "Producing and Marketing Wild Simulated Ginseng in Forest and Agroforestry Systems," Virginia Cooperative Extension, Virginia Tech, Virginia State University, 354-312 (2009): http://pubs.ext.vt.edu/354/354 -312/354-312.html.

13. Ibid.

## Chapter 13: Geo-Authentic Botanicals

1. D. E. Boufford and S. A. Spongberg, "Eastern Asian–Eastern North American Phytogeographical Relationships: A History from the Time of Linnaeus to the Twentieth Century," *Annals of the Missouri Botanical Garden* 70 (1983): http://flora.huh.harvard.edu/china /novon/eaena.htm.

2. Josef Brinckmann, interview with the author, April 1, 2014.

3. Ibid.

4. Ibid.

5. Gustavo F. Gonzales, "Of Maca and Men," in *Volume 10: Examples of The Development of Pharmaceutical Products from Medicinal Plants*, (United Nations Office for South-South Cooperation, n.d.), www.hersil.com.pe/pdf _estudio/V10_S2_OfMaca.pdf

6. Brinckmann, interview.

## Chapter 14: Postharvest Processing

1. Joachim Muller and Albert Heindl, "Chapter 17: Drying of Medicinal Plants," in *Medicinal and Aromatic Plants*, eds. R. J. Bogers, L. E. Craker, and D. Lange (Springer, 2006), 237–252; Jose Otavio Carrera Silva Junior et al., *Processing and Quality Control of Herbal Drugs and Their Derivatives: Quality Control of Herbal Medicines and Related Areas* (InTech, 2011), www.intechopen .com/books/quality-control-of-herbal -medicines-and-related-areas/processing-and -quality-control-of-herbal-drugs-and-their -derivatives; Robert Verpoorte et al, "Chapter 19: Plants as Sources of Medicines," in *Medicinal and Aromatic Plants*, eds. R.J. Bogers, L.E. Craker, and D. Lange (Springer, 2006), 261–273.

2. S. P. Cuervo-Andrade, "Quality Oriented Drying of Lemon Balm (*Melissa Officinalis L.*)" (PhD diss., Witzenhausen, 2011).

3. Sandra Patricia Cuervo-Andrade and Oliver Hensel, "Experimental Determination and Mathematical Fitting of Sorption Isotherms for Lemon Balm (*Melissa Officinalis L.*)," *Agricultural Engineering International: CIGR Journal* 15 (1) (2013): 139-145; V.B. Zagumennikov et al., "Studies of Total Ash and Moisture in Fresh *Echinacea Purpurea* Herb," *Pharmaceutical Chemistry Journal* 46, (10) (2013): 603–605; C.S. Ethmane Kane et al., "Moisture Sorption and Thermodynamic Properties of Two Mints: *Mentha Pulegium* and *Mentha Rotundifolia*," *Revue des Energies Renouvelables* 11 (2) (2008): 181–195; Ilknur Alibas, "Determination of Drying Parameters, Ascorbic Acid Contents and Color Characteristics of Nettle Leaves During Microwave, Air and Combined Microwave-air Drying," *Journal of Food Process Engineering* 33 (2010): 213–233; H. Machour et al., "Sorption Isotherms and Thermodynamic Properties of Peppermint Tea (*Mentha Piperita*) after Thermal and Biochemical Treatment," *Journal of Materials and Environmental Science*, 3 (2) (2012): 232–247.

4. Hossein Ahmadi Chernarbon et al., "Moisture Sorption Isotherms of Rosemary (*Rosmarinus Officinalis L.*) Flowers at Three Temperatures," *American-Eurasian Journal of Agricultural & Environmental Science*, 12 (9) (2012): 1209–1214.

5. Y. I. Salla et al., "Solar Drying of Whole Mint Plant under Natural and Forced Convection," *Journal of Advanced Research* (2013): 1–8.

## Chapter 16: Producing Value-Added Products

1. Melissa Matthewson, "Exploring Value-Added Agriculture," *Small Farms*, 2 (2) (2007): http://smallfarms.oregonstate.edu/sfn/su07valueadded.

2. Food and Drug Administration, "Dietary Supplement Health and Education Act (DSHEA)," Alliance for Natural Health USA, www.anh-usa.org/dshea/.

## Part Two: Herbs to Consider Growing for Market

1. David B. Hannaway and Christina Larson, "Alfalfa (*Medicago sativa L.*)," Oregon State University, http://forages.oregonstate.edu/php/fact_sheet_print_legume.php?SpecID=1.

2. Richo Cech, *Growing At-Risk Medicinal Herbs: Cultivation, Conservation, and Ecology* (Williams, Ore.: Horizon Herbs, 2002).

3. Ron L. Engeland, *Growing Great Garlic: The Definitive Guide for Organic Gardeners and Small Farmers* (Okanogan, Wash.: Filaree Productions, 1998).

4. Richo Cech, *Growing At-Risk Medicinal Herbs: Cultivation, Conservation, and Ecology.*

5. Andy Hankins, "Producing and Marketing Wild Simulated Ginseng in Forest and Agroforestry Systems," Virginia Cooperative Extension, Virginia Tech, Virginia State University 354-312 (2009): http://pubs.ext.vt.edu/354/354-312/354-312.html.

6. Ibid.

7. Epoch Times Staff, "4,000 Year-Old Ginkgo Tree Found in Guizhou, China," *The Epoch Time English Edition*, July 1, 2009, www.theepochtimes.com/n2/china-news/4000-year-old-ginkgo-tree-guizhou-china-18896.html.

8. Otto Richter and Sons Limited, "Richters ProGrowers Info, *Ginkgo biloba*," Richters, www.richters.com/progrow.cgi?search=Gingko&cart_id=.

9. Christopher Hobbs, "Ginkgo: Ancient Medicine, Modern Medicine," Christopher Hobbs's website, www.christopherhobbs.com/website/library/articles/article_files/ginkgo_01.html.

10. Richo Cech, *Growing At-Risk Medicinal Herbs: Cultivation, Conservation, and Ecology.*

# Resources

## Seed Companies

**Abundant Life Seeds**
PO Box 158
Cottage Grove, OR 97424
800-626-0866
www.territorialseed.com

**Fedco Seeds**
PO Box 520
Waterville, ME 04903
207-426-9900
www.fedcoseeds.com

**High Mowing Organic Seeds**
76 Quarry Road
Wolcott, VT 05680
802-472-6174
www.highmowingseeds.com

**Horizon Herbs**
PO Box 69
Williams, OR 97544
541-846-6704
www.horizonherbs.com

**Johnny's Selected Seeds**
955 Benton Avenue
Winslow, ME 04901
877-564-6697
www.johnnyseeds.com

**Richter's Herbs**
357 Highway 47
Goodwood ON Canada L0C1A0
905-640-6677
www.richters.com

## Nurseries & Rootstock

**Companion Plants**
7247 N. Coolville Ridge Road
Athens, OH 45701
740-592-4643
www.companionplants.com

**Elmore Roots Nursery**
631 Symonds Mill Road
Wolcott, VT 05680
802-888-3305
www.elmoreroots.com
*A resource for hardy rootstock
    for trees and shrubs*

**Fedco Seeds**
PO Box 520
Waterville, ME 04903
207-426-9900
www.fedcoseeds.com

**Horizon Herbs**
PO Box 69
Williams, OR 97544
541-846-6704
www.horizonherbs.com

**Nourse Farm**
41 River Road
Whately, MA 01093
413-665-2658
www.noursefarms.com

**One Green World**
6469 SE 134th Avenue
Portland, OR 97236
877-353-4028
www.onegreenworld.com

**Richter's Herbs**
357 Highway 47
Goodwood ON Canada
    L0C1A0
905-640-6677
www.richters.com

## Soil Amendments

**Fertrell Fertilizers**
PO Box 265
Bainbridge, PA 17502
717-367-1566
www.fertrell.com

**Gaia Green Products**
9130 Granby Road
Grand Forks, BC Canada
    V0H 1H1
250-442-3745
www.gaiagreen.com

**Harmony Farm Supply**
3244 Gravenstein
    Hightway North
Sebastopol, CA 95472
707-823-9125
www.harmonyfarm.com

**Lancaster Agriculture Products**
60 North Ronks Road
Ronks, PA 17572
717-687-9222
www.lancasterag.com

**North Country Organics**
PO Box 372
Bradford, VT 05033
802-222-4277
www.norganics.com

**Vermont Compost**
1996 Main Street
Montpelier, VT 05602
802-223-6049
www.vermontcompost.com

## Organic Suppliers

**The Ahimsa Alternative**
15 Timberglade Avenue
Bloomington, MN 55437
952-943-9449
www.neemresource.com
*A source for pure raw neem oil*

**ARBICO Organics**
PO Box 8910
Tucson, AZ 85738
800-827-2847
www.arbico-organics.com

**Johnny's Selected Seeds**
955 Benton Avenue
Winslow, ME 04901
877-564-6697
www.johnnyseeds.com

**Neptune's Harvest**
PO Box 1183
Gloucester, MA 01931
800-259-4769
www.neptunesharvest.com
*A source of hydrolyzed fish and
seaweed fertilizer*

**Organic Grower's Supply**
PO Box 520
Waterville, ME 04903
207-426-9900

www.fedcoseeds.com/ogs.htm

**Peaceful Valley Farm Supply**
PO Box 2209
Grass Valley, CA 95945
888-784-1772
www.groworganic.com

## Tools & Equipment

**Alexander Otto PhD, Solid
   Material Solutions LLC**
55 Middlesex Street, Suite 205
North Chelmsford, MA 01863
978-455-7182
alexanderotto@comcast.net

**AM Leonard**
241 Fox Drive
Piqua, OH 45356
800-543-8955
www.amleo.com

**Johnny's Selected Seeds**
955 Benton Avenue
Winslow, ME 04901
877-564-6697
www.johnnyseeds.com

**Peaceful Valley Farm Supply**
PO Box 2209
Grass Valley, CA 95945
888-784-1772
www.groworganic.com

## Greenhouses & Supplies

**Griffin Greenhouse Supplies**
1619 Main Street
Tewksbury, MA 01876
800-888-0054
www.griffins.com

**North Atlantic Specialty Bags**
929 Neversink Street
Reading, PA 19606

877-827-5270
www.northatlanticbags.com
*A source for specialized
packaging materials*

**Rimol Greenhouse Systems**
40 Londonderry Turnpike
Hooksett, NH 03106
877-746-6544
www.rimolgreenhouses.com

**Uline**
12575 Uline Drive
Pleasant Prairie, WI 53158
800-295-5510
www.uline.com
*A source for food-grade plastic
bags, paper bags, and boxes*

## Recommended Books

*Herbalism and Medicinal Uses of Plants*

Cech, Richo. *Making Plant
   Medicine.* Williams, Ore.:
   Horizon Herbs, 2000.

Gladstar, Rosemary. *Family
   Herbal: A Guide to Living
   Life with Energy, Health, and
   Vitality.* North Adams, Mass.:
   Storey Books, 2001.

Gladstar, Rosemary, and Pamela
   Hirsch, eds. *Planting the
   Future: Saving Our Medicinal
   Herbs.* Rochester, Vt.: Healing
   Arts Press, 2000.

Grieve, Maud. *A Modern
   Herbal: The Medicinal, Culi-
   nary, Cosmetic, and Economic
   Properties, Cultivation and
   Folk-lore of Herbs, Grasses,
   Fungi, Shrubs & Trees with
   Their Modern Scientific Uses.*
   Mineola, N.Y.: Dover
   Publications, 1971.

Hobbs, Christopher, and Leslie Gardner. *Grow It, Heal It: Natural and Effective Herbal Remedies from Your Garden or Windowsill*. New York: Rodale, 2013.

Hoffman, David. *The Holistic Herbal: A Herbal Celebrating the Wholeness of Life*. New York: Barnes & Noble, 1995.

Soule, Deb. *How to Move Like a Gardener: Planting and Preparing Medicines from Plants*. Rockport, Maine: Rudolf Steiner Press, 2013.

*Agriculture & Growing Practices, Permaculture & Design*

Bowman, Greg, ed. *Steel in the Field: A Farmer's Guide to Weed Management Tools*. Beltsville, Md.: Sustainable Agriculture Network, 1997.

Cech, Richo. *Growing At-Risk Medicinal Herbs: Cultivation, Conservation and Ecology*. Williams, Ore.: Horizon Herbs, 2002.

Falk, Ben. *The Resilient Farm and Homestead: An Innovative Permaculture and Whole Systems Design Approach*. White River Junction, Vt.: Chelsea Green, 2013.

Jacke, Dave, and Eric Toensmeier. *Edible Forest Gardens*. White River Junction, Vt.: Chelsea Green, 2005.

Persons, W. Scott, and Jeanine Davis. *Growing and Marketing Ginseng, Goldenseal and other Woodland Medicinals*. Fairview, N.C.: Bright Mountain Books, 2005.

Sarrantonio, Marianne. *Northeast Cover Crop Handbook*. Emmaus, Penn.: Rodale Institute, 1994.

Wiswall, Richard. *The Organic Farmer's Business Handbook*. White River Junction, Vt.: Chelsea Green, 2009.

## Herbal Educational Resources

**American Herb Association**
PO Box 1673
Nevada City, CA 95959
530-265-9552
www.ahaherb.com

**Avena Botanicals**
219 Mill Street
Rockport, ME 04856
207-594-0694
www.avenaherbs.com

**The California School of Herbal Studies**
PO Box 39
Forestville, CA 95436
707-887-7457
www.cshs.com

**Healing Spirits Herb Farm and Education Center**
61247 Route 415
Avoca, NY 14809
607-566-2701
www.healingspiritsherbfarm.com

**Jim McDonald, HerbCraft**
White Lake, MI 48383
248-238-8733
www.herbcraft.org

**LearningHerbs**
210 SE Cedar Hill Lane
Shelton, WA 98584
360-390-4590
www.LearningHerbs.com
The source for Herbs Made Simple video course, Creative Herbalist eBook, and the Herbal Remedy Kit; for children, Wildcraft board game and Herb Fairies children's program

**Northeast School of Botanical Medicine Herbal**
PO Box 6626
Ithaca, NY 14851
607-539-7172
www.7song.com

**Sage Mountain Retreat Center & Botanical Sanctuary**
PO Box 420
E. Barre, VT 05649
802-479-9825
www.sagemountain.com

**Vermont Center for Integrative Herbalism**
252 Main Street
Montpelier, VT 05602
802-224-7100
www.vtherbcenter.org

## Herbal Organizations

**American Botanical Council**
6200 Manor Road
Austin, TX 78723
512-926-4900
www.herbalgram.org

**American Herbal Products Association**
8630 Fenton Street, Suite 918
Silver Spring, MD 20910
301-588-1171
www.ahpa.org

**American Herbalists Guild**
125 South Lexington Avenue,
    Suite 101
Asheville, NC 28801
617-520-4372
www.americanherbalistsguild.com

**Northeast Herb Association**
PO Box 5480
Syracuse, NY 13220
315-458-4529
www.northeastherbal.org

**United Plant Savers**
PO Box 776
Athens, OH 45701
740-742-3455
www.unitedplantsavers.org

## Professional Services

**Flying Mammoths
    Landscape Design**
101 Frazier Road
Worcester, VT 05682
802-223-0410
www.flyingmammoths.com
A landscape design firm special-
    izing in design with native and
    edible plants

**Tiny Seed Creative**
802-962-5054
www.tinyseedcreative.org
Affordable websites for farmers
    and growers

**Todd Lynch, Ecotropy LLC**
413-320-2736
www.ecotropy.net
A design studio that integrates
    native medicinal and edible
    plants, ecology, and art to
    create outdoor spaces that
    illustrate and strengthen the
    connections shared by human
    and ecological wellness

# Index